# Green and Sustainable Advanced Materials

**Scrivener Publishing**
100 Cummings Center, Suite 541J
Beverly, MA 01915-6106

*Publishers at Scrivener*
Martin Scrivener (martin@scrivenerpublishing.com)
Phillip Carmical (pcarmical@scrivenerpublishing.com)

# Green and Sustainable Advanced Materials

## Volume 1: Processing and Characterization

**Shakeel Ahmed and Chaudhery Mustansar Hussain**

Scrivener Publishing

WILEY

This edition first published 2018 by John Wiley & Sons, Inc., 111 River Street, Hoboken, NJ 07030, USA and Scrivener Publishing LLC, 100 Cummings Center, Suite 541J, Beverly, MA 01915, USA

For more information about Scrivener publications please visit www.scrivenerpublishing.com.



**Wiley Global Headquarters**
111 River Street, Hoboken, NJ 07030, USA

For details of our global editorial offices, customer services, and more information about Wiley products visit us at www.wiley.com.



***Library of Congress Cataloging-in-Publication Data***

ISBN 978-119-40704-1

Cover image: Pixabay.Com
Cover design by Russell Richardson

Set in size of 11pt and Minion Pro by Exeter Premedia Services Private Ltd., Chennai, India

10 9 8 7 6 5 4 3 2 1

# Contents

# Preface

Sustainable development is a very prevalent concept of modern society. The concept has appeared as a critical force in marrying a special focus on development and growth by maintaining a balance of using resources between human beings and the ecosystem in which they are living. The developments of new and advanced materials are one of the most powerful efforts in establishing this concept. Overall, sustainable development is an internationally acknowledged directive and it includes green and environmental-friendly manufacturing materials and practices. Such practices orchestrate with the self-healing and self-replenishing capability of natural ecosystems. Green manufacturing encompasses synthesis, processing, fabrication, and process optimization, but also testing, performance evaluation and reliability. Similarly, future progress in these materials area will critically depend on our commitment with the sustainable exercises in research and technology. This book, divided into 2 parts provides a detailed overview of the status of advanced and sustainable materials for future of science and engineering.

Green and sustainable advanced materials are the newly synthesised material or existing modified material having superior and special properties. These fulfil today's growing demand for equipment, machines and devices with better quality for an extensive range of applications in various sectors such as paper, biomedical, food, construction, textile, and many more. Several advanced materials having novel properties have been reported such as biomaterials, nanomaterials, metal oxides, polymers etc. Some of them have natural origin such as plants, animals, minerals, ore etc. or extracted from plants and exist in different geometrical form and have flexibility to form a composite with other material for the specific application. Whereas, some are synthesised synthetically in required shape and size according to the demands, the superior properties of advanced material make them suitable for various forms. The objective of this book is to provide an overview of new developments and state-of-the-art for a variety of green and sustainable advanced materials.

To place all of the collective understanding about green and sustainable advanced materials into perspective, add a touch of reality to the concepts, and to cover extensive expansion of the green and sustainable advanced materials, the book is divided into two volumes and each volume has subdivisions of several chapters. Volume 1 mainly discusses Processing and Characterization while Volume 2 is focused on the Applications of green and sustainable advanced materials.

In the first volume, the first chapter presents an overview and characterization of green and sustainable advanced materials. The subsequent chapters encompass details of biopolymers and biocomposite materials and nanomaterials. Subsequent chapters describe biogenic approaches for $SiO_2$ nanostructures nanofabrication, polymer and composite materials, design and processing aspects of polymer and composite materials. The following chapters incorporate seaweed-based binder in wood composites, coloration and functional finishing of textile materials using natural resources. The final two chapters discuss advances in bio-nanohybrid materials, selenium nanoparticles and their biotechnological applications.

In the second volume, the first presents a critical review of green sustainability, nanotechnology and advanced materials and provides a vision for the future. Valorization of green and sustainable advanced materials from a biomedical perspective and their potential applications are detailed in the next chapters. Applications of green and sustainable advanced materials in textile technology and environmental protection are described in a very comprehensive manner in the next batch of chapters. Synthesized nanostructures alloys for optoelectronic, biochar-supercapacitors, biomedical from synthetic and natural green and sustainable advanced materials green and sustainable advanced materials are then covered. Efficiency of transition metals at the nanoscale - as heterogeneous catalysts and emerging applications of green and sustainable advanced materials in agriculture and food industry take center stage in final two of chapters.

In conclusion, both volumes incorporate in-depth technical information without compromising the delicate link between factual data and fundamental concepts or between theory and practice.

Overall, this book is planned to be a reference book for researchers and scientists who are searching for new sustainable advanced materials. The contributors are well-known researchers and scientists of materials science and engineering. We are very thankful to the chapter authors for their enthusiastic efforts in the making of this book. Finally, we extend our thanks to Wiley-Scrivener for publishing the book.

Shakeel Ahmed & Chaudhery Mustansar Hussain (Editors)
June 2018

# 1

# Green and Sustainable Advanced Materials: An Overview

**Tanvir Arfin*, Arshiya Tarannum and Kamini Sonawane**

*Environmental Materials Division, CSIR-NEERI, Nehru Marg, Nagpur, India*

***Abstract***
In today's developing and challenging environment, the need of quality material for fulfilling the requirements of various sectors is increasing. Hence, to satisfy this task there is a growing need for advanced materials. Advanced material is a material that either synthesises or modifies the existing material by using various advanced technologies to get improved properties such as physical, chemical, mechanical, and optical properties and that gives better performance than the conventional material. There are various advanced materials including polymeric materials (polystyrene, dendrimer, etc.), metal oxide ($TiO_2$, ZnO, etc.), biomaterial (dextran, cellulose, gelatine, pollulan, etc.), and nanomaterial (CNT, GO, etc.). Owing to their excellent mechanical, physical, electrical, chemical, and optical properties, ability to make composite with other materials, ease of availability, and low toxicity, they are used in various applications such as energy storage, water treatment (heavy metal, dyes, and pollutant removal), solar cell, electronics, paint, and textile industries.

***Keywords:*** Nanomaterial, heavy metal, dye, pollutant, environment

## 1.1 History

The study of advanced materials is offering a new concept in the field of material science continuously since 25 years. The primary target for the study is the interdisciplinary behaviour of materials science and the concern related to the aspects of the materials. The limitations of the advance materials and the future outlook in the upcoming generations are the main

---

**Corresponding author*: tanvirarfin@gmail.com

Shakeel Ahmed and Chaudhery Mustansar Hussain (eds.) Green and Sustainable Advanced Materials, Vol. 1 (1–34) © 2018 Scrivener Publishing LLC

topic of interest. In this chapter, the main emphasis is laid on the fundamental theories and the uses of advanced materials which on the other hand clarify about the recent growth in research field by different sources such as catalysis applicability, electrochemical, and semiconductor.

## 1.2 Biomaterials

Various types of biomaterials are described in the following subsections.

### 1.2.1 Dextran

Dextran is a material that can be produced quickly and from the cheap source of nitrogen or carbohydrates. Vegetable wastes, wheat bran, straws, and molasses can be used for the production of dextran. Different species of bacteria are required for the synthesis of dextran. In the food processing industries, dextran synthesis is carried out with the help of some strains such as *Lactobacillus plantarum*, *Leuconostoc mesenteroides*, and *Lactobacillus sanfransisco*. The fermentation process produces dextran by using energy source with sucrose as the primary energy source. In 1930, scientist Pederson and Hucker were the two who discovered the dextran production from the one specific strain 'Leuconostoc'.

There are different derivatives of dextran that can be used on the commercial level such as diethylaminoethyl cellulose (DEAE) dextran, dextran sulphate, and fluorescein labelled dextran. Among these, the dextran sulphate is the one exhibiting low molecular weight and applicable in various sectors due to its favourable properties.

#### *1.2.1.1 Chemical Structure*

The smallest glucose molecules form the polysaccharides called as dextran. Dextran has two types of structures: straight chain comprises α-1,6 glycosidic linkages; whereas, the branched structure of dextran includes α-1,4 glycosidic linkages as shown in Figure 1.1. The molecular weight is about $9 \times 10^6$–$500 \times 10^6$ Da.

#### *1.2.1.2 Properties*

(a) Completely soluble in water.
(b) Retain moisture.
(c) Excellently stable.

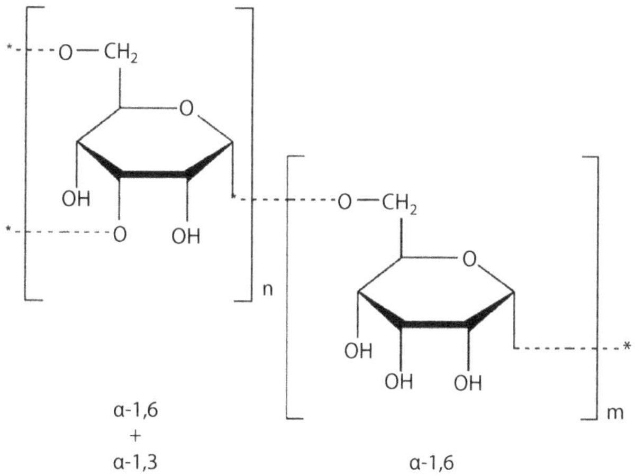


**Figure 1.1** Structure of dextran.

#### *1.2.1.3 Applications*

Dextran has its application in the waste water treatment process. By using dextran, the process becomes economic and environmental friendly, as it is biodegradable. It also has utilisation in the photographic industries to some extent but only when it shows properties such as low chloride concentration and high clarity. Dextran has various applications in different fields. The frozen product such as ice-cream contains dextran as a stabiliser. About 2–4% of dextran is present in the blend [1]. It provides viscosity and stability to frozen product and frozen dairy product.

### 1.2.2 Cellulose

Cellulose serves as the raw material. It is produced by plants in massive quantity in the world. There are various forms of cellulose namely cellulose nanocrystal [2], bacterial cellulose [3], nanofilbrillated cellulose [4], and ethyl cellulose [5]. Cellulose was discovered by Peyen in 1838. The most significant storage of organic carbon is the cellulose. The advantage is that it is a renewable resource of the polymer. The common source of cellulose is the forest area possessing wood through which the commercial production of cellulose is possible. Cotton maintains the maximum concentration of cellulose. Cellulose production by plants is about 180 billion per year, and it can serve the significant amount of carbon source. When the cellulose is present in the combined form in other compounds such as polysaccharides and lignin, it can be said to be hemicellulose. However, during the modification of cellulose, sometimes the problems arise from the naturally occurring cellulose. Cellulose was prepared on the lab scale by using the bacterial species *Gluconacetobacter xylinus* and *Acanthamoeba catellani*. In 1991, it was discovered by Tarchevsky and Marchenko.

Frequently, it serves as raw materials for the production of various products in textile and fabric. Cellulose is present in the crystalline form as crystalline I, cellulose II, and crystalline III. Among these three types, type II is the more suitable for the commercial use. Type II is also obtained from type I, and it is more stable than the other two.

#### *1.2.2.1 Chemical Structure*

It is formed by the glucose linkage of α-1,4 bonds as shown in Figure 1.2. Cellulose is a homopolymer comprising the monomers as enantiomer and regiomer. The hydrogen in cellulose is bonded together by both intermolecular and intra-molecular bonding.

#### *1.2.2.2 Properties*

(a) It is a primary constituent of plant cell wall.
(b) It helps in cell formation and protection.

#### *1.2.2.3 Application*

Cellulose has the large use in industrial products. The textile, paper and pulp industries, pharmaceutical, cosmetics, and polymer industries require cellulose. It is also used as an oxidised fabric in the pharmaceutical and medical field. It is not soluble in acid and water. Alkali serves well for complete solubilisation of oxidised cellulose material. It is employed in the pharmaceutical, agricultural, and cosmetic fields. It is used for surgical purpose but it is not directly applicable on the opened wound.

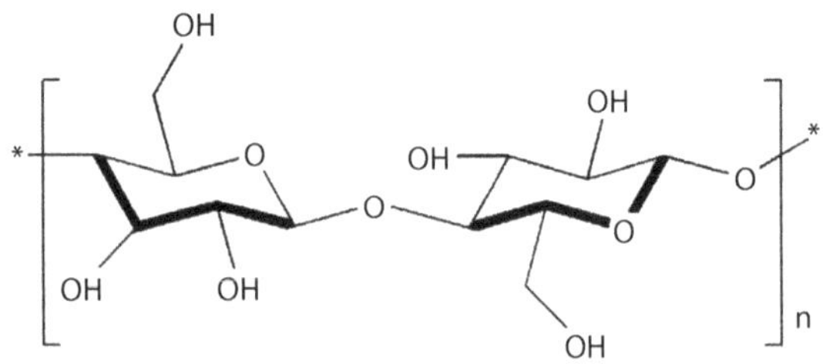


**Figure 1.2** Structure of cellulose.

Ethyl cellulose is one of the derivatives of cellulose in which some of the hydroxyl groups are directly converted into ethyl group on every repeated molecule of glucose. It is used for the antibacterial activity and to maintain mechanical stability [6, 7].

### 1.2.3 Gelatine

The primary source of gelatine is collagen in the animal. It can be extracted from various sources of the body parts of the animal such as bones, skin, or tissues. Hydrolysis is carried out for the production of gelatine. It is an anhydride form of gelatine. Gelatines are of different types based on their classification. It can be classified by source and treatment method. Gelatine is of two types: (a) Type A and (b) Type B. The 'type A' gelatine is prepared by acid hydrolysis; whereas, 'type B' gelatine can be prepared by alkali hydrolysis [8]. Other options are also available for the preparation of gelatine instead of using animals. Pectins, cellulose, xanthum gum, and the agar-agar are used for producing gelatine specifically the vegetable gelatine.

#### *1.2.3.1 Chemical Structure*

It is a polymer or protein and made up of amino acids, which is joined by the peptide bonds as shown in Figure 1.3. Its molecular weight ranges from 15,000 to 400,000 Da. The main components of gelatine are carbon, nitrogen, hydrogen, and oxygen.

#### *1.2.3.2 Properties*

(a) It appears as faint yellow.
(b) It does not have odour and taste.
(c) The chemical modification makes the gelatine more superior.
(d) It can swell in the solution.
(e) It is amphoteric.
(f) It is stable in dry form.
(g) It serves as the best medium for bacteria.
(h) It acts as a Colloid protector, especially in the electroplating industries.

#### *1.2.3.3 Application*

It can be useful in various fields. It acts as a jellifying, thickening, emulsifying, and foaming agent. Owing to the above properties, it has application

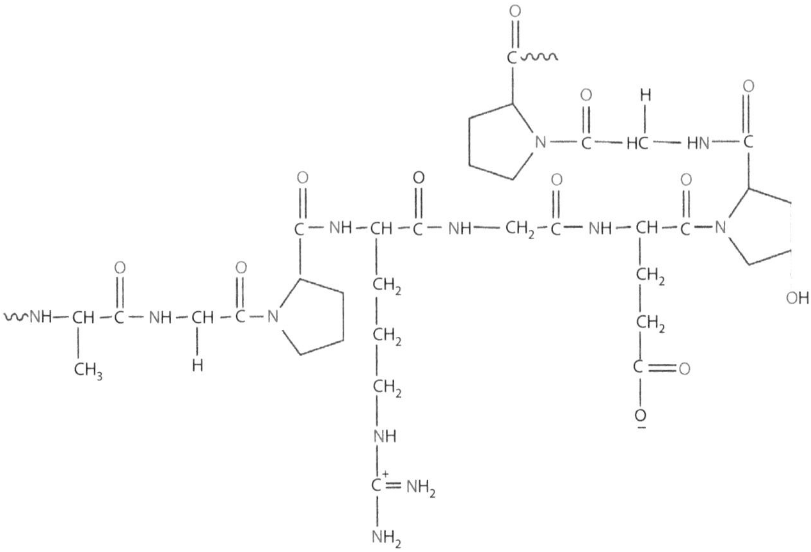


**Figure 1.3** Structure of gelatine.

in food industries, photographic field, and pharmaceutical industries. Gelatine is mainly used in the pharmaceutical industries for tablet making, granulation, coating of tablets, and encapsulation. The primary use is in pharmaceutical and food industries, where products are encapsulated to protect and preserve them from heat and moisture.

### 1.2.4 Alginate

In the nineteenth century, the alginates were discovered by the scientist E.C. Stanford. Brown seaweeds are the excellent source of alginate. About 18–35% of alginate is present in dried seaweeds. These are natural biopolymers as it is extracted from natural origin. The seaweed *Macrocystis pyrifera* is an abundant source for obtaining alginate, as this seaweed can proliferate (about two feet daily). Consequently, it gives more production of alginate. The chemical process is used for the manufacturing of alginate. During production, all the impurities including biological impurities should be removed. The hydrolysis should be proper and complete for obtaining the first functional polymer. Newtonian flow is obeyed by the small fraction of alginate.

#### *1.2.4.1 Chemical Structure*

Alginates comprises of two urinate sugars such as L-guluronic acid and mannumoric acid. These are called as M-block and G-block, as shown in

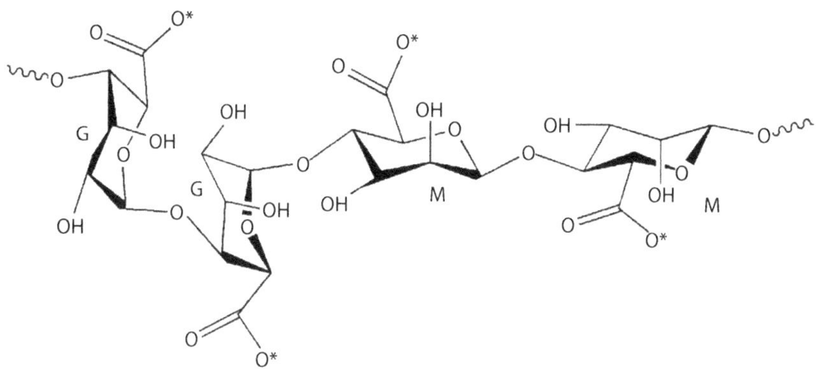


**Figure 1.4** Structure of alginate.

Figure 1.4. By the alternative arrangement of these blocks, the alginate is formed. The G-block helps to increase the ability of gel formation. However, still, the source and species affect the structure of alginates.

#### *1.2.4.2 Properties*

(a) Resistant to microorganism
(b) Hydrocolloid.
(c) Stable at pH range 5–10.
(d) Degradable.
(e) Thermo-irreversible.

#### *1.2.4.3 Application*

Alginates have high use in medical fields such as in controlling and treating diseases such as brain tumours and diabetes. It is also applied in food industries for making jelly, ice-cream, and bakery products. It can be useful in the film formation as it gives flexibility to the film by combining with the plasticiser. The film obtained will be transparent with high oxygen barrier. The sodium alginate film is prepared by using the $CaCl_2$ treatment or without $CaCl_2$ [9]. These films can be utilised in the pharmaceutical and food industries.

### 1.2.5 Chitin

It is a natural polysaccharide. It can also be called as cellulose derivative. Invertebrates and exoskeleton are the sources of chitin. It is said to be identical to cellulose even if it is not obtained from a cellulose producing organism. It is poorly soluble in water. This drawback reduces its use when compared to chitosan. The reason behind the modification of chitin is that its derivatives can be utilised more efficiently.

#### 1.2.5.1 *Chemical Structure*

It consists of β (1→4) linkage 2-acetamido-2-deoxy-β-D-glucose as shown in Figure 1.5. It is hydrophobic. Water and alcohol are not suitable for their solubility. It appears as white and hard so non elastic.

#### 1.2.5.2 *Properties*

(a) Biodegradable
(b) Compatible
(c) Suitable adsorbent
(d) Soluble in different media.

#### 1.2.5.3 *Application*

Production of cosmetic and personal care products have the most extensive area for the use of chitin. Coffee and beverage industries require chitin for the de-acidification process. It can be useful in the food, packaging, water treatment, paper and pulp industries, and textile industries. Chitin and modified chitin can be used for the preparation of fibres [10]. These fibres can be utilised for the adsorption process and wound healing operation.

### 1.2.6 Chitosan

Chitosan is a natural polymer and present in abundant quantity. It is a renewable resource. It is the derivative of chitin and has various commercial applications. As compared to cellulose, chitosan is more useful in multiple ways as it has the amine group ($NH_2$). The chemical processing makes chitosan more functional.

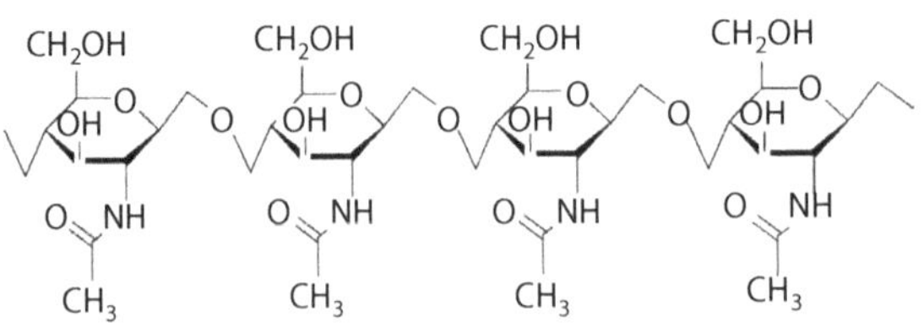


**Figure 1.5** Structure of chitin.

#### 1.2.6.1 *Chemical Structure*

The chemistry of chitosan involves α-1,4 linkages 2-amino-2-deoxy-β-D-glucopyranose structure as shown in Figure 1.6 [11].

#### 1.2.6.2 *Properties*

(a) Non-toxic.
(b) Biodegradable
(c) Renewable and biocompatible
(d) Amino groups and hydroxyl groups are reactive.
(e) Able to remove metal ions.

#### 1.2.6.3 *Application*

It can be used in the biomedical, cosmetic, and pharmaceutical fields due to their excellent properties, which includes wound healing, tissue regeneration, and differentiation. For preparing colorants, tonics, and spray for hair, chitosan serves the best. Chitosan can be used as a flocculator in water treatment. Chitosan has the advantageous effect for removal of heavy metals such as As and Hg [12].

### 1.2.7 Pullulan

Pullulan is formed by the *Aureobasidium pullulans*; Bauer observed this in 1938. Afterwards, polysaccharides were isolated by Bernier in 1958, and then Bender *et al.* [13] carried the study of the polymers, and later he named this polymer as 'Pullulan' in 1959. In 1976, pullulans was produced on the commercial scale, and it was done by Hayashibara Company.

#### 1.2.7.1 *Chemical Structure*

It is a natural and linear homopolymer. It consists of α-1,6 linkage of maltotriose units as shown in Figure 1.7. The molecular weight of pullulan is 200,000 Da.

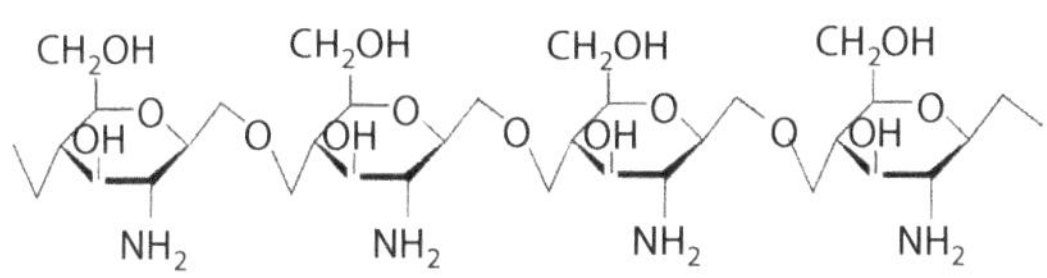


**Figure 1.6** Structure of chitosan.

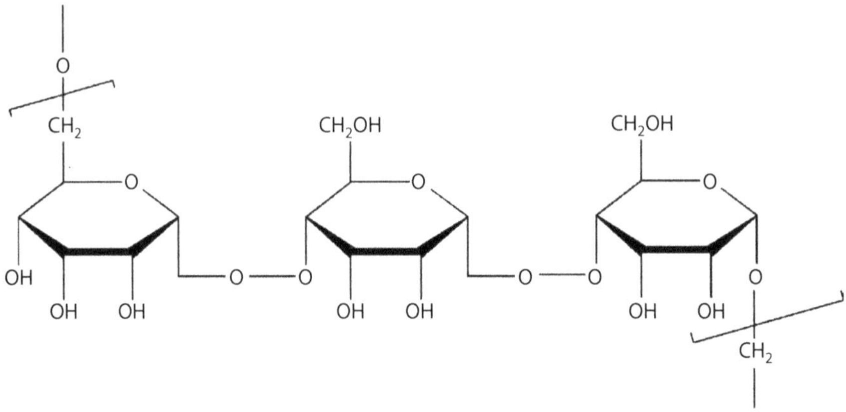


**Figure 1.7** Structure of pullulan.

#### *1.2.7.2 Properties*

(a) High water solubility.
(b) Structural flexibility.
(c) Non-ionic polysaccharide.
(d) Non-toxic
(e) Biodegradable

#### *1.2.7.3 Applications*

Pollulan is more suitable in a biomedical field. The water soluble nature of pollulan makes it more favourable in the drug delivery and tissue engineering. It acts as thickening agent, glazing agent, and helps in film formation. The significant advantage in the biomedical field is that it can serve as plasma expander [14]. Owing to its high water solubility, it is suitable for plasma expander. The molecular weight should be 60,000 Da for this purpose.

### 1.2.8 Curdlan

Curdlan is a polysaccharide and produced from bacteria; hence, it can be called as a bacterial polysaccharide. In 1966, during the succinoglucan production, curdlan was first observed. The bacteria *Alcaligenes faecalis variety myxogenes was* used for the production of curdlan polysaccharide.

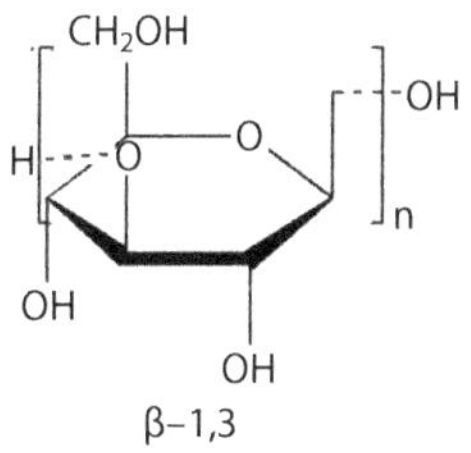


**Figure 1.8** Structure of curdlan.

### 1.2.8.1 Chemical Structure

The structure consists of repeating units α-1, 3 glycosidic linkages as shown in Figure 1.8. The molecular weight of curdlan is between $5.3 \times 10^4$ and $2 \times 10^6$ Da.

### 1.2.8.2 Properties

(a) Gel forming ability.
(b) Unaffected by thawing and freezing.
(c) Solubility in aqueous alkali solution.
(d) Insoluble in alcohol and water.
(e) Thermo-gelable.

### 1.2.8.3 Application

Curdlan is used for the formation of a gel having sturdy, firm and thermo irreversible properties. Gel formed from curdlan is stable and of high quality which can be used in the food industries. Another application is that it can be utilised as a texturising agent, stabilising agent, and thickening agent [15]. The most significant use of curdlan is in the medical field. The disease AIDS is the severe illness which does not have any permanent treatment. However, the scientist discovered the option of treating the HIV with the help of curdlan [16]. By forming curdlan sulphate by piperidine-N-sulphonic acid in DMSO, the cure for HIV is possible. This solution carries out inhibition of HIV infection.

## 1.2.9 Lignin

Candolle introduced the 'Lignin' in 1819. The name given to lignin is the Latin name of wood 'lignum' [17]. Lignin is the massive component in plants. Owing to lignin, vegetable and a huge tree grow straighter and straighter firmly. In addition, the lignin helps to restore water in plants.

#### *1.2.9.1 Chemical Structure*

It is a polymer formed by joining the different linkages of phenyl propane. A phenylpropanoid consists of the subunits such as p-coumaryl alcohol, conniferyl, and synapyl as shown in Figure 1.9. The concentration of these subunits varies according to the type of trees such as angiosperm, gymnosperm, and graminaceous plants. The complex structure is formed owing to the attached functional groups. The structure of lignin is still unknown.

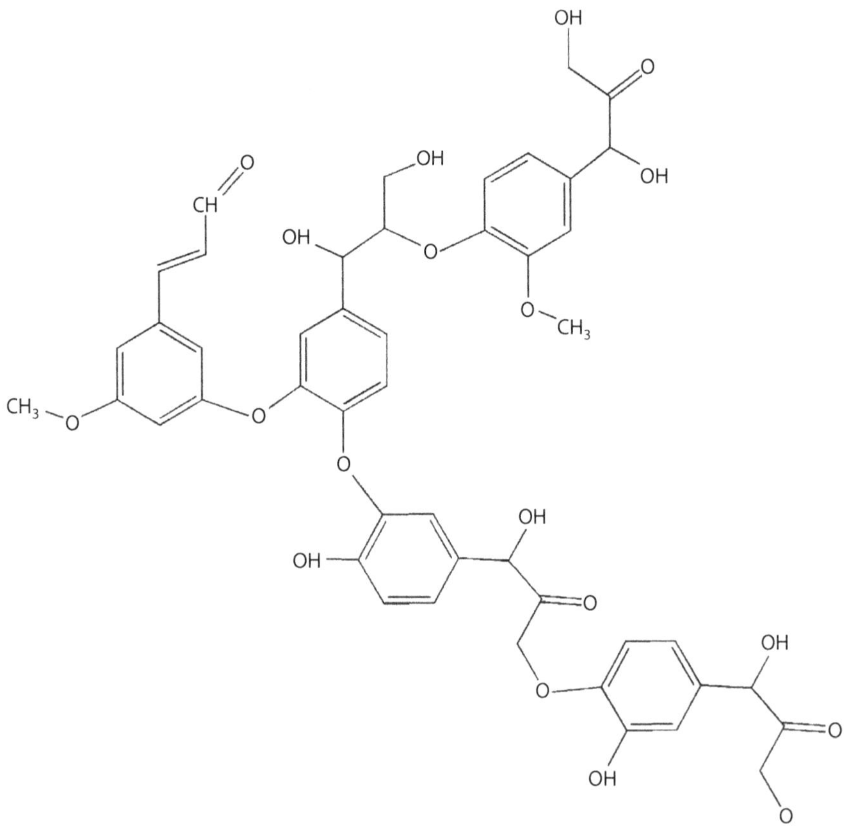


**Figure 1.9** Structure of lignin.

### 1.2.9.2 Properties

(a) Amorphous
(b) Natural defender of plants against degradation.
(c) Highly reactive owing to different functional groups.

### 1.2.9.3 Application

Lignin is used to produce the biofuel. The fuel generation from food waste and then with lignin creates the first and second generation fuels [18]. The waste generated in the early fuel generation is avoided in the second fuel production; thus, it is more efficient. The second fuel production is done with the lignocellulosic processing.

## 1.2.10 Xanthan Gum

The xanthan gum is produced in the United States research laboratory. The polysaccharide also produced by *Xanthomonas campestris*. During the

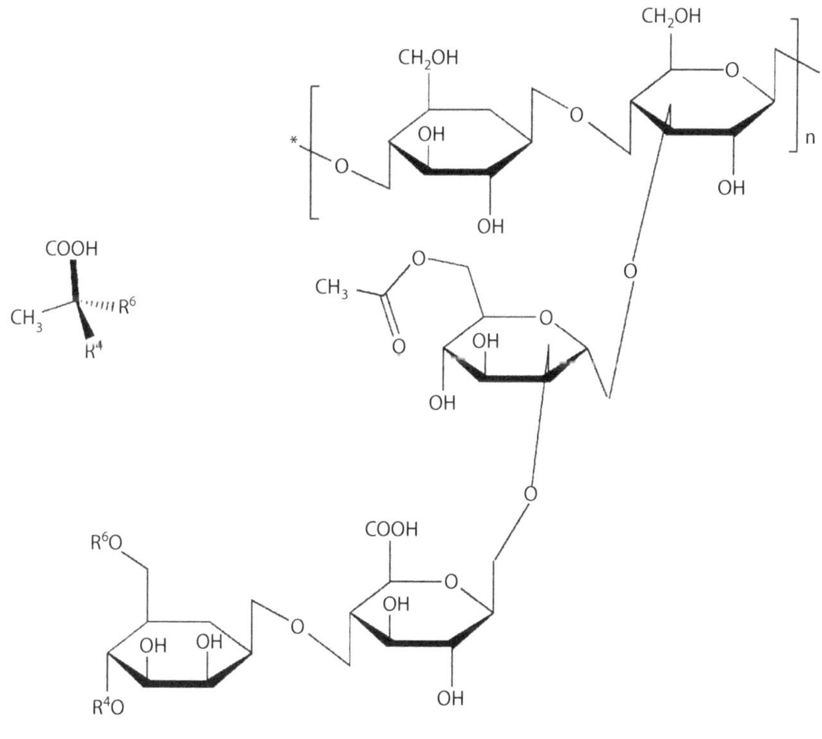


**Figure 1.10** Structure of xanthan gum.

microbial biopolymer application, it was observed about that the polysaccharide protects the bacterium. This is xanthan gum.

#### *1.2.10.1 Chemical Structure*

There are two groups of glucuronic acid and mannose attached to the repeating structure which then constitutes the pentasaccharide structure of xanthan gum as shown in Figure 1.10.

#### *1.2.10.2 Properties*

(a) Present in powder form with no taste.
(b) Provides proper viscosity by dissolution.
(c) Acts as a stabilising agent.
(d) Thermally stable

#### *1.2.10.3 Applications*

It can be used in the various fields such as food and pharmaceutical industries. This is utilised for the preparation of toothpaste. Xanthan gum helps to pump and flow out the tube and also to stay stable on the surface of the brush. Xanthan gum is used in food industries involving different purposes such as beverages, food additives, and bakery products. When the xanthan gum is added to beverages consisting of some fruit juice, the quality of the drinks is maintained by xanthan gum [19].

### 1.2.11 Hydrogels

The cross-linked polymers are the hydrogels. They are a merely hydrophilic compound which can soak the aqueous medium present around the polymers. Swelling of the polymers can be done with the help of this hydrogel depending on its properties. There are various sources for producing hydrogels such as polymers and copolymers, esters, polysaccharides, and gelatine. Thus, the different types of hydrogels are prepared according to the use. The material, shape, and sizes can vary. It can be made as nanoparticles, a film, a coating material, or micro or macro particles. Wherever the hydrogels need to be applied, it attains the shape of the surface. This type of materials is used in the biomedical and pharmaceutical fields.

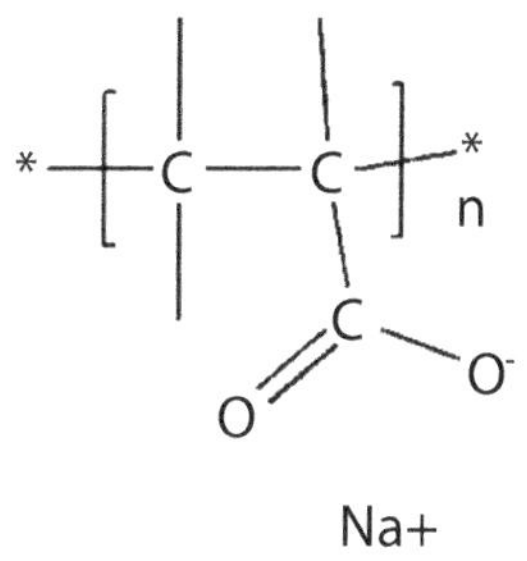


**Figure 1.11** Structure of hydrogel.

### *1.2.11.1 Chemical Structure*

The crosslinked 3D networks of hydrophilic polymer chains are able to grasp maximum quantity of water because of the hydrophilic structure as shown in Figure 1.11.

### *1.2.11.2 Properties:*

(a) Chemically stable.
(b) Swelling
(c) Completely degradable.
(d) Have porous structure.
(e) Biocompatible.

### *1.2.11.3 Application*

It can be used in the biosensor making. This is done by enzyme immobilisation technique. The fructose determination can be done by forming the hydrogels with chemicals such as polycarbamoyl suphonate and polyethyleneimine [20]. The D-fructose dehydrogenase enzyme is immobilised for sensing the fructose.

## 1.2.12 Xylan

The versatile material in the environment is the polymers. These can be synthesised biologically but not like cellulose. By using these polymers and polysaccharides, the micro-particles can be prepared. The advanced form of the micro-particle can be made using xylan. It is also called as hemicellulose. Different parts of plants consist of varying concentration of xylan.

As it is present in a significant amount, it is preferable to use it for the processing in the industrial pharmaceutical sector.

#### *1.2.12.1 Chemical Structure*

They are present with lignin and cellulose. Moreover, 20–30% of hemicellulose, which is currently available in plants, is essential for rigidity as shown in Figure 1.12.

#### *1.2.12.2 Properties*

(a) Thickening agent
(b) Provides adhesion
(c) Biodegradable
(d) Biocompatible

#### *1.2.12.3 Application*

They can be used to prepare the coating materials, films, micro particles, and nanoparticles, and for delivery of drugs [21]. It can be applied in the

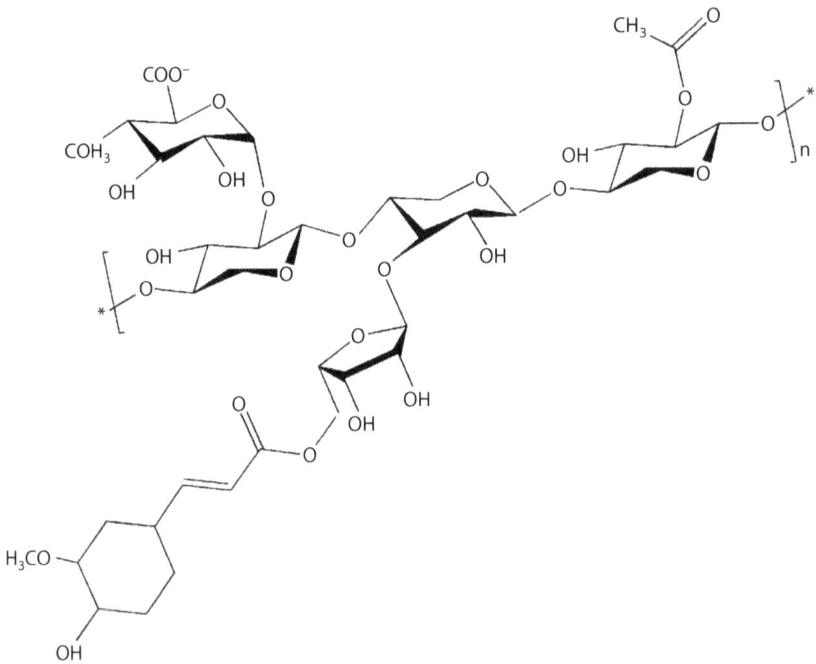


**Figure 1.12** Structure of xylan.

biomedical, automobile, textile, and pharmaceutical fields. The glycosidic bonds degrade in the sugar units. For this, the colonic drug delivery system is required. The production of this scheme can be prepared by using the xylan as a raw material [22].

### 1.2.13 Arabic Gum

It is well known gums which are widely employed in the food industries for increasing the quality of food owing to the availability of functional groups within the structure. The Arabic gum obtained from the trees is usually found in Africa. As is supplied from the Arabian ports, it got its name gum Arabic. The trees required for the extraction of Arabic gum Acacia are Senegal and Acacia seyal. Because of this reason, it is also called as acacia gum.

The Arabic gums are of different types, and they are classified on different basis such as follows:

(i) Source
(ii) Physical nature and structure
(iii) Chemical structure

#### *1.2.13.1 Chemical Structure*

It is a complex polysaccharide and consists of a hydroxyl group and salts of some compounds such as calcium, magnesium, and potassium. The structure involves side chain 1,3-β-D-galactopyranosyl units connected to the main chain as 1, 6 linkages as shown in Figure 1.13. Sometimes they have branched polymers as well. Its composition mostly consists of carbohydrates and lower amount of proteins. It is neutral but sometimes behaves slightly acidic.

#### *1.2.13.2 Properties*

It is a compound of dual nature: hydrophobic and hydrophilic. Thus, the reactions with water and without water can perform well.

(a) Less viscous than other gums.
(b) Highly water soluble.
(c) Acts as antioxidants.
(d) Controls blood glucose concentration.
(e) Acts as an emulsifier.

### *1.2.13.3 Applications*

Because of its properties, it can better act as an emulsifier, stabiliser, and most probably encapsulating agent. Pharmaceutical and food industries have more uses rather than textile and cosmetics. It is used mostly in the food as a dietary material. Even though it is added in the food products to reduce the fat and calories, it imparts the taste as well. Its creamy texture remains the same [23].

## 1.3 CdS

It is an inorganic compound. It occurs naturally in two crystalline structural forms in the mineral: hawleyite and greenockite, but mostly found as an impurity in zinc ore Wurtzite and Sphalerite. It appeared as yellow colour solid and is used as a yellow pigment in pigment industry. At present, it has gained importance in nanotechnology as one of the nanomaterials has various applications. CdS is soluble in acid but insoluble in water. It acts as a semiconductor owing to the presence of n-type conductivity exerted by the presence of sulphur vacancies and excess of Cd atoms [24]. Energy band gap in the bulk of CdS is found to be 2.42 eV at a temperature of 300 K, and it shows maximum absorption at 515 nm [25]. It exists in three kinds of crystalline substance: zinc blende, Wurtzite, and rock-salt. Wurtzite has hexagonal geometry; zinc blende has cubic structure, whereas rock-salt possesses octahedral geometry. Zinc blende is a stable one. Properties of CdS depend upon its size. CdS nanoparticles show excellent physical, chemical, and electrical properties when compared to the bulk form. Unique properties of CdS-NP make it suitable for a wide range of technological application. It is most commonly used as a pigment in paint industries owing to its thermal stability.

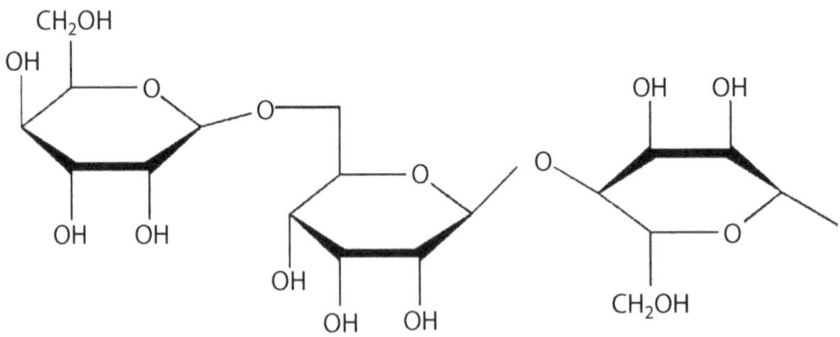


**Figure 1.13** Structure of Arabic gum.

Yang *et al.* [26] developed hybrid nanofibers by depositing CdS-NPs on nanofibers of bacterial cellulose which was used as a substrate. Hybrid nanofibers were utilised for photocatalysis application. CdS nanocrystals having hexagonal structure were prepared by the hydrothermal reaction between cadmium chloride and thiourea at low temperature, and the prepared nanocrystals were deposited on the surface of nanofibers of bacterial cellulose. The hybrid nanofibers CdS/BCF show very high photocatalysis efficiency giving 82% degradation of MO after irradiation for 90 min. CdS/BCF hybrid can also be recycled. Efficient photocatalysis for degradation of organic dyes can be done using CdS/BCF hybrid.

## 1.4 Carbon Nanotube

CNT is a carbon allotrope with cylindrical structure possessing their sizes on the nanometre scale as shown in Figure 1.14. It was discovered in 1991 by a Scientist while producing fullerene via an arc discharge method. CNT possess excellent physical, mechanical and electrical properties. It is much stronger than steel as the bonding between each carbon atom in CNT is very strong. It is exhibited as an excellent conductor of heat as well as electricity. There are two types of CNT: SWNT and MWNT. SWNT possesses single layer or wall just like a strand, whereas MWNT possesses multiple nested tubes with increasing diameter. Each tube in SWNT is placed at a finite distance from its adjacent tube and linked together by an interatomic force. CNT occurs in various structure forms based on thickness, length, and number of layers. The way of rolling the graphene sheet to form the tubes determines the characteristics of CNT whether it will be metallic or semiconductor. It is a low weight and flexible materials which can be added to other materials to make the composite. CNT has a wide range of applications in various fields such as ceramic, pharmaceuticals, forensic, industrial and manufacturing, fibre analysis, and earth science.

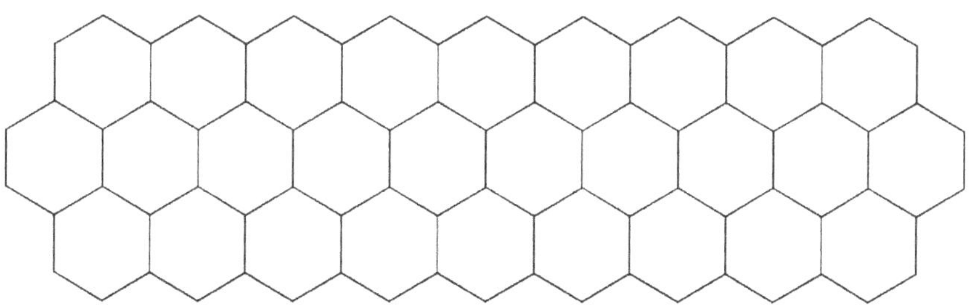

**Figure 1.14** Structure of CNT.

Jung *et al.* [27] synthesised Ag/CNT hybrid NPs for antimicrobial filtration of air. It was produced by using aerosol nebulisation followed by thermal evaporation/condensation. The aerosol method used for generating Ag/CNT hybrid is straightforward and can provide Ag/CNT continuously. CNT possesses unique physical, mechanical, and optical properties, whereas Ag is known for its antimicrobial properties. The hybrid NPs of both show higher antimicrobial activity than CNT and Ag NPs. Hence, it can be used for application related to public health as well as biomedical engineering.

## 1.5 Fe Containing Nanomaterial

Fe-containing nanomaterial is mainly consisting of ZVI and found in the form of iron oxide. Iron oxide is a collective term used for oxides, oxy-hydroxides, or hydroxide of Fe. Iron oxide is sometimes referred to as SPIONS which is an oxide form of a transition metal. Around 16 phases of iron oxides are known so far. Trivalent iron is found in iron oxide which is less soluble and imparts intense colour. The essential properties of iron oxides include small size, biocompatibility, low toxicity, biodegradability, and large surface area [28]. It is a cost effective material. The essential characteristics such as low toxicity as well as biodegradability make it advantageous over another nanomaterial. SPIONS consist of two parts; central part is occupied with iron oxide (either magnetite or maghemite) with a coating on the outer side. Size of SPIONS determines its category: MION (10–30 nm), SSPIONS (10–50 nm), USPIONs (10–15 nm), and micron sized SPIONS (300 nm–3.5 μm). SPIONS are used as a contrasting agent for malignant tumour treatment owing to its higher magnetic susceptibility than paramagnetic material. As iron oxide-NPs have a magnetic property, it is easy to separate. Owing to the presence of antimicrobial property, it finds application in the biomedical field. It is also used in water treatment as well as for removal of heavy metals and dye, drug delivery, therapeutics, soil remediation, etc. [29].

Owing to its excellent physiochemical property and easy separation by application of magnetic field, it was used as an adsorbent for water treatment to remove heavy metals such as As and Cd. It is easy to regenerate and reuse which makes it a cost-effective adsorbent and also decreases the economic burden [30].

## 1.6 Graphene

It is a carbon allotrope having two-dimensional structure as shown in Figure 1.15. It was discovered in 1962 by Boehm and rediscovered by Geim

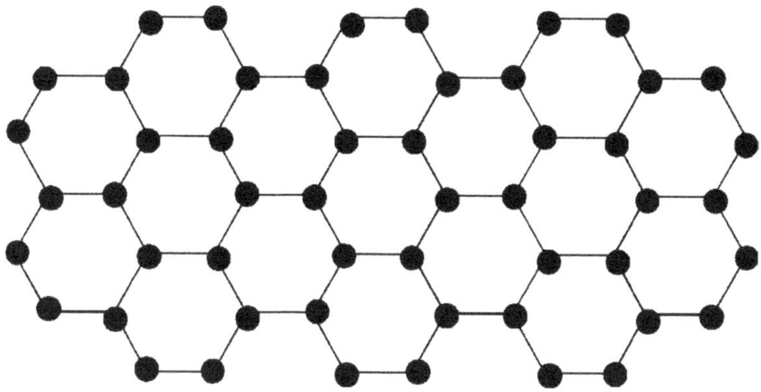

**Figure 1.15** Structure of graphene.

and Novoselov in 2004 for which they get the Nobel Prize. The carbon atom in graphene is bonded together in a hexagonal lattice and is a primary unit for other carbon allotropes such as fullerene, charcoal, CNT, and graphite. It is the thinnest, lightest, and most durable material ever measured in the universe.

It possesses novel mechanical, physical, and optical properties. It is 200 times stronger than steel having the same thickness and has a thermal conductivity more than double of that like a diamond. Being thinnest material, it is optically transparent and highly flexible material. Graphene can be synthesised by a number of methods including arc discharge, mechanical exfoliation, CVD, chemical exfoliation, epitaxy, and pyrolysis. Because of its extraordinary properties, it has a wide range of applications. Displays and electronic devices, photonics, composite materials, paints and coatings, energy generation and storage, sensors, biomedical, and water treatment are among the many application areas in which graphene could be used. Graphene's large surface area and atomic thickness make it a suitable candidate for creating devices useful for detection and diagnosis of microbes [31].

Gupta *et al.* [32] developed a green method for synthesising graphene from cane sugar and use it in water purification application. They immobilised the cane sugar on sand without using the binder. The resultant material was called as GSC. Raman Spectroscopy confirmed that the resultant material is graphene. It was used for water purification. To demonstrate its application, they used R6G dye and CP pesticide; it was found that it effectively removes R6G and CP from water by physical adsorption process with an adsorption capacity of 55 and 46 mg/g, respectively. The result proves that it is much better than activated carbon and can be regenerated also. Hence, it might be a cost effective adsorbent for water purification.

## 1.7 Graphene Oxide

GO is an oxidised form of graphene or can be called as a graphene derivative as shown in Figure 1.16. It is a thin atom sheet having oxygen containing functional group on its surface. It can also be produced from graphite which is an abundantly available material. It was first developed in 1859 by a chemist Brodie from graphite using fuming $HNO_3$ and $KC1O_3$. Later, in 1957, Hummer and Offeman developed a new efficient and straightforward method, which is known as Hummer's method, to produce GO. They used the mixture of $NaNO_3$, $KMnO_4$, and $H_2SO_4$. Hummer's approach is a commonly used method for manufacturing GO. Because of the presence of functional group 'O', GO is hydrophilic and can disperse in water and various solvents as well. This property makes it easy to process [33]. It can also be used for graphene synthesis by reducing it. Flake size of GO can be tuned from nm to mm according to their application. The ability to optimise both its chemical composition as well as size, allows it to be used in a various form such as a biosensor, medicine, biomedical engineering, and environment friendly energy devices [34].

Liu *et al.* [35] prepared 3D GO sponge and used it in water treatment application for the removal of MB and MV dyes which are the critical contaminants in waste water from textile and dye manufacturing industries.

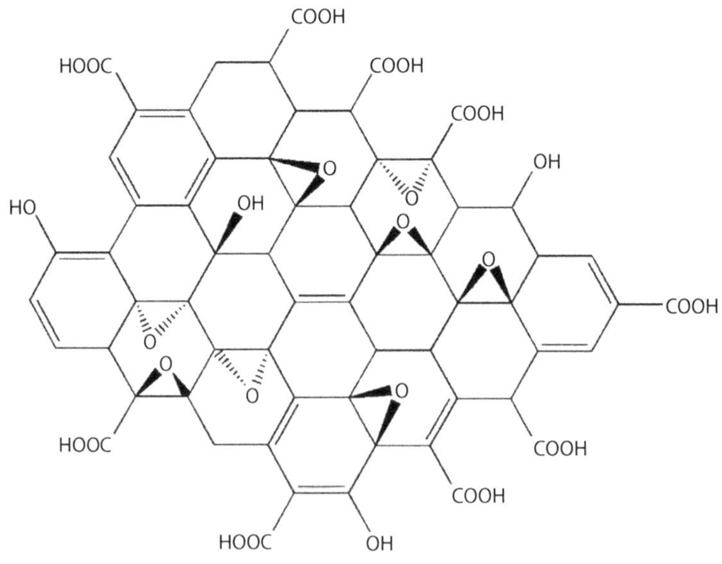


**Figure 1.16** Structure of graphene oxide.

They prepared the GO sponge by a simple method of Centrifugal vacuum evaporation. GO sponge removes the dyes by an adsorption process with high efficiency in a concise time of 2 min. It has very high adsorption capacity: 397 mg/g for MB and 467 mg/g for MV. After the adsorption process, GO sponge can be easily recovered by vacuum filtration. Hence, it proves to be an efficient and rapid adsorbent for dye removal.

## 1.8 Inulin

It is a polysaccharide of natural origin. It is produced by various plants naturally. Commercially, it is extracted from chicory. It belongs to the class of fructan which is a dietary fibre. It was first discovered by Valentine Rose, a German scientist in 1804 in Inula Helenium roots; hence, it was named as Inulin. The other names of inulin are helenin, meniantin, and alatin. It is used by around 36,000 plant species for energy storage purpose. It is a natural storage for carbohydrate in plants but does not store its other form like starch. It is made up of a heterogeneous group of fructose polymers. The polymeric chain consists of repetitive units of a fructosyl moiety with glucosyl moieties as a chain terminator. Monomeric units are linked together by β (2,1) bonds as shown in Figure 1.17. The DP of polymeric chain in inulin varies from 2 to 60. The molecule having DP < 10 is termed as oligosaccharide, whereas particles having high DP are termed as inulin. It is indigestible in human being due to its β (2,1) bonds. On mixing with liquid, it forms gel alike fat. It enhances the stability of emulsion as well

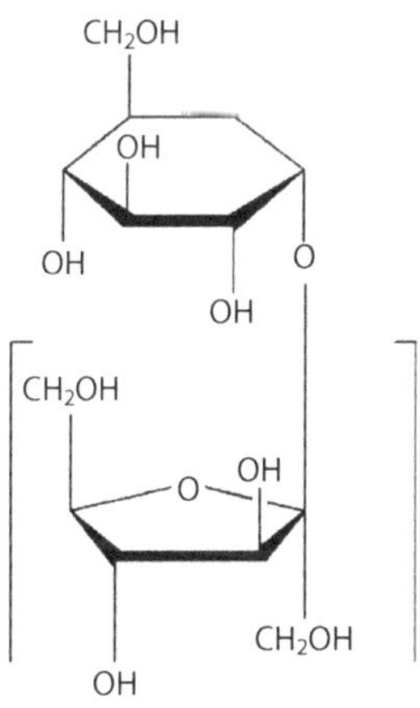


**Figure 1.17** Structure of inulin.

as foams. It is used as an ingredient in functional foods and also used in determining the filtration rate of glomerulus [36].

Zhang *et al.* [37] made the use of inulin in drug delivery application for injury of spinal cord. Modification of inulin with ibuprofen was carried out by direct esterification in which carboxylic acid was *in situ* activated with N,N′-carbonyldiimidazole. Pyrene was used as a fluorescence probe to determine the aggregation concentration of modified inulin. Copolymer of inulin and ibuprofen and methylprednisolone were used to prepare the Methylprednisolone loaded NPs having spherical shape with a diameter of 132 nm. Methylprednisolone loaded NPs possess great potential in the treatment of injured spinal cord.

## 1.9 Pectin

It is a heteropolysaccharide generally found in the cell wall of terrestrial plants. It is the main constituent of middle lamella. In the middle lamella, it helps in binding the cell together. It helps in the growth of plant. It is rich in galacturonic acid. It is mainly composed of D-galacturonic acid which are joined together by $\alpha \rightarrow 1(1 \rightarrow 4)$ glycosidic bonds and varying degree of carboxyl functional group methylesterified as shown in Figure 1.18. Henri Braconnot was the first who isolated it in 1825. Commercially, it is removed from citrus fruit and produced as light brown or white powder. It is found in various plants and its properties vary with source to source. The main property of pectin is its gel forming ability which enables it to use in food and pharmaceutical products. Pectin is generally used in food as gel forming and thickening agent in food such as jams and jellies. It is also used as stabiliser in food and cosmetics. It has application such as drug delivery agent and wound healing agent in biomedical field. Moreover, it is also used in medicine for increasing the stool viscosity for the treatment of constipation as well as diarrhoea.

Rebeiro *et.al.* [38] developed pectin coated LDH beads and used it as DDS for the treatment of colon disease. 5ASA, which is used as NSAID for

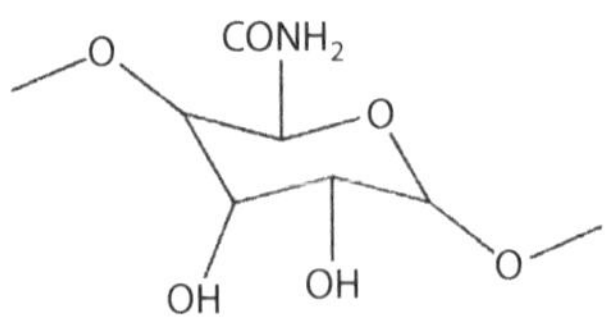


**Figure 1.18** Structure of pectin.

Crohn's and colitis disease, was used for delivery. The hybrid material and 5ASA was prepared via co-precipitation method by intercalating them in layered double-hydroxide form of $Mg_2Al$ and then incorporated into chitosan to utilise the benefit of its mocoadhesiveness. The compound obtained was processed into the beads and it was coated with pectin. Coating provides them the stability at the acidic pH of gastric fluid. The coated composite swells in water procuring the controlled release of 5ASA along its passage. The composite beads are a promising candidate to deliver 5ASA drug for colon disease treatment with minimal side-effects.

## 1.10 Metal Oxide

Metal oxides are significant and it undergoes the task of environmental remediation and serves as the best for electronics owing to their ability to liberate charge carriers on stimulation within the specific quantity of energy. Different types of metal oxide are explained in the upcoming part:

### 1.10.1 $TiO_2$

$TiO_2$ is also called as Titania and is naturally found in mineral form: rutile, ilmenite, and anatase. $TiO_2$ can be refined by different methods. EU uses sulphate method for the extraction of 70% of total $TiO_2$ while remaining 30% is extracted by chloride method. It can be manufactured in various sizes ranging from micrometre to nanometre scale. Arrangement of atoms in $TiO_2$ determines its crystal structure. It has the capability to reflect and scatter the incident light. It also possesses very high diffraction index. These properties make it suitable for use in pigment industry as a white pigment. It is also used in various industries such as paint, varnishes, plastics, textile, paper, and dyes. It holds E-number E171, hence it is used as an additive in food, cosmetics, toothpaste, and drugs. Production of $TiO_2$-NPs is less than 1% and is not used as an additive in food. Instead, it is used as an additive in suncreams, wood preservative, and textile fibres. It has good protection against UV-radiation which is harmful for skin hence used in suncream. The application of suncream containing $TiO_2$-NPs is easier than that containing microscale $TiO_2$. $TiO_2$ can be coated with other material to enhance its property. $TiO_2$-NPs possess very high catalytic activity owing to its high surface to volume ratio, hence used in photocatalysis process. It is hydrophilic in nature and forms thin film of water forms on its surface which carries away the pollutants along with it. Owing to this property, it is used for self-cleaning in tiles and paints and as an anti-fogging agent on glass pane coated with $TiO_2$. $TiO_2$-NPs also have application in dye-sensitised solar cell [39].

Today, because of intensive use of pesticides and chemicals as a warfare agent, a terrorist attack is more common and that demands quality and affordable protective material and equipment which having improved properties to protect against toxic chemicals. Keeping this in mind, Zeljko *et al.* [40] developed a 'smart textile' having ability to decontaminate the toxic chemical by incorporating the $TiO_2$-NPs onto the textile substrates. $TiO_2$-NPs were used because it can decompose toxic chemicals and also can be synthesised easily at a low price. It solves the problem and enables the textile to be more comfortable, self-cleaning, and highly durable.

### 1.10.2 ZnO

It is an inorganic material and appears as a white coloured powder. It is naturally found in mineral form 'Zincite' which contains other impurities and also imparts red or yellow colour to ZnO. Mostly, it is produced by synthetic methods and used as an additive in various products such as pigments, rubber, adhesive, food, lubricants, cement, and glass. It is a semiconductor material having wide band-gap of 3.3 eV at room temperature [41]. Owing to the presence of oxygen group, n-type doping is carried out in ZnO. It crystallises in two geometrical forms; cubic form is Zincblende, and hexagonal structure is Wurtzite which is the more stable one. It possesses various properties including wide band gap, luminescence, excellent transparency, and electron mobility. These features make it favourable for applications, including electronics (LED, transistors) and energy saving. It is synthesised in various nanostructural forms including nanorods, nanowires, nanoparticles, nanobelts, and tetrapods.

Davar *et al.* [42] synthesised ZnO-NPs using a green method in which zinc acetate, lemon juice, and sucrose were used as a precursor. They used these NPs as a photocatalyst for degradation of organic dyes including MB, Mr, and MO. ZnO-NPs synthesised by this method had a particle size of 21.5 nm and showed improved photo-catalytic activity. It can be used as a promising photocatalyst in textile industries for the degradation of organic dyes

### 1.10.3 $CeO_2$

It is also known as ceria or ceric oxide or cerium dioxide. It is an oxide form of cerium and appears as a pale yellow powder. It is obtained naturally from ores such as monazite and bastnaesite in which cerium is present in the mixture form with rare earth metal. Cerium oxide is obtained by adding oxidant to the mixture. As other rare earth metals in the mixture are resistant to oxidation, cerium gets separated as cerium oxide. $CeO_2$ can also be prepared by the calcination of cerium hydroxide or cerium oxalate. It has excellent optical and electrical properties with good thermal as well as chemical stability. It is used in semiconductor devices as a luminescent material. It exhibits excellent ionic conductivity for oxygen, high absorption for UV radiation, and excellent storage capability for oxygen [43]. It possesses unusual catalytic activity. It has applications in bio-field as biomedicine, drug delivery agent, bioscaffolding, and bioanalysis. $CeO_2$ at the nanoscale has various biological as well as engineering applications such as pharmacological agents, solar cell, fuel cell, and catalytic material [44].

Dudek [45] reported the synthesis of cerium oxide based electrolytes for the application in solid oxide fuel cell filled with solid carbon or gaseous hydrogen. Hydrothermal and Pechini methods were used for the synthesis of ceria based $CeO_2$–$Gd_2O_3$–SrO but the composite prepared by hydrothermal method gave better results. The prepared $CeO_2$ based electrolyte reduced the sintering temperature from 1500 °C to 125 °C. Addition of strontium in the composite electrolyte $Ce_{0.8-x}Gd_{0.2}SrxO_{2-\delta}$ improved its ionic conductivity which is more than that of $Ce_{0.8}Gd_{0.2}O_{2.}$

## 1.11 Polymer

The polymer can be defined as an essential chemical substance composed of a variety of repeated units. There are three different structural forms of polymer namely 1D, 2D, and 3D networks [46]. Types of polymer are explained in the upcoming section:

### 1.11.1 Polystyrene

Bonastre first classified polystyrene in 1831 resulting from distillation of Storax produced from 'Tree of turkey' as shown in Figure 1.19 [47]. Polystyrene has an ability to react easily with a variety of metal compounds namely nickel arsenate [48], titanium arsenate [49], cobalt arsenate [50], titanium-vanadium phosphate [51], and calcium phosphate [52] framing the metallic membrane. It has various fundamental properties such as ion conductivity [53], potential [54], cytotoxic [55], impedance [56], DC conductivity [57], thermodynamic [58], dielectric [59], and antibacterial [60].

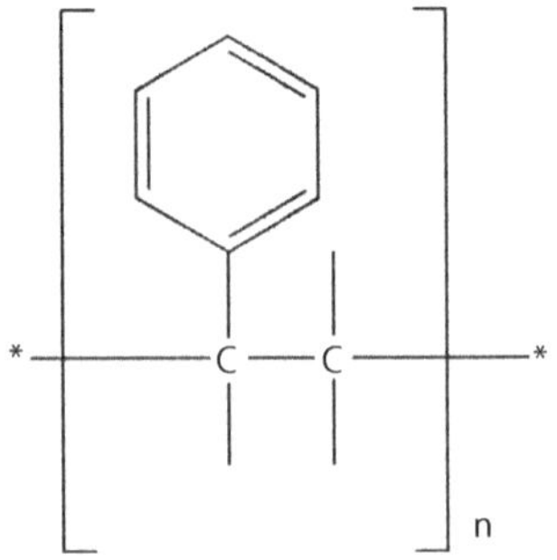


**Figure 1.19** Structure of polystyrene.

### 1.11.2 PANI

In 1835, PANI was considered as an 'aniline black' used for the fabrication via aniline oxidation as shown in Figure 1.20. PANI has been the interesting topic for scientist since it is recognised as a novel material in the technological fields due to processibility and stability. PANI has gathered the interest of the people from various parts of the country because of the variety of properties namely proton dopability, low cost, ease of synthesis, excellent redox recyclability, variable ionic conductivity [61], and photocatalytic [62].

### 1.11.3 Starch

Starches are polysaccharides and are basic sources of food which is derived from parts of the plants. It is the source of carbohydrate for human being as they consume it in daily to day life process as it contains 70–80% of calories. The structure of starch is given below in Figure 1.21. The starches undergo chemical modification and along with the reactive sites they transmit biologically active compound in the form of biocompatible carriers. They are even metabolised in the body of human being. The study of starch chemistry is on the pace of development as its technological aspects elaborate the capability of starch isolated from plants sources. They are commonly used for the environmental friendly applications [63].

### 1.11.4 Dendrimer

Tomalia in 1985 suggested about the word dendrimer originally derived from Greek term 'Dendron' which means 'the tree'. Dendrimer consists of branched synthetic polymer with spherical shape having the diameter that

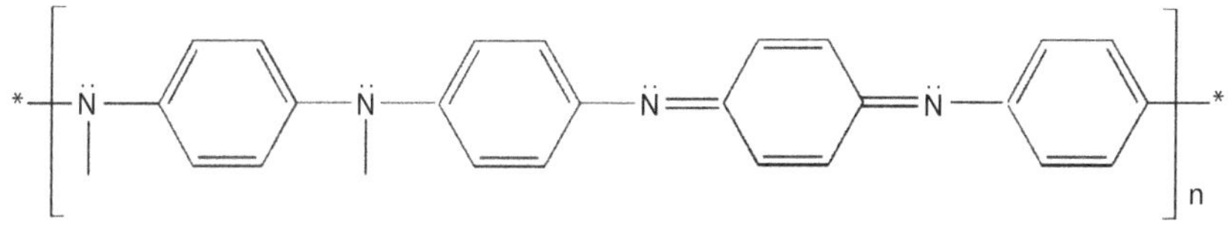


**Figure 1.20** Structure of PANI.

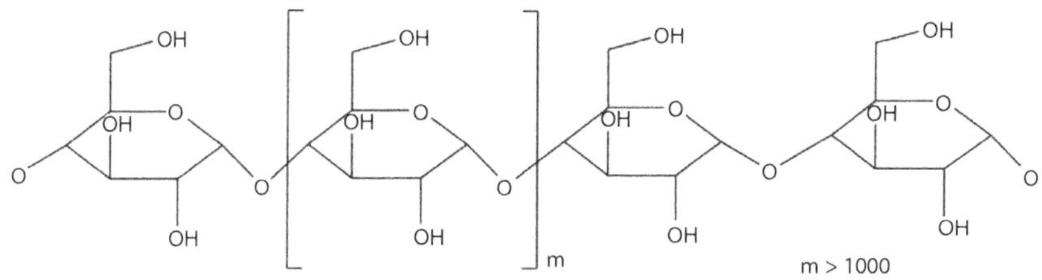


**Figure 1.21** Structure of starch.

ranges from 1.1 nm in the first generation to 9 nm in the eighth generation [64].

## 1.12 Bentonite

All the definition the literature related to bentonite clarifies that it is a rock term applied for explaining clay mineral, montmorillonite which occurs naturally and are finely grained. Different uses of bentonites are there in the industries which are resulted from swelling ability when it comes in contact with the water. When the large quantity of bentonite is added to the comparatively less quantity of water, the mixture formed from these addition forms the sticky lumps possessing the property of adhesiveness [65].

## 1.13 Conclusion

Advanced materials are the newly synthesised material or existing modified material having superior properties. It fulfils today's growing need for equipment with better quality for an extensive range of applications in various sectors such as paper, biomedical, and textile [66]. Several advanced materials having novel properties have been reported till date such as biomaterials, nanomaterials, metal oxide, and polymer to serve in the various fields. Some of them have natural origin such as plants, animals, minerals, and ore or extracted from plants and exist in different geometrical forms and have flexibility to form a composite with other materials for the specific application. In contrast, some are synthesised synthetically in required shape and size according to the demands. The superior properties of advanced material make them suitable for various forms.

## References

1. Kothari, D., Das, D., Patel, S., Goyal, A., Dextran and food application, in: *Polysaccharide,* pp. 1–16, Springer International Publishing, Switzerland, 2014.
2. Athar, S., Bushra, Arfin, T., Cellulose nanocrystals and PEO/PET hydrogel material in biotechnology and biomedicine: current status and future prospects, in: *Nanocellulose and nanohydrogel matrices: biotechnological and biomedical applications*, M. Jawaid., F. Mohammad (Eds.), pp. 139–173, Wiley-VCH, Germany, 2017.
3. Borkar, R., Waghmare, S.S., Arfin, T., Bacterial cellulose and polyester hydrogel matrices in biotechnology and biomedicine: current status and future prospects, in: *Nanocellulose and nanohydrogel matrices: biotechnological and biomedical applications,* M. Jawaid., F. Mohammad (Eds.), pp. 21–46, Wiley-VCH, Germany, 2017.
4. Khan, A.U., Malik, N., Arfin, T., Nanofibrillated cellulose and copoly(amino acid) hydrogel matrics in biotechnology and biomedicine, in: *Nanocellulose and nanohydrogel matrices: biotechnological and biomedical applications,* M. Jawaid., F. Mohammad (Eds.), pp. 331–352, Wiley-VCH, Germany, 2017.
5. Mohammad, F., Arfin, T., Al-Lohedan, H.A., Sustained drug release and electrochemical performance of ethyl cellulose-magnesium hydrogen phosphate composite. *Mater. Sci. Eng.* C 71, 735, 2017.
6. Arfin, T., Mohamamd, F., Electrochemical, antimicrobial and anticancer effects of ethyl cellulose-nickel (II) hydrogen phosphate. *Innovations Corrosion Mater. Sci.,* 6, 10, 2016.
7. Arfin, T., Mohamamd, F., Electrical conductivity, mechanical stability, antibacterial and anticancer activities of ethyl cellulose-tin (II) hydrogen phosphate. *Adv. Mater. Lett.,* 6, 1058, 2015.
8. Athar, S., Arfin, T., Commercial and prospective applications of gelatin, in: *Natural polymers: derivatives, blends and composites,* S. Ikram, S. Ahmed (Eds.), Vol. II, pp. 199–216, Nova Science Publishers, New York, 2017.
9. Rhim, J.-W., Physical and mechanical properties of water resistant sodium alginate films. *LWT-Food Sci. Technol.,* 37, 323, 2004.
10. Arfin, T., Chitosan and its derivatives: overlook of commercial application in diverse field, in: *Chitosan: derivatives, composites and applications,* S. Ahmed, S. Ikram (Eds.), pp. 115–150, Wiley-Scrivener Publishing, Massachusetts, 2017.

11. Arfin, T., Mohammad, F., Chemistry and structural aspects of chitosan towards biomedical applications, in: *Natural polymers: derivatives, blends and composites*, S. Ikram, S. Ahmed (Eds.), Vol. I, pp. 265–280, Nova Science Publishers, New York, 2016.
12. Weltrowski, M., Martel, B., Morecellet, M., Chitosan N-benzyl sulfonate derivatives as sorbents for removal of metal ions in an acidic medium. *J. Appl. Polym. Sci.*, 59, 647, 1996.
13. Bender, A.E., Palgrave, J.A., Doell, B.H., A collaborative test of Moore and Stein's resin-chromatographic method for determining amino acids. *Analyst*, 84, 526, 1959.
14. Rekha, M.R., Sharma, C.P., Pullulan as a promising biomaterial for biomedical applications: a perspective. *Trends Biomater Artif Organs*, 20, 116, 2007.
15. Zhang, R., Edgar, K.J., Properties, chemistry, and applications of the bioactive polysaccharide curdlan. *Biomacromolecules*, 15, 1079, 2014.
16. Yoshida, T., Hatanaka, K., Uryu, T., Kaneko, Y., Suzuki, E., Miyano, H., Mimura, T., Yoshida, O., Yamamoto, N., Synthesis and structural analysis of curdlan sulfate with a potent inhibitory effect *in vitro* of AIDS virus infection. *Macromolecules*, 23, 3717, 1990.
17. Verma, S.R., Dwivedi, U.N., Lignin genetic engineering for improvement of wood quality: applications in paper and textile industries, fodder and bioenergy production. *S. Afr. J. Bot.*, 91, 107, 2014.
18. Naik, S.N., Goud, V.V., Rout, P.K., Dalai, A.K., Production of first and second generation biofuels: a comprehensive review. *Renew. Sustain. Energy Rev.*, 14, 578, 2010.
19. Katzbauer, B., Properties and applications of xanthan gum. *Polymer Degrad. Stab.*, 59, 81, 1998.
20. Kwan, R.C.H., Chan, C., Renneberg, R., An amperometric biosensor for determining amino acids using a bienzymatic system containing amion acid oxidase and protease. *Biotechnol. Lett.*, 24, 1203, 2002.
21. Oliveira, E.E., Silva, A.E., Nagashima Jr, T., Gomes, M.C.S., Aguiar, L.M., Marcelino, H.R., Araujo, I. B., Bayer, M. P., Ricardo, N. M. P. S., Oliveira, A. G., Egito, E. S. T., Xylan from corn cobs, a promising polymer for drug delivery: Production and characterization. *Bioresour. Technol.*, 101, 54025406, 2010.
22. Saha, B.C., Alpha-L-arabinofuranosidases: Biochemistry, molecular biology and application in biotechnology. *Biotechnol. Adv.*, 18, 403, 2000.
23. Ibrahim, O.B., Osman, M.E., Hassan, E.A., Characterization and simple fractionation of Acacia Senegal. *J. Chem. Acta.*, 2, 11, 2013.
24. Onwudiwe, D.C., Arfin, T., Strydom, C.A., Synthesis, characterization, and dielectric properties of N-butyl aniline capped CdS nanoparticles. *Electrochim. Acta.*, 116, 217, 2014.
25. Dumbrava, A., Badea, C., Prodan, G., Ciupina, V., Synthesis and Characterization of Cadmium Sulfide Obtained at Room Temperature. *Chalcogenide Lett.*, 7, 111, 2010.

26. Yang, J., Yu, J., Fan, J., Sun, D., Tang, W., Yang, X., Biotemplated preparation of CdS nanoparticles/bacterial cellulose hybrid nanofibers for photocatalysis application. *J. Hazard Mater.*, 189, 377, 2011.
27. Jung, J.H., Hwang, G.B., Lee, J.E., Bae, G.N., Preparation of airborne Ag/CNT hybrid nanoparticles using an aerosol process and their application to antimicrobial air filtration. *Langmuir*, 27, 10256, 2011.
28. Mohammad, F., Arfin, T., Thermodynamics and electrochemical characterization of core-shell type gold-coated superparamagnetic iron oxide nanoparticles. *Adv. Mater. Lett.*, 5, 315, 2014.
29. Mohammad, F., du Plessis L., Arfin, T., X-ray analysis of metal oxide-metal core shell nanoparticles, in: *X-ray diffraction: Structure, principles and applications*, K. Shih(Ed.), pp. 161–181, Nova Science Publishers, New York, 2013.
30. Dave, P.N., Chopda, L.V., Application of iron oxide nanomaterials for the removal of heavy metals. *J. Nanotechnol.*, 2014, 398569, 2014
31. Arfin, T., Rangari, S.N., Graphene oxide-ZnO nanocomposite modified electrode for the detection of phenol. *Anal. Methods*, 10, 3, 347
32. Gupta, S.S., Sreeprasad, T.S., Maliyekkal, S.M., Das, S.K., Pradeep, T., Graphene from sugar and its application in water purification. *ACS Appl. Mater. Interfaces*, 4, 4156, 2012.
33. Stankovich, S., Dikin, D.A., Piner, R.D., Kohlhaas, K.A., Kleinhammes, A., Jia, Y., Wu, Y., Nguyen, S.T., Ruoff, R.S., Synthesis of graphene-based nanosheets via chemical reduction of exfoliated graphite oxide. *Carbon* 45, 1558, 2007.
34. Arfin, T., Bushra, R., Mohammad, F., Electrochemical sensor for the sensitive detection of o-nitrophenol using graphene oxide-poly (ethyleneimine) dendrimer-modified glassy carbon electrode. *Graphene Technol.*, 1, 1, 2016.
35. Liu, F., Chung, S., Oh, G., Seo, T.S., Three-dimensional graphene oxide nanostructure for fast and efficient water-soluble dye removal. *ACS Appl. Mater. Interfaces*, 4, 922, 2012.
36. Roberfroid, M.B., Functional foods: concepts and application to inulin and oligofructose. *Br. J. Nutr.*, 87, S139, 2002.
37. Zhang, L., Li, Y., Wang, C., Li, G., Zhao, Y., Yang, Y., Synthesis of methylprednisolone loaded ibuprofen modified inulin based nanoparticles and their application for drug delivery. *Mater. Sci. Eng. C*, 42, 111, 2014.
38. Ribeiro, L.N., Alcântara, A.C., Darder, M., Aranda, P., Araújo-Moreira, F.M., Ruiz-Hitzky, E., Pectin-coated chitosan–LDH bionanocomposite beads as potential systems for colon-targeted drug delivery. *Int. J. Pharm.*,463, 1, 2014.
39. Rezvani, F., Parvazian, E., Hosseini, S.A., Dye-sensitized solar cells based on composie $TiO_2$ nanoparticle-nanorod single and bi-layer photoelectrodes. *Bull. Mater. Sci.*, 39, 1397, 2016.
40. Senić, Ž., Bauk, S., Vitorović-Todorović, M., Pajić, N., Samolov, A., Rajić, D., Application of $TiO_2$ nanoparticles for obtaining self-decontaminating smart textiles. *Sci. Techn. Rev.*, 61, 63, 2011.

41. Onwudiwe, D.C., Arfin, T., Strydom, C.A., Surfactant mediated synthesis of ZnO nanospheres at elevated temperature, and their dielectric properties. *Superlattices Microstruct.,* 81, 215, 2015.
42. Davar, F., Majedi, A., Mirzaei, A., Green synthesis of ZnO nanoparticles and its application in the degradation of some dyes. *J. Am. Ceram Soc.,* 98, 1739, 2015.
43. Babitha, K.K., Sreedevi, A., Priyanka, K.P., Sabu, B.,Varghese, T., Structural characterization and optical studies of CeO2 nanoparticles synthesized by chemical precipitation. *Indian J. Pure Ap. Phy.,* 53, 596, 2015.
44. Mohammad, F., Arfin, T., Al-Lohedan, H.A., Enhanced biological activity and biosorption performance of trimethyl chitosan-loaded cerium oxide particles. *J. Ind. Eng. Chem.,* 45, 33, 2017.
45. Dudek, M., Ceramic electrolytes in the $CeO_2$–$Gd_2O_3$–SrO system-preparation, properties and application for solid oxide fuel cells. *Int. J. Electrochem. Sci.,* 7, 2874, 2012.
46. Arfin, T., Tarannum, A., Polymer materials: from the past to the future, in: *Green polymeric materials: advances and sustainable development,* S. Ahmed, Annu, S. Ikram (Eds.), pp. 35–52, Nova Science Publishers, New York, 2017.
47. Arfin, T., Mohammad, F., Yusof, N.A., Applications of polystyrene and its role as a base in industrial chemistry, in: *Polystyrene: synthesis, characteristics and applications,* C. Lynwood (Ed.), pp. 269–280, Nova Science Publishers, New York, 2014.
48. Arfin, T., Rafiuddin., Transport studies of nickel arsenate membrane. *J. Electroanal. Chem.,* 636, 113, 2009.
49. Arfin, T., Rafiuddin., Electrochemical properties of titanium arsenate membrane. *Electrochim. Acta,* 54, 6928, 2009.
50. Arfin, T., Yadav, N., Impedance characteristics and electrical double-layer capacitance of composite polystyrene–cobalt–arsenate membrane. *J. Ind. Eng. Chem.,* 19, 256, 2013.
51. Arfin, T., Jabeen, F., Kriek, R.J., An electrochemical and theoretical comparison of ionic transport through a polystyrene based titanium–vanadium (1: 2) phosphate membrane. *Desalination,* 274, 206, 2011.
52. Arfin, T., Fatima, S., Conductometric studies with polystyrene calcium phosphate membrane. Asian J. Adv. Basic Sci., 2, 1, 2013.
53. Arfin, T., Rafiuddin, Thermodynamics of ion conductivity of alkali halides across a polystyrene-based titanium arsenate membrane. *Electrochim. Acta,* 55, 8628, 2010.
54. Arfin, T., Rafiuddin, An electrochemical and theoretical comparison of ionic transport through a polystyrene-based cobalt arsenate membrane. *Electrochim. Acta,* 56, 7476, 2011.
55. Mohammad, F., Arfin, T., Cytotoxic effects of polystyrene–titanium–arsenate composite in cultured H9c2 cardiomyoblasts. *Bull. Environ. Contam. Toxicol.,* 91, 689, 2013.

56. Arfin, T., Yadav, N., Impedance characteristics and electrical double layer capacitance of polystyrene based nickel arsenate membrane. *Anal. Bioanal. Electrochem.*, 4, 135, 2012.
57. Arfin, T., Mohammad, F., DC electrical conductivity of nano-composite polystyrene–titanium–arsenate membrane. *J. Ind. Eng. Chem.*,19, 2046, 2013.
58. Arfin, T., Rafiuddin., Metal ion transport through a polystyrene-based cobalt arsenate membrane: Application of irreversible thermodynamics and theory of absolute reaction rates. *Desalination*, 284,100, 2012.
59. Arfin, T., Mohammad, F., Electrochemical, dielectric behaviour and *in vitro* antimicrobial activity of polystyrene-calcium phosphate. *Adv. Ind. Eng. Manag.*, 3, 25, 2014.
60. Arfin, T., Fatma, S., Synthesis, Influence of Electrolyte Solutions on Impedance Properties and *In-vitro* Antibacterial Studies of Organic-inorganic Composite Membrane. *Adv. Ind. Eng. Manag.*, 3, 19, 2014.
61. Arfin, T., Bushra, R., Kriek, R.J., Ionic conductivity of alkali halides across a polyaniline-zirconium (IV)-arsenate membrane. *Anal. Bioanal. Electrochem.*, 5, 206, 2013.
62. Bushra, R., Arfin, T., Oves, M., Raza, W., Mohammad, F., Khan, M.A., Ahmad, A., Azam, A., Muneer, M., Development of PANI/MWCNTs decorated with cobalt oxide nanoparticles towards multiple electrochemical, photocatalytic and biomedical application sites. *New J. Chem.*, 40, 9448, 2016.
63. Mogarkar, P.R., Arfin, T., Chemical and structural importance of starch based derivative and its applications, in: *Natural polymers: derivatives, blends and composites,* S. Ikram, S. Ahmed (Eds.), Vol. II, pp. 73–87, Nova Science Publishers, New York, 2017.
64. Arfin, T., Mohammad, F., Dendrimer and its role for the advancement of nanotechnology and bioengineering, in: *Advances in materials science research,* M.C. Wythers (Ed.), Vol. 21, pp. 157–174, Nova Science Publishers, New York, 2015.
65. Arfin, T., Applications of bentonite and its role as a base of engineering chemistry, in: *Bentonite: characteristics, uses and implications for the environment,* A.K. Mishra (Ed.), pp. 127–136, Nova Science Publishers, New York, 2017.
66. Mohammad, F., Arfin, T., Saba, N., Jawaid, M., Al-Lohedan, H.A., Electrical conductivity and biological efficacy of ethyl cellulose and polyaniline-based composites, in: *Electrically conductive polymers and polymer composites,* A. Khan, M. Jawaid, A.A.P. Khan, A.M. Asiri (Eds.), pp. 181–198, Wiley-VCH, Germany, 2018.

2

# Characterization of Green and Sustainable Advanced Materials

**Pintu Pandit*** **and Gayatri T. Nadathu**

*Institute of Chemical Technology, Department of Fibres and Textile Processing Technology, Mumbai, MH, India*

***Abstract***

Advanced materials refer to all new materials and modifications to existing materials to obtain superior performance in one or more characteristics that are critical for the application under consideration. They can also exhibit completely novel properties. Advanced materials typically have properties that are superior to and outperform conventional materials in their applications. Advanced materials are designed for targeted properties. Both completely new materials such as graphene or high-temperature superconductors and those that are next-level iterations on traditional materials such as alloys or composites may be described as advanced materials. Such materials show novel or improved structural (strength, hardness, flexibility) and/or functional properties (electronic, magnetic, optical). Advanced Characterization and Testing of advanced materials explores developments in physical and chemical testing and specific high-performance analysis of advanced materials. Details of specific application properties and associated testing for each characteristic are discussed. This chapter reflects a critical, comparative, and condensed overview of major analytical techniques for materials characterization with emphasis on industrial applications.

***Keywords:*** Physical characterization, chemical characterization, graphene, nanocomposite, advanced materials

**Corresponding author*: pintupanditict@gmail.com

Shakeel Ahmed and Chaudhery Mustansar Hussain (eds.) Green and Sustainable Advanced Materials, Vol. 1 (35–66) © 2018 Scrivener Publishing LLC

## 2.1 Introduction

The advanced materials industry encompasses a full life cycle from material extraction, primary production, process development, and materials characterization to product fabrication, testing, use and end-of-life waste management, and recycling. Advanced materials are nowadays used in various industrial and scientific contexts, owing to their particular and sometimes unique properties. In many cases, those properties are an output of the composition of the materials. An integral part of the characterization of advanced materials is therefore to determine their precise elemental composition as well as to detect possible contaminants. This knowledge allows the suitable optimization of the production as well as the properties of the final products [1]. New applications ranging from energy storage, optoelectronics, photocatalysis harvesting and conversion, the miniaturization of existing systems, improved energy-conversion efficiencies, and the development of eco-friendly technologies necessarily require that material properties to be improved and tuned to a significant level.

Physical and chemical characterization can be used to measure the materials structure and properties. It is a fundamental process in the field of materials science that helps to clear our scientific understanding of engineering of advanced materials. The scope of the physical or chemical characterization often differs as per end application of different advanced techniques that reflects the study of the microscopic structure and properties of advanced materials while others include macroscopic techniques that encompass mechanical testing, thermal analysis, and density calculation. The scale of the structures observed in advanced material characterization ranges from angstrom to centimeters [2–5].

Many advanced characterization techniques have been practiced since long, like basic optical microscopy, though new techniques and methodologies are constantly emerging. Advancement in the field of the electron microscope and secondary ion mass spectrometry has revolutionized the field and allowed the imaging and analysis of advanced structures and compositions on much smaller scales than was previously possible. This led to a huge increase in the level of understanding and clearer queries regarding the behavior and issues related to different advanced materials that show varying properties and behaviors [5]. Recently, atomic force microscopy has further increased the maximum possible resolution for analysis of certain advanced materials is possible [6, 7]. The constant evolution of material science and technology challenge existing methods of chemical science metrology and lead to an evolving development and application of improved analytical methods. Measurement capabilities for chemical compositional characterization are applied to detect elemental signatures, quantify elemental mass, and map its physical distribution.

In the field of advanced fibrous materials, the ability to better control the quality and performance of output product came thanks to the evolution of raw materials available for textile manufacturing ranging from natural fibers with very high inherent variability to synthetic fibers produced with commodity polymers to high performance fibers and gain in precision of production techniques and equipment [8–12]. This gave rise to large efforts to develop test methods that could characterize the performance of textile properties including physical, mechanical, chemical, barrier, and thermal properties as well as comfort, flammability, and durability. Test methods also exist to assess the nature of the functional additive used and its efficiency as antibacterial, insect-repellent, stain-free, and ultraviolet (UV) protective additive and/or finish [10–20]. As sustainable manufacturing has become a concern for all the advanced materials, characterization methods have been developed to quantify the sustainability and also test the biodegradable product and perform its lifecycle assessment [10–12, 18].

The advancement in characterization and materials may allow products designed to protect us from mechanical hazards, chemicals, biological agents, extreme temperatures, fire, impacts, electric shocks, drowning, or falls for instance. They may also help identify, see, locate, or on the contrary conceal. Characterization of advanced materials like graphene, which represents a member of an increasingly populated class of carbon allotropes, was motivated by developments in preparing carbon nanotubes as rolled up sheets of graphene. These materials include graphene nanoribbons, graphene nanosheets, graphene nanoflakes, and solar cells (refer Figure 2.1). They are most effective for electrically conductive composites [21–25].

Research into phase relations within multi-component systems, determination of crystal structures and defects, studies of the mechanisms and kinetics of the chemical reactions that take place during the synthesis of advanced materials, studies of the phase transformations, and microstructure development of polycrystalline materials, as well as the morphology development of nanostructured materials reveal the relations between these parameters and the resulting properties of the materials. Such knowledge could lead to new functional properties and consequently to the development of new applications. The essential requirements for such investigations are the tools of characterization and manipulation of the advanced materials with control of structure, size, and shape at the micro/nano level.

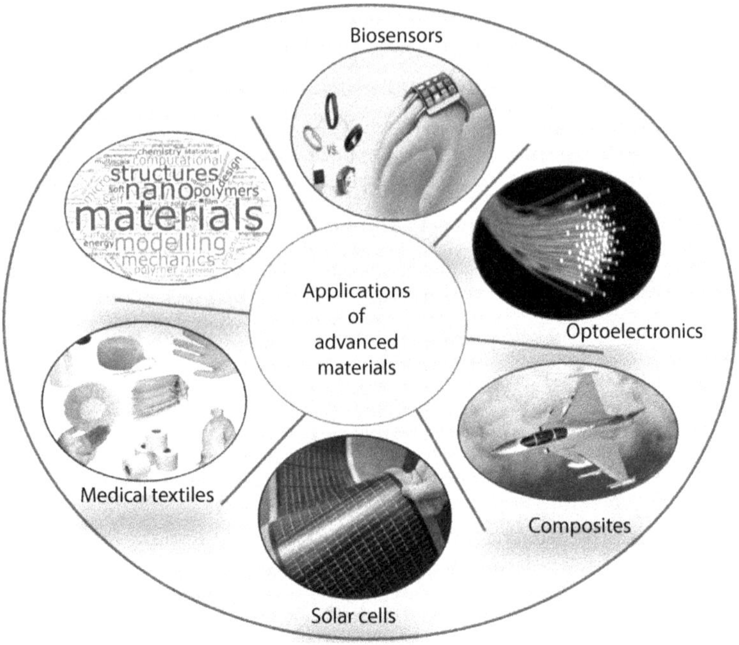


**Figure 2.1** Applications of advanced materials.

Recently, standards have rapidly been developed to guide manufacturers in characterizing advanced materials in a consistent manner. In order to differentiate the performance of the materials, adapted test methods are needed that allow demonstration of their benefits during end application activities and thus help manufacturers prove that these are not mere marketing claims but also benefit the environment by informing the customer in a lucid manner. The test methods assess the interactions between materials and end application with respect to the user [26, 27].

## 2.2 Characterization of Advanced Materials

Characterization of advanced materials has been classified into two parts: (a) physical characterization and (b) chemical characterization in order to understand its properties and important applications. This usually means measuring the Performance properties in terms of physical/mechanical and chemical process (refer Figure 2.2). Characterization needs to be accurate enough to fulfill criteria of quality control such as consistency and research & development.

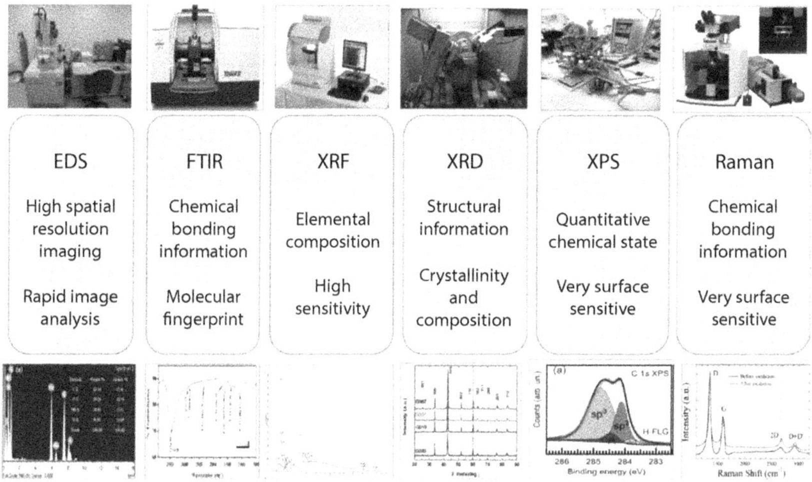


**Figure 2.2** Physico-chemical characterization of advanced materials (graphene).

Quality control is necessary for materials, and it defines a set of activities or techniques whose purpose is to ensure that all quality requirements are being met by monitoring of processes and solving performance problems through inspecting and testing. Testing of materials is normally carried out according to a "STANDARD." The standard follows the instruction and methods such as American Association for textile Chemists and Colorists (AATCC), American Society for Testing and Materials (ASTM), American National Standard Institute (ANSI), British Standard Institute (BSI), ISO, BIS, Japanese Industrial Standard (JIN), and Canada Standard Association (CSA).

## 2.3 Physical Characterization of Advanced Materials

Physical methods of characterization techniques are a category that probes and maps the surface and sub-surface structure of materials. These techniques can use photons, electrons, ions, or physical cantilever probes to gather data about a sample's structure on a range of length scales. It is done with help of microscopy such as optical microscope, scanning electron microscope (SEM), transmission electron microscope (TEM), field ion microscope (FIM), scanning tunneling microscope (STM), scanning probe microscopy (SPM), atomic force microscope (AFM), and X-ray diffraction topography (XRT).

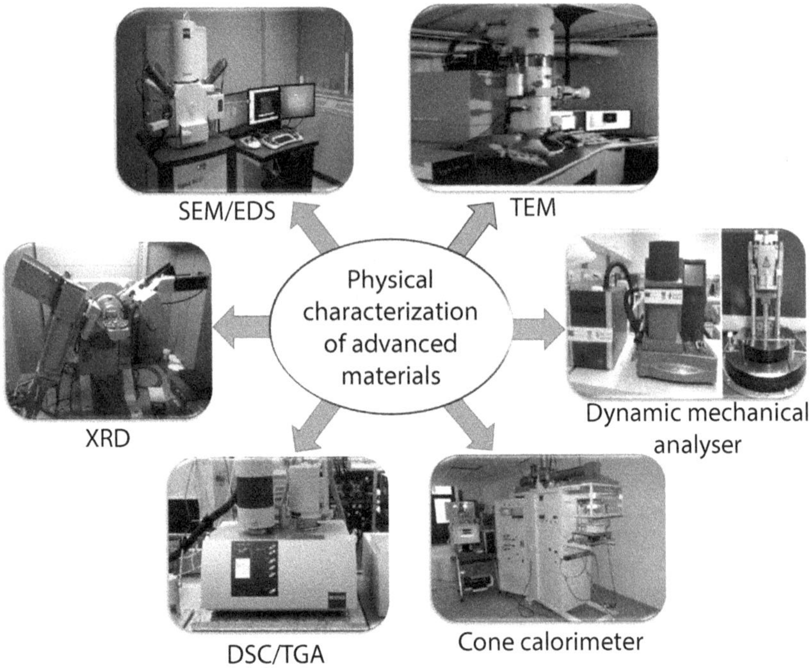


**Figure 2.3** Physical characterization testing instruments for advanced materials.

Spectroscopy can also be used as techniques for characterization based on principles that reveal the chemical composition, crystal structure composition variation, and photoelectric properties of materials. Some common instrumentation methods (refer Figure 2.3) include energy-dispersive X-ray spectroscopy (EDX), wavelength dispersive X-ray spectroscopy (WDX), X-ray diffraction (XRD), mass spectrometry, nuclear magnetic resonance spectroscopy (NMR), secondary ion mass spectrometry (SIMS), electron energy loss spectroscopy (EELS), Auger electron spectroscopy, X-ray photoelectron spectroscopy (XPS), Fourier transform infrared spectroscopy (FTIR), photoluminescence (PL), thermoluminescence (TL), photon correlation spectroscopy/dynamic light scattering (DLS), terahertz spectroscopy, small-angle X-ray scattering (SAXS), small-angle neutron scattering (SANS), and X-ray photon correlation spectroscopy (XPCS).

Macroscopic characterization techniques are used to characterize various properties of materials, for example, by mechanical testing, which includes tensile, compressive, torsional, creep, fatigue, toughness, and hardness testing. Other macroscopic characterizations include differential thermal analysis (DTA), dielectric thermal analysis, thermogravimetric analysis (TGA), differential scanning calorimetry (DSC), impulse excitation technique (IET), ultrasound techniques including resonant ultrasound spectroscopy, and time domain ultrasonic testing methods.

### 2.3.1 Scanning Electron Microscopy

SEM produces images of a sample by scanning the surface with a focused beam of electrons. In the sample electrons interact with atoms, producing various signals from sample's surface topography and composition that contain information about the surface of the sample. The electron beam is scanned in a raster scan pattern and the beam's position is combined with the detected signal to produce an image. SEM can have resolution better than 1 nm. Specimens can be observed in high vacuum in conventional SEM or in low vacuum or wet conditions in variable pressure or environmental SEM and at a wide range of cryogenic or elevated temperatures with specialized instruments [28].

The most common SEM mode is the detection of secondary electrons emitted by atoms excited by the electron beam. The number of secondary electrons that may be detected depends on specimen topography. By scanning the sample and collecting the secondary electrons that are emitted using a special detector, an image displaying the topography of the surface is created. For conventional imaging in SEM, specimens must be electrically conductive at least at the surface and electrically grounded to prevent the accumulation of electrostatic charge. Non-conducting materials are usually coated with an ultrathin coating of electrically conducting material deposited on the sample either by low-vacuum sputter coating or by high-vacuum evaporation. Conductive materials in current use for specimen coating include gold, gold/palladium alloy, platinum, iridium, tungsten, chromium, osmium [29], and graphite. For SEM, a specimen is normally required to be completely dry, since the specimen chamber is under high vacuum. Hard, dry materials such as wood, bone, and feathers can be examined with little further preparation.

### 2.3.2 Energy-Dispersive X-Ray Spectroscopy

Energy-dispersive X-ray spectroscopy (EDS) is an analytical technique used for the elemental analysis or chemical characterization of a sample. It relies on an interaction of some source of X-ray excitation and a sample. Its characterization capabilities are due in large part to the fundamental principle that each element has a unique atomic structure allowing a unique set of peaks on its electromagnetic emission spectrum [30] (which is the main principle of spectroscopy). To stimulate the emission of characteristic X-rays from a specimen, a high-energy beam of charged particles such as electrons or protons, or a beam of X-rays, is focused into the sample being studied. At rest, an atom within the sample contains ground state (or unexcited) electrons in discrete energy levels or electron shells bound to the nucleus. The incident beam may excite an electron in an inner shell, ejecting it from the shell while creating an electron–hole where the electron was. An electron from an outer, higher-energy shell then fills the hole, and the difference in energy between the higher-energy shell and the lower energy shell may be released in the form of an X-ray. The number and energy of the X-rays emitted from a specimen can be measured by an energy-dispersive spectrometer. As the energies of the X-rays are characteristic of the difference in energy between the two shells and of the atomic structure of the emitting element, EDS allows the elemental composition of the specimen to be measured [31].

EDS can be used to determine the chemical elements that are present in a sample and can be used to estimate their relative abundance. The accuracy of this quantitative analysis of sample composition is affected by various factors. Many elements will have overlapping X-ray emission peaks (e.g., Ti Kβ and V Kα, Mn Kβ and Fe Kα). The accuracy of the measured composition is also affected by the nature of the sample. The likelihood of X-ray escaping the specimen and thus its availability to be detected and measured depend on the energy of the X-ray and the composition, amount, and density of material it has to pass through, to reach the detector. Owing to this, accurate estimation of the sample composition from the measured X-ray emission spectrum requires the application of quantitative correction procedures, referred to as matrix corrections [32, 33]. Plasma treated surfaces, nanocoatings on different base materials, etc. have all been examined by SEM/EDS [8].

### 2.3.3 Transmission Electron Microscopy

TEM is a microscopy technique in which a beam of electrons is transmitted through a specimen to form an image. The specimen is most often an ultrathin section less than 100 nm thick or a suspension on a grid. An image is formed from the interaction of the electrons with the sample as the beam is transmitted through the specimen. The image is then magnified and focused onto an imaging device such as a fluorescent screen, a layer of photographic film, or a sensor such as a charge-coupled device. TEMs find application in cancer research, virology, and materials science as well as pollution, nanotechnology, and semiconductor research. At lower magnifications, TEM image contrast is due to differential absorption of electrons by the material due to differences in composition or thickness of the material. At higher magnifications, complex wave interactions modulate the intensity of the image, requiring expert analysis of observed images. Alternate modes of use allow for the TEM to observe modulations in chemical identity, crystal orientation, electronic structure, and sample induced electron phase shift as well as the regular absorption based imaging. High-quality samples will have a thickness that is comparable to the mean free path of the electrons that travel through the samples that may be only a few tens of nanometers. Preparation of TEM specimens is specific to the material under analysis and the type of information to be obtained from the specimen. Materials that have dimensions small enough to be electron transparent, such as powdered substances, small organisms, viruses, or nanotubes, can be quickly prepared by the deposition of a dilute sample containing the specimen onto films on support grids [34, 35].

### 2.3.4 X-Ray Diffraction

XRD is a technique used for determining the atomic and molecular structure of a crystal, in which the crystalline atoms cause a beam of incident X-rays to diffract in many specific directions. By measuring the angles and intensities of these diffracted beams, a crystallographer can produce a three-dimensional picture of the density of electrons within the crystal. From this electron density, the mean positions of the atoms in the crystal can be determined as well as their chemical bonds, their disorder, and various other information.

As many materials such as salts, metals, minerals, and semiconductors as well as various inorganic, organic, and biological molecules can form crystals, X-ray crystallography has been fundamental in the development of many scientific fields. In its first decades of use, this method determined the size of atoms, the lengths and types of chemical bonds, and the atomic-scale differences among various materials, especially minerals and alloys. The method also revealed the structure and function of many biological molecules, including vitamins, drugs, proteins, and nucleic acids such as DNA. X-ray crystallography is still the chief method for characterizing the atomic structure of new materials and in discerning materials that appear similar to others. X-ray crystal structures can also account for unusual electronic or elastic properties of a material, shed light on chemical interactions and processes, or serve as the basis for designing pharmaceuticals against diseases.

X-ray crystallography is related to several other methods for determining atomic structures. Similar diffraction patterns can be produced by scattering electrons or neutrons that are likewise interpreted by Fourier transformation. If single crystals of sufficient size cannot be obtained, various other X-ray methods can be applied to obtain less detailed information; such methods include fiber diffraction, powder diffraction, and (if the sample is not crystallized) SAXS. If the material under investigation is only available in the form of nanocrystalline powders or suffers from poor crystallinity, the methods of electron crystallography can be applied for determining the atomic structure [36, 37].

### 2.3.5 Ultraviolet Protection

The sun supports life on our planet, but its life-giving rays also pose dangers in the form of UV radiation. Artificial sources such as welding machines, tanning beds, and lasers can also produce UV radiation. There are three types of UV radiation. UV-C is absorbed by the ozone layer and does not present any threat. However, UV-A and UV-B radiation can have long and short term negative effects on the skin and visual ability.

The ultraviolet protection factor (UPF) values of the textile and polymeric materials were measured using a spectrophotometer in the range of 280–400 nm. The UPF value of sample was determined from the total spectral transmittance. The instrument on which this test could be done is Shimadzu UV-2600 spectrophotometer based on AS/NZS 4399:1996 method for textile materials. The procedures have to be done as per ASTM D6544 (preparation), AATCC TM183 (testing), and ASTM D6603 (labeling) [10, 15].

| Ultraviolet protection factor (UPF) Range | Effective UV transmission | Protection category |
|---|---|---|
| 15–24 | 6.7–4.2% | Good |
| 25–39 | 4.1–2.6% | Very Good |
| 40–50, 50+ | ≤2.5% | Excellent |

### 2.3.6 Thermal Characterization (TGA, DTA, DSC, Cone Calorimetry)

Thermal characterization is essential to understand the mechanism of thermal degradation, flame, and fire formation in order to protect materials so that the rate of fire development can be minimized. Heating a material at steady rate can produce chemical changes such as oxidation and degradation and/or physical changes such as the glass transition in polymers, conversion in ceramics, and phase change in metals. The various techniques of thermal analysis measure one or more physical properties of a sample as a function of temperature. There are generally three basic methods of thermal analysis, namely TGA, DTA, and DSC. They can apply programmed heating and cooling but usually operate at a slowly rising temperature. The sample chamber may contain air, oxygen, nitrogen, argon, etc. or be evacuated. Recently developed methods have extended the range of thermal analysis, and other aspects of behavior can now be studied. For instance, dynamic mechanical thermal analysis (DMTA) mechanical as well as structural information can be obtained on the viscoelastic response of a polymeric sample to tensile, bend, or shear stresses during heating.

Standard methods can evaluate the burning behavior of both untreated and treated samples. For the determination of flammability, limiting oxygen index (LOI) analysis, IS 13501:1992, for textiles test procedure was used. As per standard, flame contact time was kept as 30 s; specimen size was 6 × 4 $cm^2$, oxygen and nitrogen flow meter in $cm^3$/min was set as per LOI value with a 38 mm length of the flame. For LOI of 25, oxygen flow meter was set as 77 (4275 $cm^3$/min) and nitrogen flow meter was set as 108 (12825 $cm^3$/min). The results are expressed as follows:

$$\text{LOI}(\%) = \frac{O_2}{O_2 + N_2} \times 100$$

In vertical flammability, the different parameters were measured as per ASTM D 6413-09 standard test method for flame resistance of textiles. For example, specimen size was 30 × 7.6 $cm^2$ and flame height 38 mm. The cut edge of the fabric on the bottom was exposed to a controlled flame for 12 s. After exposure to the flame, after flame, afterglow, and char length were measured [10, 11, 16].

#### 2.3.6.1 *Thermogravimetric Analysis*

Thermogravimetric analysis or thermal gravimetric analysis (TGA) is a method of thermal analysis in which the mass of a sample is measured over time as the temperature changes. This measurement provides information about physical phenomena such as phase transitions, absorption, and desorption, as well as chemical phenomena including chemisorption, thermal decomposition, and solid–gas reactions (e.g., oxidation or reduction) [38].

In a balance, mass of the sample is continuously determined and recorded while the sample is being slowly heated. Temperature up to at least 1000 **°C** is available. It has been applied to the decomposition of rubbers, kinetic studies of metallic oxidation, glass transitions, and softening in polymers. Equilibrium is not attained within the sample, and the method is insensitive to the more subtle solid-state changes.

In a desired temperature range, if a species is thermally stable, there will be no observed mass change. Negligible mass loss corresponds to little or no slope in the TGA trace. TGA also gives the upper use temperature of a material. Beyond this temperature, the material will begin to degrade. TGA is used in the analysis of ceramics and thermally stable polymers. Ceramics usually melt before they decompose as they are thermally stable over a large temperature range, thus TGA is mainly used to investigate the thermal stability of polymers. Most polymers melt or degrade before 200 °C. However, there is a class of thermally stable polymers that are able to withstand temperatures of at least 300 °C in air and 500 °C in inert gases without structural changes or strength loss, which can be analyzed by TGA. For example, a combustion reaction could be tested by loading a sample into a thermogravimetric analyzer at normal conditions. The thermogravimetric analyzer would combust the sample by heating it beyond the ignition temperature of a sample. The resultant TGA curve plotted with the y-axis as a percentage of initial mass would show the residue at the final point of the curve. Thermogravimetric kinetics may be explored for insight into the reaction mechanisms of thermal decomposition. Activation energies of the decomposition process can be calculated using the Kissinger method [39–42].

Measurements are used primarily to determine the composition of materials and to predict their thermal stability at temperatures up to 1200 °C. The technique can characterize materials that exhibit weight loss or gain due to decomposition, oxidation, or dehydration. Gases mostly used are nitrogen and helium that often provides the best baseline but will make furnace work hard at high temperature. Air can sometimes improve resolution because of differences in the oxidative stability (versus thermal stability) of components. Copper oxalate can be used to detect any oxygen contamination.

The thermal methods usually employed are DSC, DTA, TGA, and derivatographic analysis. For a comparative study of thermal behavior of related polymer or simple molecules, each molecule is analyzed by any one or more of these methods of analysis under identical experimental condition. For example, TGA is carried out in the air and in oxygen-free nitrogen [40–42]. It is carried out at different heating rates. The amount and particle size of the material being examined, influence the nature of the thermogram. The speed of the recorder noting the change in weight and the shape of the sample container also influence the thermogravimetric results. The rate of heating the sample and the ambient atmosphere during analysis are very important factors to be controlled during thermal analysis. The information furnished by TGA and DTA is to some extent complementary. From the results of DTA and TGA, it is possible to determine the temperature up to which the material does not lose weight. It is also possible to know the temperature at which material starts decomposing and whether the decomposition occurs in one or multiple stages [39–42].

### *2.3.6.2 Differential Thermal Analysis*

DTA is a thermoanalytical technique that is similar to differential scanning calorimetry. In DTA, the material under study and an inert reference are made to undergo identical thermal cycles (i.e., same cooling or heating programme) while recording any temperature difference between sample and reference. This differential temperature is then plotted against time, or against temperature (DTA curve, or thermogram). Changes in the sample, either exothermic or endothermic, can be detected relative to the inert reference. Thus, DTA curve provides data on the transformations that have occurred, such as glass transitions, crystallization, melting, and sublimation. The area under a DTA peak is the enthalpy change and is not affected by the heat capacity of the sample.

DSC measures the energy required to keep both the reference and the sample at the same temperature; whereas, DTA measures the difference in temperature between the sample and the reference when the same amount of energy has been introduced into both. DTA reveals changes during the heating of a sample that involve evolution or absorption of energy. Small samples can be used to give sharper, narrower peaks, provided that they are fully representative of the source material. DTA is generally regarded as a semi-quantitative or qualitative method. It has been used in studies of the glass transition in polymers.

DTA curve can be used as a fingerprint for identification purposes, but usually the applications of this method are the determination of phase diagrams, heat change measurements, and decomposition in various atmospheres. DTA is widely used in the pharmaceutical and food industries. DTA may be used in cement chemistry, mineralogical research, and environmental studies. DTA curves may also be used to date bone remains or to study archaeological materials [43–45].

#### *2.3.6.3 Differential Scanning Calorimetric Analysis*

DSC is a thermoanalytical technique in which the difference in the amount of heat required to increase the temperature of a sample and reference is measured as a function of temperature. Both the sample and reference are maintained at nearly the same temperature throughout the experiment. Generally, the temperature program for DSC analysis is designed such that the sample holder temperature increases linearly as a function of time. The reference sample should have a well-defined heat capacity over the range of temperatures to be scanned. The result of a DSC experiment is a curve of heat flux versus temperature or versus time. There are two different conventions: exothermic reactions in the sample shown with a positive or negative peak, depending on the kind of technology used in the experiment. This curve can be used to calculate enthalpies of transitions. This is done by integrating the peak corresponding to a given transition.

In this method, unlike DTA the sample and reference body have separate resistive heaters. When a difference in temperature develops between the sample and reference an automatic control loop heats the cooler of the two until the difference is eliminated. It has been mostly used for protective materials using textile, polymers, film, and advanced fibers. Using this technique, it is possible to observe fusion and crystallization events as well as glass transition temperatures (Tg). DSC can also be used to study oxidation, as well as other chemical reactions. Glass transitions may occur as the temperature of an amorphous solid is increased. These transitions appear as a step in the baseline of the recorded DSC signal. This is due to the sample undergoing a change in heat capacity; no formal phase change occurs [46, 47].

#### *2.3.6.4 Cone Calorimetry*

A cone calorimeter is a modern device used to study the fire behavior of small samples of various materials in the condensed phase. It is widely used in the field of fire safety engineering. It gathers data regarding the ignition time, mass loss, combustion products, heat release rate, and other parameters associated with the sample's burning properties. The principle for the measurement of the heat release rate is based on Huggett's principle that the gross heat of combustion of any organic material is directly related to the amount of oxygen required for combustion. This is based on the fact that the heat output from many combustible materials including most natural and synthetic plastics, rubbers, and textiles is almost always a constant 13.6 kJ per gram of oxygen consumed. Other gas measurements, such as carbon monoxide and carbon dioxide concentrations, are made together with measurement of smoke density [48, 49].

Oxygen consumption calorimetry has made the measurement of heat release rate of a fire a routine part of fire testing for both types of research and for regulatory compliance. Heat release rate is a primary metric of fire size that is foundational in modern fire protection engineering. The device allows a sample to be exposed to different heat fluxes over its surface. Its name comes from the conical shape of the radiant heater that produces a nearly uniform heat flux over the surface of the sample under study. It allows characterization of the fire properties of samples of materials (approx. $100 \times 100$ $mm^2$). The fire characteristics of a material can be determined from several different standard models of the cone calorimeter that can be used to evaluate different aspects of the flammable materials. In order to maintain a safe environment, it is important to know and understand the flammability, heat of combustion, ignitability, heat release, and smoke production of many materials all of which can be measured using a calorimeter. The cone calorimeter is a reduced-scale apparatus. Scale effects must be considered when using cone results to predict real-world fires. Cone calorimeter test standards are as follows: ASTM E 1354, ASTM D 5485, CAN/ULC-S135, ISO 5660-1, NFPA 271.

### 2.3.7 Characterization for Mechanical Properties of Advanced Materials

The basic mechanisms operating in materials during creep, fatigue, brittle fracture, and their combinations in relation to microstructural and structural development may be investigated by tests. The mechanical tests used include creep tests, fatigue tests, tensile tests, fracture tests as well as combinations of some of them, for example, combined creep/fatigue tests. Materials would be subjected to similar strain and stress histories as they encounter in critical locations of components and structures in the transport and energy production industries. The changes of the mechanical response are recorded, and the modification of the internal structure and fatigue damage introduced by simulated complex loading situations are studied using transmission and scanning electron microscopy and atomic force microscopy. The data gathered help to improve the resistance of new advanced materials and predict their fatigue life under most severe external conditions. This is applied to the case of nickel-based super alloys [50] or light weight TiAl intermetallic alloys that are often subjected to complex loading conditions due to external load transfer, abrupt changes in geometry, temperature gradients, and material imperfections. Developments in the testing system for thermo-mechanical biaxial tests of metallic materials with surface layers could be used for testing wires and cylindrical specimens made of shape-memory materials. This would be essential to identify the proper temperature and strain parameters of shape-memory actuators, stents, and other components of advanced medical devices based on the shape-memory effect [50–52].

Dynamic mechanical analysis better known as DMA is a powerful tool for understanding the thermal and mechanical properties of polymeric materials. Using deformation as the main variable for evaluation, DMA provides information about a material's glass transition temperature and deterioration limit. DMA measures the modulus (stiffness) and damping (energy dissipation) of materials as they are deformed under periodic stress. It can be used to assess the effects of temperature on material and characterize its physical properties. Dynamic mechanical analysis can be used to determine fundamental transition temperatures, such as glass transition and how that value changes with material formulation and conditioning. The mechanical response of different materials being considered for an application can be compared over the temperature range anticipated in service. For more advanced applications, DMA can include stress–relaxation behavior and creep testing.

Material property combinations and ranges have been extended by the development of composite materials. In a specific case of Nd–Fe–B/epoxy composite materials, the mechanical and magnetic properties of compression molded Nd–Fe–B magnets with different content of magnetic powder in the epoxy matrix were studied. Mechanical properties were investigated at ambient temperature according to ASTM standard D 3039-00. The data may be interpreted for tailoring of properties of epoxy bonded magnetic composites in general, especially for automobile and information technology applications where relatively brittle metallic permanent magnets are not usable [54].

The durability of the textile materials has been carried out by ASTM D4158: guide for abrasion resistance of textile fabrics (uniform abrasion). ASTM D4966: test method for abrasion resistance of textile fabrics (Martindale abrasion tester method). AATCC TM93: (abrasion resistance of fabrics) accelerator method.

## 2.4 Chemical Characterization of Advanced Materials

The well-established techniques of crystallography, transmission electron microscopy, and spectroscopy are the modern methods of scanning probe and tunneling spectroscopies [55–56] elucidate the physical topography and structure. A technique like extended X-ray absorption fine structure (EXAFS) synchrotron radiation spectroscopy is an extremely useful way to determine the chemical state of practically important species that occur in very low abundance or concentration. Surface chemical and compositional analysis via X-ray photoelectron spectroscopy (XPS), Auger Electron Spectroscopy (AES), Secondary Ion Mass Spectroscopy (SIMS), Inductively Coupled Plasma Mass spectroscopy (ICP-MS) and Inductively Coupled Plasma Optical Emission Spectroscopy (ICP-OES), chemical/structural analysis of single and mixed phase materials with NMR, and chemical analysis with LC/GC/FTICR-MS have provided powerful tools for the characterization of nanostructures (refer Figure 2.4). Materials chemistry applies the insights of chemical thermodynamics, kinetics, and quantum mechanics to problems in material science. It involves the

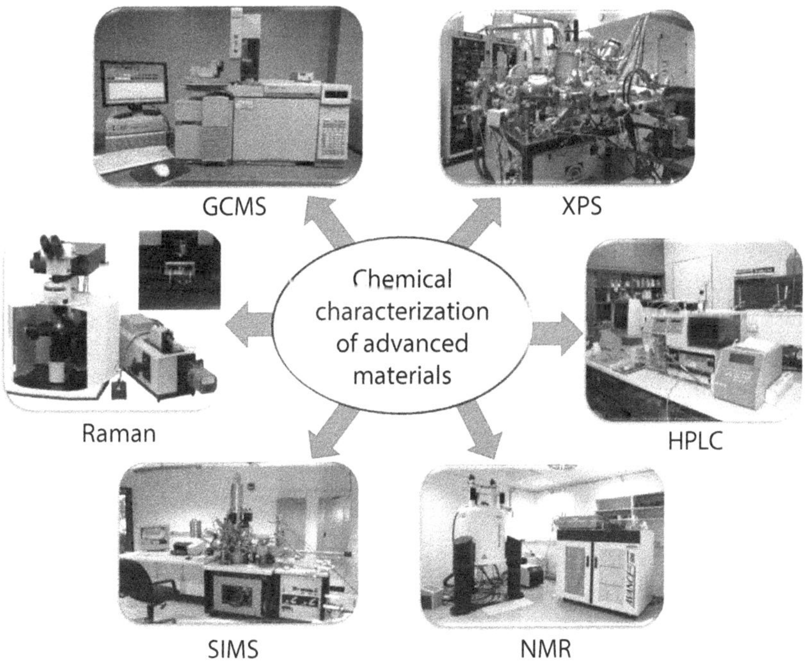


**Figure 2.4** Chemical characterization testing for advanced materials.

synthesis and characterization of thin films, fullerenes and other carbon materials, ceramics, catalysts, fibers, and polymeric materials, covering a broad spectrum of applications.

### 2.4.1 EXAFS, XPS, and AES

Commonly used surface analysis techniques such as AES and XPS are utilized in different modes to obtain information on the near-surface composition and together with ion sputtering or other methods to remove surface layers, to determine the composition of the material as a function of depth from the original surface. For advanced materials, for example, semiconductor devices, magnetic storage media, and new classes of nanostructures and nanomaterials, the materials are fabricated with critical dimensions on the nanometer scale and there is little distinction between surface, bulk, thin film, and interface properties. This generates the need to determine composition as a function of position, in the neighbouring regions of surfaces and interfaces for materials that may have complex morphologies.

EXAFS Spectroscopy is a powerful technique for probing the local structure around almost any specific element in the periodic table (except the lightest) giving information on the number and chemical identities of near neighbours and the average interatomic distances up to 5–6 Å. EXAFS studies can be performed at trace levels (<1000 ppm), which is useful, for example, for studying metal complexes on surfaces, catalysis, and metal sites in bioinorganic samples. XAFS is an accurate measure of the X-ray absorption coefficient as a function of incident X-rays, in an energy range that is below and above the absorption edge of the selected element of material under investigation. The X-ray source is synchrotron radiation from a synchrotron storage ring in which the electrons have more than 1 GeV energies. The synchrotron radiation consists of only continuous X-rays and no characteristic X-rays, allowing the EXAFS technique to be used for atoms in any aggregation state (solid, liquid or gas) in all kinds of environment: crystalline solids, glasses, amorphous phases, liquids, and solutions. A monochromatic beam of X-rays of the energy E, which passes through a homogeneous sample of the thickness x, is attenuated. The EXAFS signal, therefore, is related to the terms of the local configuration environment and to the complex function that describes the scattering of the photoelectron from the surrounding atoms. The fine structure is oscillatory due to the sinusoidal dependence upon inter-atomic distance R. EXAFS is a local structure probe due to the effect of the photoelectron mean free path, $\lambda$, in the exp $[-2R_j/\lambda(k)]$ term [55].

This fine structure is intrinsically a quantum mechanical phenomenon that is based on the X-ray photoelectric effect, in which an X-ray photon incident on an atom within a sample is absorbed and liberates an electron from an inner atomic orbital (e.g., 1 s). The "photoelectron" wave scatters from the atoms around the X-ray absorbing atom, creating interferences between the outgoing and scattered parts of the photoelectron wave function. These quantum interference effects cause an energy-dependent variation in the X-ray absorption probability, which is proportional to the X-ray absorption coefficient, a measurable quantity. This fine structure contains detailed local structural information and chemical information specific to the scattering from atomic shells. An atomic shell is a group of atoms of the same species at the same distance from the absorbing atom. EXAFS region (~1000 eV above the absorption edge) has been used to reveal structural information such as the bond distance between an atom of interest and its neighbouring atoms and also coordination numbers around such atoms.

*In situ* X-ray Absorption Spectroscopy (XAS) results of $MnO_2$/N-$rGO_{ae}$ ($MnO_2$ nanosheets and N-doped reduced graphene oxide aerogel) supercapacitor [56] may be practically used to improve performance in terms of high power and energy applications. In studies on the fabrication of carbon-based metal ferrite composite nanofibers for battery applications, the structure and novel electrochemical properties of carbon/$MFe_2O_4$ (where M = Mn, Ni, and Cu) composite nanofibers — produced by a combination of electrospinning and carbonization were researched. From XAS results, the structure of $MFe_2O_4$ showed cubic inverse spinel ferrite structure with the oxidation states of $Fe^{3+}$ and $Mn^{3+}$, $Ni^{2+}$ and $Cu^{2+}$ ions in structure, respectively [56, 57].

X-ray photoelectron spectroscopy (XPS) is used for investigating the surface chemistry of electrically conducting and non-conducting samples. It provides the following information from the top 10 nm of a sample with a spatial resolution between 3 and a few hundred μm.

- Elemental composition with up to 0.1% sensitivity
- Chemical state information
- Thickness measurement of over layers of up to 8 nm on a substrate
- Surface chemical imaging with a resolution of 3 μm
- Angle-resolved XPS for thickness and depth-distribution of chemical species
- Depth profiling using an ion gun to sputter away the surface during analysis.

A surface is irradiated with X-rays (commonly Al Kα or Mg Kα) in a vacuum. When an x-ray photon hits and transfers this energy to a core-level electron, it is emitted from its initial state with a kinetic energy dependent on the incident X-ray and binding energy of the atomic orbital from which it originated. The energy and intensity of the emitted photoelectrons are analyzed to identify and determine the concentrations of the elements present. These photoelectrons originate from a depth of <10 nm; therefore, the information obtained from within this depth reveal chemical state quantification, oxidation states, chemical environment, and functional groups.

Most XPS instruments include an ion gun that produces monatomic argon ions ($Ar^+$) for this purpose. A range of beam energies is usually available, typically from a few hundred eV up to a few keV, allowing operators to choose to etch rates depending on the thickness of the layers to be analyzed. Spectra are collected after each cycle of material removal by ion milling to generate a depth profile, typically displayed as atomic concentration, which shows the variation in the chemistry with depth into the surface. The process works particularly well for most inorganic materials and keeps the chemical structure intact as layers are removed by ion bombardment. The types of layer structures that are now being developed for applications such as biosensors, display screens, and energy generation depend on mixtures of polymers, metals, and oxides that can be damaged by interaction with a monatomic ion beam, changing the material's chemistry and distorting test data.

An ion source that sputters the sample surface using large, singly charged gas clusters with both Ar+ and Arn+ ($50 \leq n \leq 2000$) gas cluster sputtering, enables both surface cleaning and depth profiling of the growing class of advanced materials built from mixtures of organic and inorganic compounds. The technique helps to negate the low-energy sputtering threshold by making the projectile much heavier. By using a weakly bound cluster of several thousand gas atoms, the material can still be removed, and the beam energy can spread across the whole cluster. Upon impact, the cluster removes surface material, but also breaks apart, minimizing penetration of the projectile into the surface, and instead disperses the energy in the beam laterally. Such a low energy per atom reduces damage to the remaining surface so that the resultant spectra represent the sample's actual chemistry. An application of XPS investigated the film of copper phthalocyanine (CuPc), obtaining depth profiles in the CuPc FET (Field effect transistor) based on silicon.

AES is a method of analyzing the type and amount of elements on a solid surface to a depth of several nanometers by measuring the energy and signal intensity of Auger electron generated when a finely focused beam of the electron (2–10 KeV) irradiates the surface. The Auger effect is a process by which electrons with characteristic energies are ejected from atoms in response to a downward transition by another electron in the atom and in Auger spectroscopy, the vacancy is produced by bombardment with high energy electrons. Light elements tend to have a higher Auger yield in comparison to heavier atoms. Heavy atoms otherwise have a higher X-ray yield. Therefore, AES is an ideal technique to examine the surface of thin films made from light elements. AES has been adapted to allow unambiguous and high-throughput determination of thickness up to six graphene layers and detection of defect and dopant in graphene films on almost any substrate [58]. It has also been used to analyze thin sample processed by a focused ion beam, enabling measurement free from charge-up even if sample is composed of insulating material and achieving depth profiling with good resolution at interface sites deep within the sample [59].

### 2.4.2 ICP-MS, ICP OES, and SIMS

General applicability of X-ray based techniques is sometimes limited by insufficient sensitivities.

For the mass spectrometric techniques, where sensitivities are usually higher, low sample throughput and high instrument costs are major disadvantages. Moreover, the need for appropriate matrix matched standards or certified reference materials for quantitative determinations limits the applicability of some of the above-mentioned techniques. Routine analysis of trace elements in advanced materials is therefore usually performed with atomic absorption spectrometry, inductively coupled plasma optical emission spectroscopy (ICP-OES), or inductively coupled plasma mass spectrometry (ICP-MS); the advantage lies with ICP-based techniques with their wide range of accessible elements and large dynamic range, compared to single element Atomic absorption spectroscopy (AAS) methods [1]. Solid sampling techniques in combination with ICP-OES or ICP-MS have therefore been employed for the direct analysis of advanced materials, circumventing the problems associated with sample dissolution. Such sampling techniques include slurry nebulization (SN), electrothermal vaporization (ETV), and laser ablation (LA). LA provides access to compact solid samples, even though quantitative analysis is much more challenging. However, it provides information about the distribution of the analyte within the sample that can be obtained (i.e., elemental mapping and depth profile analysis). When deciding on the choice of the appropriate detection system, typical parameters that should be considered are sensitivity, selectivity (spectral interferences), acquisition speed, and susceptibility to matrix effects.

The significant advantage of ICP-MS over ICP-OES is its higher detection power. The use of ICP-MS as a detection system allows those samples that could not or only partially examined with ICP-OES. In addition to the determination of impurities at the ultra-trace level, ICP-MS also allows the analysis of samples with restricted sample amounts such as single crystals or thin film coatings. Another benefit of higher detection power is improvements in depth profiling and laterally resolved analysis of advanced materials, since the sensitivity of analysis determines the achievable resolution. However, although ICP-MS provides in most cases better detection limits, under certain circumstances ICP-OES offers comparable or even higher sensitivity. Sensitivity of ICP-MS analysis is also affected by the presence of spectral interferences. Modern instrumentation offers measurements with increased mass resolution or the application of collision/reaction cells to overcome most isobaric and/or polyatomic interferences. However, improvements achieved in selectivity are always accompanied by a decrease in the detection power. In contrast to ICP-MS, where flexibility of analysis is limited by the number of existing nuclides per element, detection with ICP-OES provides a multitude of emission lines for each element. Existing spectral interferences may be overcome since usually analyte wavelengths can be selected in accordance with the needs in selectivity and/or sensitivity of the investigated sample. In the direct analysis of advanced materials, an example is the determination of calcium traces in strontium titanate ($SrTiO_3$) – a commonly used solid-state electrolyte. In ICP-MS analysis, all calcium nuclides are interfered by doubly charged strontium ions or argon ions, hampering accurate determination. With the use of ICP-OES, analysis of Ca in $SrTiO_3$ could be conducted without interference. A general advantage of ICP-OES compared to ICP-MS is the larger tolerance toward matrix load of the inductively coupled plasma, hence more rugged in relation to samples. Generally, ICP-MS is the more sensitive method, compared to ICP-OES, when in combination with ETV sampling. Especially for heavy elements, sensitivity is significantly better for ICP-MS than for ICP-OES, often by two to three orders of magnitude (Cd, Pb, Sb). This also holds true for elements that are difficult to analyze by means of ICP-OES; such an example is Laser Ablation with ICP-OES/MS detection has been used for elemental mapping of sodium in photovoltaic modules to determine degradation process, of Au, Cr, Pt, Ti in microelectronic devices for impurity mapping and device testing; and on bones with Mg-based implants for degradation studies [1].

Secondary ion mass spectrometry enables material surfaces to be analyzed with great chemical sensitivity and excellent resolution in depth, based on the phenomenon of sputtering. The target surface generates secondary ions when bombarded with a focused beam of primary ions that has been accelerated with a potential of 1–30 kV within a high-vacuum chamber ($10^{-5}$–$10^{-10}$ torr). The ion beam generates a series of collision cascades in a shallow surface layer 0.5–5 nm deep causing neutral atoms and, to a lesser extent, secondary ions to be sputtered from the specimen surface. The secondary ions, for a metallic oxide sample, may be a source of M, O, $M^+$, $O^+$, etc. species. They are either monoatomic or clustered, positive or negative, and are directed in a mass spectrometer to be sorted and identified as per their mass/charge ration. High elemental sensitivities to ppm or ppb level are achievable. All elements in the Periodic table can be analyzed, and it is possible to distinguish isotopes. However, SIMS is usually semiquantitative, with dependence on calibration with standard samples of known composition, as the physical process of ion ejection is difficult to fit in rigorous theoretical terms. It complements other surface analysis methods. The range of beam diameters is 1 μm to several mm and among the various types of ion beam available with $Ar^-$, $Cs^+$, $^{16}O^-$, positively charged beams are a common choice. SIMS laser beam may be used instead of ion beams if samples are insulating. SIMS is available in dynamic, static, and imaging modes to investigate materials such as metals, ceramics, polymers, catalysts semiconductors, and composites. SIMS imaging has greater sensitivity than EDX in SEM analysis to provide maps of elemental distribution. Dynamic mode uses a high beam current and is used to determine the distribution and low concentration of dopants in semiconductors, wherein beam scans a raster 100–500 μm in size, eroding sample surface. Static mode uses smaller beam current to provide final spectra with chemical data on top few atomic layers of the surface while doing little damage to the surface and is applied in polymers [60].

### 2.4.3 LC/GC/FTICR-MS

The coupling of high-performance liquid chromatography (HPLC) and Gas chromatography separation methods with mass spectrometry detection methods was first demonstrated in the 1970s [61]; nevertheless, it was with the development and commercialization of atmospheric pressure ionization sources (ESI, APCI) that for the first time the combination of liquid chromatography and mass spectrometry entered the realm of routine analysis [62]. Concurrent with the development of soft ionization methods, suitable to ionize fragile and non-volatile compounds of biological origin, several innovations in mass analyser technology, such as the introduction of high-field and superfast magnets, as well as the improvements in the Time of flight (TOF) and Fourier transform ion cyclotron resonance

(FTICR) enhanced the sensitivity and the upper mass range. Generally, a mass spectrometer is composed of five components: inlet system, ion source, a mass analyzer, an ion detector, and data system. All mass spectrometers are equipped with a vacuum system in order to maintain the low pressure (high vacuum) required for operation. This high vacuum is necessary to allow ions to reach the detector without undergoing collisions with other gaseous molecules. In the ion source, the gas-phase analytes are ionized and transferred into the mass analyzer where they are separated according to their mass-to-charge ratios (m/z). Ion detection can be accomplished by electron multiplier systems that enable m/z and abundance to be measured and displayed by means of an electric signal perceived by the data system, which also controls the equipment [63]. Unlike other mass spectrometers (e.g., sector instruments, time-of-flight, quadrupole) where mass analysis and ion detection are spatially separated events, in FTICR all analytical steps are made on the same spatial place but separated in time. The masses are not resolved in space or time as with other techniques but only by the ion cyclotron resonance (rotational) frequency that each ion produces as it rotates in a magnetic field. Thus, the different ions are not detected in different places as with sector instruments or at different times as with time-of-flight instruments, but all ions are detected simultaneously during the detection interval. This provides an increase in the observed signal-to-noise ratio owing to the principles of Fellgett's advantage [64]. In FTICR-MS, resolution can be improved either by increasing the strength of the magnet (in teslas) or by increasing the detection duration. High-resolution ESI-MS spectra may be obtained by FTICR-MS with complete resolution of the isotope envelope of the charged ions. The mass analyzers nowadays are capable of performing high resolution and high accuracy mass measurements on a routine basis, wherein spectra are acquired in a couple of minutes. They have end applications in laser desorption for materials and their surface characterization.

### 2.4.4 NMR

The strength of NMR nuclear magnetic resonance spectroscopy comes from the local nature of the underlying interactions and its high selectivity. Therefore, it is equally suited for highly ordered and disordered materials or amorphous materials. Methods are equally applicable to organic and inorganic materials or even organic–inorganic hybrid materials. It may be used to examine the structure and molecular dynamics of several classes of materials, from polymers to liquid crystals to building materials. Such advances in NMR techniques as multinuclear and multidimensional scanning allow investigation of the structure and adsorption behavior of metal–organic framework compounds. Researchers have examined $^{63}Cu$ and $^{17}O$ NMR chemical shift data to study cuprates, one class of high-temperature superconductor [65]. NMR generally lacks the sensitivity required to study the novel states of matter that exist in thin films and interfaces. Recently, a technique was developed in which an NMR signal is detected through radioactive decay of an implanted probe nucleus, such as $^{8}Li^{+}$. The radioactive detection scheme, combined with the high intensity, low energy ion beam, enable depth-resolved NMR measurements in crystals, thin films, and multilayers on depth scales of 2–200 nm. The method has roughly 10 orders of magnitude greater sensitivity than conventional NMR and is well suited to probing electronic properties of thin films, interfaces, and heterostructures [66]. Some other examples of the use of NMR spectroscopy in materials science research include the following:

- Characterizing the surface features of silicon nanoparticles [67].
- Analyzing the structure of lignin polymers [68].
- Characterizing graphite oxide, from which graphene can be prepared [69].
- Validating the structure of ultrastrong and stiff layered polymer nanocomposites [70].

### 2.4.5 FTIR and Raman Spectroscopy

Infrared (IR) spectroscopy has always been a powerful tool for the identification of organic materials. With the development of FTIR spectroscopy, it has become a more popular method for the quantitative analysis of complex mixtures, as well as the investigation of surface and interfacial phenomena. An FTIR spectrometer simultaneously collects high-spectral-resolution data over a wide spectral range. This confers a significant advantage over a dispersive spectrometer, which measures intensity over a narrow range of wavelengths at a time. The term Fourier-transform infrared spectroscopy originates from the fact that a Fourier transform (a mathematical process) is required to convert the raw data into the actual spectrum. The currently used instrumentation and sampling modules allow its application to gas, liquid, bulk and powdered solid samples, and thin films using both transmission and reflectance methods, including specular, diffuse, and attenuated total reflectance (ATR) techniques. The optimization of instrumental parameters, sample handling techniques, signal processing, and data analysis allow definitive investigations into the composition of materials. The characteristic vibrational absorption bands used for the identification of key chemical structures as well as methods for quantitative measurement of chemical composition are applied in such problems as

- Deformulation of polymers, rubbers, and other materials through thermogravimetric infra-red (TGA-IR) or gas chromatography infra-red (GC-IR) analysis
- Microanalysis of small sections of materials to identify contaminants
- Analysis of thin films and coatings
- Monitoring of automotive or smokestack emissions
- Failure analysis.

In the functionalization of graphene oxide nanosheets with polyurethane acrylates wherein the NCO groups bond with the hydroxyl groups, FTIR spectra confirm the bond formation in the polymer/graphene nanocomposites [71].

Raman spectroscopy is a form of vibrational spectroscopy, like infrared (IR) spectroscopy—the difference being IR bands arise from a change in the dipole moment of a molecule due to an interaction of light with the molecule, whereas Raman bands arise from a change in the polarizability of the molecule due to the same interaction. When the energies of these transitions are plotted as a spectrum, they can be used to identify the molecule as they provide a "molecular fingerprint" of the molecule being observed. Certain vibrations that are allowed in Raman are forbidden in IR; whereas, other vibrations may be observed by both techniques although at significantly different intensities; thus, these techniques can be considered as complementary. A Raman spectrum is a plot of the intensity of Raman scattered radiation as a function of its frequency difference from the incident radiation (usually in units of wavenumbers, $cm^{-1}$). Because it is a difference value, the Raman shift is independent of the frequency of the incident radiation. Raman spectroscopy can be used for both qualitative and quantitative applications. As Raman detects fundamental vibrations, Raman bands have a good signal-to-noise ratio and are non-overlapping. The spectra are very specific, and chemical identifications can be performed by using search algorithms against digital databases. Raman bands are inherently sharper than their infrared counterparts; isolated bands are often present in the spectrum for more straightforward quantitative analysis, since band areas and heights are proportional to the concentration of species. As Raman involves only illuminating a sample with a laser and collecting the scattered photons, it requires little sample preparation and is non-destructive. The

Raman scattering from a compound (or ion) adsorbed on or even within a few Angstroms of a structured metal surface can be $10^3$–$10^6\times$ greater than in solution. This surface-enhanced Raman scattering is strongest on silver but is observable on gold and copper as well for common excitation sources. Surface-enhanced Raman spectroscopy (SERS) is now widely used as a rapid and inexpensive tool for chemical/biochemical analysis [72]. The method can give enormous increases in the intensities of the Raman signals of low-concentration molecular targets if they are adsorbed on suitable enhancing substrates, which are typically composed of nanostructured Ag or Au. It can be used as a powerful probe of the surface chemistry of any nanostructured material that can provide SERS enhancement as it is the surface chemistry that controls how these materials interact with their local environment and, in real applications, this may be more important than commonly measured properties such as morphology or plasmonic absorption. SERS can be used to study either wet or dry sample in the open laboratory, or as components within electrochemical cells. The measurements are rapid and may be carried out at micrometer lateral resolution using widely available microscope-based systems or at nanometer resolution using more-specialized tip-enhanced approaches. SERS data consist of the appearance of the characteristic vibrational spectrum of a compound in the SERS signal of a material indicating that the compound, a particular molecule or ion, is located within the enhancing region, which extends a few nanometers at most from the material's surface. They also give information on the structure of the adsorbed molecules or ions. Areas that benefit from Raman spectroscopy include the following:

- Identification of materials constituting nanostructures
- Number of layers for graphene and transition metal dichalcogenide (TMDC) layers [73]
- Thickness of heterostructure layers
- Diameter and chirality of carbon nanotubes
- Stress/strain characterization
- Electronic properties of materials (metallic/semiconductive, doping)
- Localization of separated carbon nanotubes or quantum dots

Chemical characterization techniques applied to advanced materials allow the further synthesis and fabrication of materials by design that is the ability to create materials with specific properties to order through the ability to control molecular assemblies on the atomic level.

## 2.5 Conclusions

This chapter has evaluated the techniques for characterization of advanced materials through their classification into the properties (physical and chemical) being measured. The research on wonder advanced materials such as graphene, plasma, biomaterials, dendrimers for drug delivery, gene transfection, photodynamic therapy, light harvesting, biomimics, and catalysis is still in its early stage and requires greater work in characterization to discover its true potential. The existing technology for characterization of advanced materials helps to understand and expand the core science of advanced materials. Thus, the adoption of advanced characterization of new materials would help to develop superior quality products while addressing the commercial issues in terms of its application successfully.

## References

1. Limbeck, A., Bontaa, M. and Nischkauera, W., Improvements in the direct analysis of advanced materials using ICP-based measurement techniques. *J. Anal. At. Spectrom.*, 32, 212, 2017.
2. Zhang, S. and Zhao, D. (Eds.), *Aerospace Materials Handbook*, CRC Press, 2012.
3. Leng, Y., *Materials characterization: introduction to microscopic and spectroscopic methods.* John Wiley & Sons, 2009.
4. Zhang, S., Li. L. and Kumar, A. *Materials characterization techniques.* CRC Press, 2008.
5. Flewitt, P.E.J and Wild, R.K. *Physical methods for materials characterisation.* Taylor & Francis, 2015.
6. Bennig, G.K., Atomic force microscope and method for imaging surfaces with atomic resolution. U.S. Patent No. 4724318, 1988.
7. Greco, A., Lionetto, F., Maffezzoli, A. Orientation of graphene nanoplatelets in thermosetting matrices. *IEEE Trans. Nanotechnol*, 15(6), 877, 2016.
8. Teli, M.D., Samanta, K.K., Pandit, P., Basak, S., and Chattopadhyay, S.K., Low-temperature dyeing of silk fabric using atmospheric pressure helium/nitrogen plasma. *Fibers Polym.* 16 (11), 2375, 2015.
9. Teli, M.D., Samanta, K.K., Pandit, P., Basak, S. and Chattopadhyay, S.K., Low temperature dyeing of silk using atmospheric plasma treatment. *Ind. J. Nat. Fibers*, 1, 1, 2014.
10. Teli, M.D. and Pandit, P., Novel method of ecofriendly single bath dyeing and functional finishing of wool protein with coconut shell extract biomolecules. *ACS Sustain. Chem. Eng.*, 5(9), 8323, 2017.
11. Teli, M.D. and Pandit, P., Development of thermally stable and hygienic colored cotton fabric made by treatment with natural coconut shell extract. *J. Ind. Text.*, 2017. https://doi.org/10.1177/1528083717725113.

12. Teli, M.D. and Pandit, P., A Novel Natural Source Sterculia foetida Fruit Shell Waste as Colorant and Ultraviolet Protection for Linen. *J. Nat. Fibers*, 15 (3), 337, 2018.
13. Teli, M.D., Samanta, K.K., Pandit, P., Basak, S. and Gayatri, T.N., Hydrophobic silk fabric using atmospheric pressure plasma. *Int. J. Bioresour. Sci.*, 2, 15, 2015.
14. Teli, M.D., Pandit, P., Samanta and K.K., Application of atmospheric pressure plasma technology on textile. *J. Text. Assoc.*, 75, 422, 2015.
15. Teli, M.D. and Pandit, P., Multifunctionalised silk using delonixregia stem shell waste. *Fibers Polym.*, 18 (9), 1679, 2017.
16. Teli, M.D., Pandit, P. and Basak, S., Coconut shell extract imparting multifunction properties to ligno- cellulosic material. *J. Ind. Text.*, 47, 6, pp. 1261–1290, 2017.
17. Basak, S., Samanta, K.K., Chattopadhyay, S.K., Pandit, P and Maiti, S., Green flame retardant finishing and combined dyeing of proteinous wool fabric. *Color. Technol.*, 132, 135, 2016.
18. Pandey, R., Patel, S., Pandit, P., Nachimuthu, S and Jose, S. Colouration of textiles using roasted peanut skin-an agro processing residue. *J. Clean Prod.*, 172, 1319, 2018.
19. Teli, M.D and Pandit, P., Application of *Sterculia foetida* fruit shell waste bio-molecules on silk for aesthetics and wellness properties. *Fibers Polym.* 19 (1), 41, 2018.
20. Teli, M.D. and Pandit, P., A Novel natural source sterculia Foetida fruit shell waste as colorant and antibacterial agent for cotton. *J. Text. Assoc.*, 76, 293, 2017.
21. Yetisen, A. K., Qu, H., Manbachi, A., Butt, H., Dokmeci, M.R., Hinestroza, J.P., Skorobogatiy, M., Khademhosseini, A. and Yun, S. H., Nanotechnology in textiles. *ACS nano*, 10, 3042, 2016.
22. Zeng, Wei, Lin Shu, Qiao Li, Song Chen, Fei Wang and Xiao-Ming Tao, Fiber-based wearable electronics: a review of materials, fabrication, devices, and applications. *Adv. Mater.*, 26, 5310, 2014.
23. Geim, A.K. and Novoselov, K.S., The rise of graphene. *Nat. Mater.*, 6, 183, 2007.
24. Wallace, P.R., The band theory of graphite. *Phys. Rev.*, 71, 622, 1947.
25. Novoselov, K.S., Geim, A.K., Morozov, S.V., Jiang, D., Zhang, Y., Dubonos, S.V., Grigorieva, I.V.and Firsov, A.A., Electric field effect in atomically thin carbon films. *Science*, 306, 666, 2004.
26. Dolez, P., Vermeersch, O., Izquierdo, V., *Advanced Characterization and Testing of Textiles A volume in The Textile Institute Book Series*, 2018.
27. Serio, M.A., Gruen, D.M.and Malhotra, R. (Eds.). *Synthesis and Characterization of Advanced Materials*, American Chemical Society, 1997.
28. Stokes, D. *Principles and practice of variable pressure: environmental scanning electron microscopy* (VP-ESEM), John Wiley & Sons, 2008.
29. Suzuki, E. High-resolution scanning electron microscopy of immunogold-labeled cells by the use of thin plasma coating of osmium. *J. Microscopy*, 208 (3), 153, 2002.

30. Goldstein, J.I., Newbury, D.E., Michael, J R., Ritchie, N.W., Scott, J. H.J., and Joy, D. C. *Scanning electron microscopy and X-ray microanalysis*, Springer, 2017.
31. Malick, L.E., Wilson, R.B. and Stetson, D., Modified thiocarbohydrazide procedure for scanning electron microscopy: routine use for normal, pathological, or experimental tissues. *Stain Technol.*, 50 (4), 265, 1975.
32. McMullan, D., Scanning electron microscopy 1928–1965. *Scanning*, 17(3), 175,1995.
33. Corbari, L., Cambon-Bonavita, M.A., Long, G.J., Grandjean, F., Zbinden, M., Gaill, F. and Compère, P., Iron oxide deposits associated with the ectosymbiotic bacteria in the hydrothermal vent shrimp Rimicaris exoculata. *Biogeosci. Discus.*, 5(2), 1825, 2008.
34. Kawase, N., Kato, M., Nishioka, H. and Jinnai, H., Transmission electron microtomography without the missing wedge for quantitative structural analysis. *Ultramicroscopy*, 107(1), 8, 2007.
35. Williams, D. B., and Carter, C. B. *The transmission electron microscope. In Transmission electron microscopy*, pp. 3–17, Springer, Boston, MA, 1996.
36. Guinier, A., Sainte-Marie Lorrain, D., X-ray Crystallography. (Book Reviews: X-Ray Diffraction in Crystals, Imperfect Crystals, and Amorphous Bodies). *Science*, 142, 1564, 1963.
37. Baron, A.Q., Introduction to high-resolution inelastic x-ray scattering. *arXiv preprint arXiv:1504.01098*, 2015.
38. Coats, A.W. and Redfern, J.P., Thermogravimetric analysis. A review. *Analyst*, *88*(1053), 906, 1963.
39. Liu, X., Yu, W., Evaluating the thermal stability of high performance fibers by TGA. *J. Appl. Polym. Sci.*, 99(3), 937, 2006.
40. Marvel, C.S., Synthesis of thermally stablep polymers. *ARIZONA UNIV TUCSON*, 1972.
41. Zhang, T., Jin, J., Yang, S., Hu, D., Li, G. and Jiang, J., Synthesis and characterization of fluorinated PBO with high thermal stability and low dielectric constant. *J. Macromolecular Sci., Part B*, 48(6), 1114, 2009.
42. Voitovich, V.B., Lavrenko, V.A., Voitovich, R.F. and Golovko, E.I., The effect of purity on high-temperature oxidation of zirconium. *Oxidation of metals*, 42 (3), 223, 1994.
43. Ramos-Sanchez, M. C., Rey, F. J., Rodriguez, M. L., Martin-Gil, F. J., Martin-Gil, J., DTG and DTA studies on typical sugars. *Thermochimica Acta*, 134, 55, 1988.
44. Ferrer, S., Borras, J., Martin-Gil, J. Martin-Gil, F.J., Thermal studies on sulphonamide derivative complexes: Part 1. Thermal behaviour of acetazolamide11 Acetazolamide (Acm) is 5-acetamido-1, 3, 4-thiadiazole-2-sulphonamide.-Copper complexes. *Thermochim. Acta*, 147(2), 321, 1989.
45. Berger, K.G. and Akehurst, E.E., Some applications of differential thermal analysis to oils and fats. *Int. J. Food Sci. Technol.*, 1(3), 237, 1966.
46. O'neill, M.J., The analysis of a temperature-controlled scanning calorimeter. *Anal. Chem.*, 36 (7), 1238, 1964.

47. Pungor, E., *A practical guide to instrumental analysis*, Boca Raton, Florida, 181, 1995.
48. Huggett, C., Estimation of rate of heat release by means of oxygen consumption measurements. *Fire Mater., 4*(2), 61, 1980.
49. Beyler, C., Croce, P., Dubay, C., Johnson, P. and McNamee, M., Oxygen consumption calorimetry, William Parker: Di Nenno Prize. *Fire Sci. Reviews*, 6(1), 1, 2016.
50. Kunz, L., Lukas, P., Konecna, R., High-cycle fatigue of Ni-base superalloy Inconel 713 LC. *Int. J. Fatigue*. 32(6), 908, 2010.
51. Chlup, Z., and Salamon, D. Properties of porous multi-layered freestanding ceramic microchannels. *Scripta Mater.* 63(6), 597, 2010.
52. Dobes, F., Pesicka, J. and Kratochvil, P., Creep of Fe-18Al-4Cr alloy with zirconium addition. *Intermetallics*, 18(7), 1353, 2010.
53. Grujić, A., Talijan, N., Stojanović, D., Stajić-Trošić, J., Burzić, Z., Balanović, Lj.an Aleksić, R., Mechanical and magnetic properties of composite materials with polymer matrix. A. *J. Min. Metall. Sect. B-Metall.*, 46 (1) B, 25, 2010.
54. Jortner, J., Rao, C.N.R., Nanostructured advanced materials. Perspectives and directions. *Pure Appl. Chem.*, 74 (9), 1491, 2002.
55. Gaur, A., Shrivastava, B.D. and Nigam, H.L., X-Ray Absorption Fine Structure (XAFS) Spectroscopy–A Review. *Proceedings of the Indian National Science Academy*, 79(Part B), 921, 2013.
56. Kumar, R., Shukla, S.K., Pandey, A., Srivastava, S.K. and Dikshit, A., Copper oxide nanoparticles: an antidermatophytic agent for trichophyton spp. *Nanotechnol. Rev.*., 4(5), 401, 2015.
57. Xu, M., Fujita, D., Gao, J. and Hanagata, N., Auger electron spectroscopy: a rational method for determining thickness of graphene films. *ACS Nano.*, 4(5), 2937, 2010.
58. Sato, M., Shirai, S., Seyama, Y. and Itakura, T., 2010. Analysis of insulating materials and deep interfaces by Auger electron spectroscopy. *Fujitsu Scient. Techn. J.*, 46(3), 257, 2010.
59. Smallman, R.E. and Ngan, A.H.W. *Physical metallurgy and advanced materials*, Butterworth-Heinemann, 2011.
60. Dass, C., *Fundamentals of contemporary mass spectrometry*. John Wiley & Sons, 16, 2007.
61. Whitehouse, C.M., Dreyer, R.N., Yamashita, M. and Fenn, J.B., Electrospray interface for liquid chromatographs and mass spectrometers. *Anal. Chem.*, 57(3), 675, 1985.
62. Madeira, P.J.A., Alves, P.A. and Borges, C.M., High resolution mass spectrometry using FTICR and orbitrap instruments. *In Fourier Transform-Materials Analysis. InTech*, 2012.
63. Marshall, A.G., Hendrickson, C.L. and Jackson, G.S., Fourier transform ion cyclotron resonance mass spectrometry: a primer. *Mass Spectrum. Rev.*, 17(1), 1, 1998.

64. Haase, J., Slichter, C.P. and Williams, G.V., Evidence for two electronic components in high-temperature superconductivity from NMR. *J. Phys.: Condens. Matter*, 21(45), 455702, 2009.
65. MacFarlane, W.A., Implanted-ion βNMR: A new probe for nanoscience. *Solid State Nucl. Magn. Reson.*, 68, 1, 2015.
66. Faulkner, R.A., DiVerdi, J.A., Yang, Y.; Kobayashi, T., Maciel, G.E., The surface of nanoparticle silicon as studied by solid-state NMR. *Materials, 6*, 18, 2013.
67. Wen, J.L., Sun, S.L., Xue, B.L. and Sun, R.C., Recent advances in characterization of lLignin polymer by solution-state nuclear magnetic resonance (NMR) methodology. *Materials*, 6, 359, 2013.
68. Cai, W., Piner, R.D., Stadermann, F.J., Park, S., Shaibat, M.A., Ishii, Y.,... and An, J., Synthesis and solid-state NMR structural characterization of 13C-labeled graphite oxide, *Sci.*, 321 (5897), 1815, 2008.
69. Podsiadlo, P., Kaushik, A.K., Arruda, E.M., Waas, A.M., Shim, B.S., Xu, J., Nandivada, H., Pumplin, B.G., Lahann, J., Ramamoorthy, A. and Kotov, N.A., Ultrastrong and stiff layered polymer nanocomposites. *Science*, 318 (5847), 80, 2007.
70. Qingzhou, X., Advanced Materials, Technology and Application: Proceedings of the 2016 International Conference on Advanced Materials, Technology and Application (AMTA2016). World Scientific Publishing Company, 2016.
71. Dick, S., Konrad, M.P., Lee, W.W., McCabe, H., McCracken, J.N., Rahman, T., Stewart, A., Xu, Y. and Bell, S.E., Surface-enhanced Raman spectroscopy as a probe of the surface chemistry of nanostructured materials. *Adv. Mater.*, 28(27), 5705, 2016.
72. Du, J., Yang, Z., Wang, X., Qi, C., Li, Y., Mao, W., Qiao, H., Yu, Z., Ren, T. and Qiao, Q., Fabrication of multilayered-sandwich $MoS_2$/c architectures with advanced lithium storage properties. *Electrochim. Acta*, 250, 238, 2017.

3

# Green and Sustainable Advanced Biopolymeric and Biocomposite Materials

**T.P. Mohan* and K. Kanny**

*Composite Research Group (CRG), Department of Mechanical Engineering, Durban University of Technology, Durban, South Africa*

***Abstract***
Much energy, resources, waste, and by-products have been generated in developing synthetic materials causing severe strain to the environment and eco-system. In past decades, there has been a growing concern to develop the materials in an environmentally friendly way. Significant progress was made in the development of green materials based on polymer and composites. However, several technical and application issues of these materials need to be addressed for effective usage of these materials and products in future. This chapter reviews the green polymers and composites that are competitive to their synthetic counterpart materials. The materials under scrutiny in this chapter are biopolymers, biofillers, natural fibers and biocomposites. The merits and demerits of these green materials are discussed, and their possible way for improving their properties (mechanical, thermal, and physical) are discussed as well.

***Keywords:*** Green materials, biopolymers, natural fibers, natural fillers biocomposites

## 3.1 Introduction

Nowadays, research has started to lean toward developing more environmentally friendly and reusable materials; biocomposites match these criteria. Biocomposites have various uses ranging from the automotive industry to the medical field. Recent research has found that adding fillers improves various mechanical properties [1].

**Corresponding author*: mohanp@dut.ac.za

Shakeel Ahmed and Chaudhery Mustansar Hussain (eds.) Green and Sustainable Advanced Materials, Vol. 1 (67–92) © 2018 Scrivener Publishing LLC

The use of another type of composite material, called nanocomposites, has increased, because they have a fine surface that allows more bonding at the interface as opposed to the more traditional composites, which have a coarser surface [2].

The application of synthetic materials such as nanoclays, glass, and carbon fiber for the manufacturing of nanocomposites have been of strong interest prior but these materials became problematic [3]. The biggest problem that has risen is that of the cost and the lack of reusability [4]. Owing to this, research has been undertaken to find sustainable materials.

Using polymers from renewable resources over synthetic materials has been the solution to the environmental problem because they reduce the amount of plastic waste [5]. These materials have allowed more options in the economic sense owing to their sustainability [6].

Research in agricultural resources to produce green materials has increased owing to the ability to use original products and waste materials [7]. The application of these biocomposites is expected to improve the rate of production while staying environmentally friendly [8].

## 3.2 Classification of Green Materials

Natural fibers have increased in use in recent decades owing to their high strength-to-weight ratio, and researchers have invested interest in the use of these as reinforcements owing to their desirable properties, cheap costs, and recyclability.

The structural flaws, inconsistencies, and chemical composition of the fibers are of concern, and these variables dictate the overall mechanical properties [9]. Some significant natural fibers are listed in Figure 3.1.

Research is currently geared toward developing fully biodegradable composites by combining natural fibers and biodegradable resins. The benefits of using green composites are that they are environmentally friendly because they can be disposed as compost.

Biodegradable polymers are divided into four categories: biomass products, microorganisms, biotechnology, and fossil. Categories 1–3 are renewable, while category 4 is of fossil origin. Many of these biodegradable polymers (Biopolymers) are commercially available, showing various properties and they can compete with non-biodegradable polymers in various fields [11]. The classification of significant biopolymers is shown in Figure 3.2.

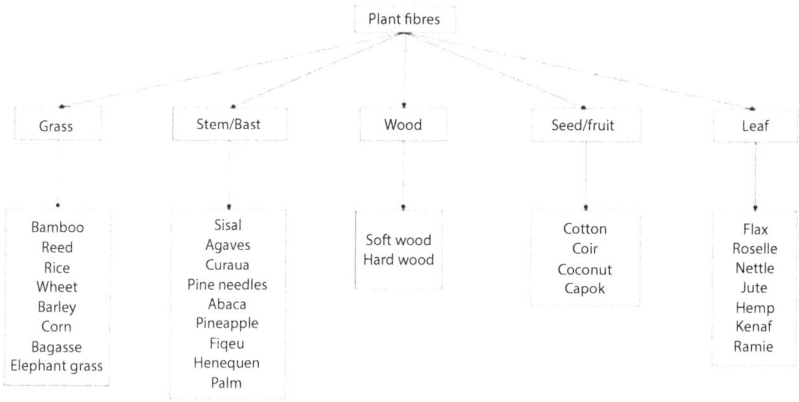


**Figure 3.1** Classification of Significant Natural Fibers.

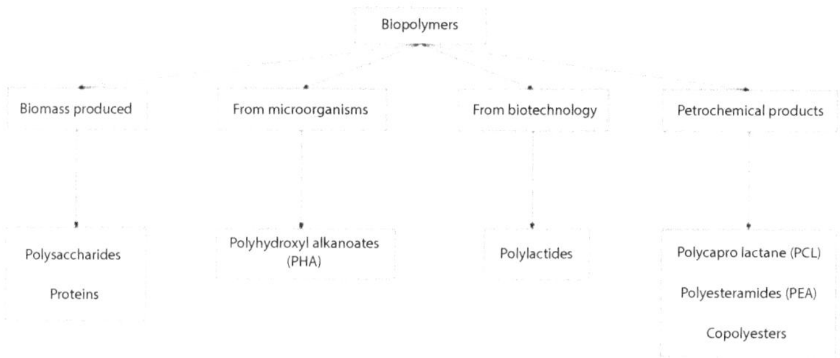


**Figure 3.2** Classification of Significant Biopolymers.

## 3.3 Biopolymers

Biopolymers have two principal categories: those formed by biological schemes; and those that are formed chemically although they are derived from biological materials such as amino acids [12].

Biopolymers degrade naturally via microorganisms, moisture, and heat. These are environmentally friendly because they can be made from waste starch from crops that was already used in food. Biomass is used to make biopolymers. Biomass refers to fuel that is made from organic materials, which is a renewable (sustainable) source of energy used to generate electricity [13]. Increased use of this would decrease the dependence on fossil fuels.

A wide range of emerging uses depend on these production techniques. Biopolymers are being developed for various applications such as medical materials, clothing fabrics, water treatment chemicals, and biosensors [14].

Biopolymer fibers are infused in a biopolymer matrix. This matrix dictates the composite structure and geometry, mechanically transfers the shear forces between the fibers, and guards them against radiation. Usually, polymers (excluding elastomer type) are split into two categories. These categories are thermosets and thermoplastics. Both categories are suitable as a matrix for construction materials that are made from biocomposites [15].

Polymers that originates naturally (starch and cellulose, for example) must be modified either physically or chemically to make them suitable for processing as thermoplastic resins. The structure of starch, for example, can be made thermoplastic by using adjuvant substances such as glycerol and water [16].

Several biopolymers such as cellulose and starch have been under growing interest owing to various financial and environmental implications. Biocomposites have at least one contributing material of natural origin. This could include plant fibers like cotton and hemp from reused wood. To truly progress in the engineering of true biocomposites, the matrix component should be made mainly from renewable resources. A few biopolymers have been applied directly or after modification, to substitute the conventional fillers allowing partial biodegradation [17].

## 3.4 Natural Fillers

Composites contain two (or more) components, which, when merged, result in a material with new properties from the contributing parts [18]. The two most important factors that affect the performance of composites are filler–matrix interface and the filler's architecture.

The filler's architecture is affected by the extraction process, and the aspect ratio (length to diameter) is one of the major influences in the strength of the composite [19]. The interface between filler and matrix is also critical; the interface works to transfer externally applied loads to the reinforcement via shear stresses over the interface. Controlling the strength of the interface is critical. Strong bonding at the interface allows true reinforcing to take place [20].

Fillers are widely utilized in thermoplastics, elastomers, and thermosets. They may be inert materials that serve to reduce resin cost [21]. Fillers are utilized in polymers for numerous reasons such as regulating density and flame retardancy, and improved characteristics such as tear resistance and hardness [22].

Some polymers catch aflame easily. Polymers such as acrylates, which are mostly used for extruded sheets, paints, and shellacs, are highly flammable and very difficult to reduce flame retardance, even if large volumes of conventional Flame Retardant (FR) agents are added, like phosphorus and inorganic materials. Recently, several groups have reported that addition of a small amount of organoclay can majorly decrease the heat release and mass loss rate [23].

Clays have long been utilized as fillers in polymer systems owing to lower cost and greater mechanical properties of the resulting polymer composites. If all other parameters are equal, the efficiency of a filler to improve the physical and mechanical properties of a polymer system is sensitive to its degree of dispersion in the polymer matrix [24].

The treatment of Montmorillonite (MMT) with amino acids permitted a dispersion of the individual 1-nm-thick silicate layers of the clay in polyamide on a molecular scale. The hybrid material produced by them presented major improvement in mechanical strength, even at very low clay content. This has led to further development in various materials and manufacturing methods available [25].

The most widely available fiber that is renewable in nature is cellulose, which contributes structural rigidity to plants. Nevertheless, they can be made from agricultural waste that makes it more feasible than most fiber sources currently in use. Cellulose fiber offers a good spread of excellent mechanical properties, a low density, and safer working conditions in comparison to synthetic fibers [26]. Various fibers have been applied as reinforcements for composites and nanocomposites. A list of some significant natural fibers is provided with their properties in Table 3.1.

**Table 3.1** List of Natural Fibers and their Properties [27].

| **Fiber source** | **Treatment** | **Cellulose content (%)** | **Tensile strength (MPa)** | **Young's modulus (MPa)** | **Elongation (%)** |
|---|---|---|---|---|---|
| Cotton stalk bark | Alkali | 79 | 377 | 19 | 3 |
| Velvet leaf | Alkali | 69 | 325–500 | 18–38 | 2–3 |
| Grass leaves | Alkali | 61 | 715 | 31 | 2 |
| Grass stems | Alkali | 68 | 351 | 9 | 7 |
| Ramie | Untreated | 69–76 | 560 | 25 | 3 |
| Soybean straw | Alkali | 85 | 351 | 12 | 4 |
| Hop stem | Alkali | 84 | 84 | 20 | 3 |
| Corn stalk | Alkali | 52 | 286 | 16 | 2 |

## 3.5 Natural Fibers

Nowadays, there is a lot of focus regarding the research and development of natural fiber composite because it has been established that they are less costly, readily available, eco-friendly, and renewable.

Natural fiber is one of the renewable sources that are used as reinforcements and supplements that are highly used in polymer-based materials. The increase in environmental awareness has brought forward natural fiber as a topic of discussion. Strong natural fibers are one of the rising materials that substitutes in for synthetic materials and its related products where a strength-to-weight ratio is of concern. They are adept in the avenue of energy conservation.

Automotive and aircraft industries are moving forward by using various natural fibers, mainly with sisal and bio-resin systems for interior components. High specific properties sold for lower prices in natural fiber composites are making it attractive for various applications [28].

Choosing the appropriate fibers is determined by the critical values, such as stiffness and tensile strength of the composite. Further considerations for the selection of suitable reinforcing fibers include the elongation at failure, price and processing costs, and the long-term behavior [29].

Areca appears between all the natural fiber-enhancing materials; its husk contains sufficient yield to become self-sustaining, owing to its small amounts of water and ash content [30]. Banana fibers are one of the most widely available materials because approximately 70 million tons of bananas are produced annually [31].

Bamboo has many advantages over other plant fibers, for example, a low density, high mechanical strength, and high rate of production. One of the desirable properties is that of its ability to fix atmospheric carbon dioxide. It also has disadvantages in various applications such as a high moisture content and difficulty of extracting consistent fiber. Bamboo has traditionally been used in construction owing to its high strength-to-weight ratio [32].

Hemp, one of the most environmentally friendly fibers and the oldest, requires a government license for cultivation, and this imposes disadvantages compared to other fibers. It has many applications like for ropes, ships sail's, and paper. Fiber extraction is done through maceration and the separation of the fibers from the straw is then performed, this can be carried out mechanically [33].

Kenaf is cultivated in multiple continents. It has various fiber lengths and the geometry is not consistent. The stem produces a coarser fiber in the outer layers, and a finer fiber is found in the core. The main application of Kenaf is for producing ropes, paper, and as material inside the door of cars [34].

Coconut fiber is extracted from the outer husk of a coconut. There are two kinds of coconut fibers: brown fiber obtained from ripe coconuts and extracted from matured coconut; and white fibers obtained from immature coconuts. These different types of fibers have different applications, and the brown fiber is mostly used in engineering. These are resistant to fungi and rot. They are also excellent insulators, and their use in construction is limited owing to its inconsistent properties. Fiber dimensions vary with individual cells and on the type of species, maturity of the coconut, and even the extraction process [35].

## 3.6 Biocomposites

### 3.6.1 Thermoplastic Starch Based Composites

Prior research in thermoplastic starch-based composites focused on the use of plasticized starch as matrix for green nanocomposites. The use of thermoplastic starch to manufacture composites by melt intercalation was done via twin screw extruders. The composites were prepared with regular plasticized corn starch with glycerol and reinforced with hydrated kaolin. The study noted found a major increase in the tensile strength from 5 to 7.5 MPa, for the composite with 50 phr clay composition.

The modulus of elasticity had also seen an increase from 120 to 290 MPa and the tensile strain at break decreased from 30% to 14%. The highest modulus of elasticity and the tensile strength corresponded to the maximum clay content that was infused in the matrix. One other noteworthy fact was that the higher clay content made the composite more fragile [36].

The addition of plasticizers was varied to find the effect of plasticizers on the mechanical and structural properties of resultant composites using solution method [37]. It was assumed that the sequence of adding components such as: starch, plasticizer, glycerol, and clay had a significant effect on the nature of composites. Consequently, the properties were changed with no well-established method to obtain the appropriate processing technique for the composites.

The filler dispersal was found to be highly heterogeneous and the more brittle when starch was plasticized before filling with clay. The modulus of composites had increased significantly for all compositions relative to the unfilled matrix, irrespective of the preparation method. An improvement in all the properties of the composites could be obtained by improved dispersion of the clay filler, not necessarily by varying the plasticizers content.

### 3.6.2 Polylactic Acid (PLA) Based Composites

Studies on nanocomposites using PLA showed that nanocomposites prepared with PLA had enhanced mechanical, thermal, and impact properties. Commercial PLA was also applied as the matrix with cellulose whiskers treated with anionic surfactant as reinforcement [38].

The lack of improvement in mechanical properties of the composite was due to poor adhesion between the fiber and the matrix that is directly related to the nature of the fiber and the type of mechanical treatment given [39]. The tensile modulus of the composite had an increase from 62.5% to 169.5% as compared to prior research where no improvement was recorded.

Recently, nanocomposites made with PLA with a compatibilizer and cellulose fibrils have been recognized [40]. Bleached wood pulp was used as the reinforcement and commercial PLA as the matrix. A chemo-mechanical method was used to prepare cellulose nanofibrils discrete consistently in an organic solvent. Polyethylene Glycol (PEG) was added to the matrix and had improved the tensile strength and elongation percentage when compared with pure PLA.

It was found that PEG covers the surface of the cellulose nanofibrils and acts as not only just a plasticizer for PLA to improve its elongation, but also as a compatibilizer between the hydrophobic PLA and the hydrophilic cellulose nanofibrils. It was also noted that PEG also prevents aggregation of the nanofibrils, allowing the cellulose nanofibrils to disperse in the PLA matrix homogeneously to form a network structure.

### 3.6.3 Cellulose Based Composites

Cellulose from agricultural products has been identified as a source of biopolymer that can replace petroleum polymers. Green nanocomposites have been successfully produced from Cellulose Acetate (CA), Triethyl Citrate (TEC) plasticizer, and organically modified clay via melt compounding [41].

The cellulosic plastic with 80 wt.% pure cellulose acetate and 20 wt.% triethyl citrate plasticizers was utilized as the polymer matrix for nanocomposite production. Mechanical properties of the composites were determined. Tensile strength and modulus of cellulosic plastic reinforced with 10% (on mass) organoclay improved strength to 180%. Thermal stability of the cellulosic plastic had also seen an increase [42].

A dynamic antimicrobial packaging material has been developed using Methyl Cellulose (MC) as the base material with Montmorillonite (MMT)

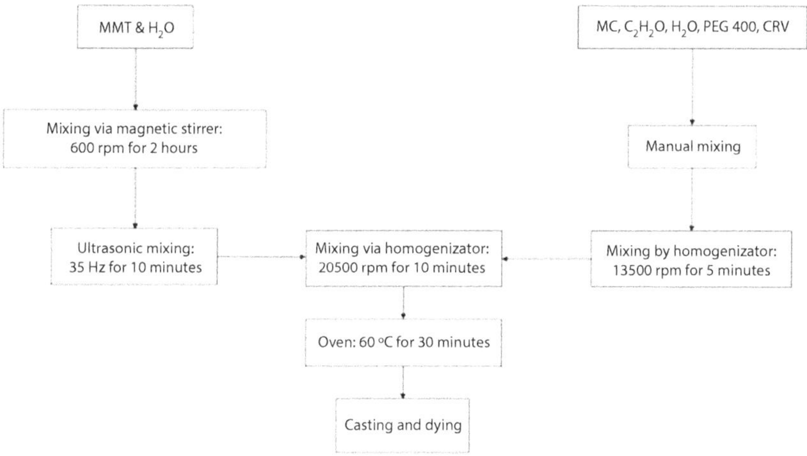


**Figure 3.3** MC/CRV/MMT Nanocomposite Film Procedure.

as reinforcement [43]. MC and MMT were industrially prepared and mixed with Carvacrol (CRV) to make nanocomposites as shown in Figure 3.3.

### 3.6.4 Plant Oil Based Composites

Bio-based nanocomposites have been developed from epoxidized soybean oil, Diglycidyl Ether of Bisphenol-A (DGEBA), and organically enhanced Montmorillonite (OMM) [46]. Soy-based epoxy resin was mixed with different organophilic Montmorillonite clay (as reinforcement) at various concentrations to make nanocomposites. The tensile strength of OMM reinforced soy-based epoxy nanocomposites had increased. Impact strength and flexural properties of the nanocomposites showed significant enhancement. A maximum increase of 25% was noted for flexural strength of the nanocomposites [45].

Nanocomposites reinforced by exfoliated clay were found to exhibit higher storage moduli. Scanning Electron Microscope (SEM) images of the composites revealed relatively uniform dispersion of organoclay layers in addition to the strong interface of the material. The improvement was due to the plasticizing effect of the long aliphatic chains in the soy-based epoxy polymer matrix and reinforcement effect of the nanoclay layers [46].

The use of mango puree-based edible films for nanocomposites were developed by adding Cellulose Nanofibers (CNF) at different concentrations as a nano-reinforcement to the mango puree-based edible films for packaging applications. Edible films can be used as a matrix for biocomposites, given that the tensile strength had improved from 4.06 MPa to 8.09 MPa and Young's modulus from 19.85 MPa to 322.05 MPa. This was due to the nanofibers seeming to have presented positive interactions with the mango puree. They were found to disperse in the matrix without significant clustering [47].

Soy-based polyurethane can be used as a matrix to obtain bio-based nanocomposites. The Halloysite Nanotubes (HNT) nanoparticles that were used to reinforce polyurethane were utilized to make E-glass-reinforced composites using low-cost vacuum-assisted resin transfer molding process then characterized for ultimate compressive strength, flexural strength, and modulus [48].

### 3.6.5 Polymer—Polymer Blends-Based Composites

The blending of two or more polymers to obtain a polymer that is biodegradable has also drawn research interest. These polymers have been tested for their degradability and mechanical properties and thus are recommended for use as degradable polymers for composite applications. Such polymer mixtures include Starch/PLA mixtures, starch/modified polyester blends, and thermoplastic starch/polyester amide mixtures [49]. The blends were prepared using a single-step extrusion process.

The transition temperatures for TPS/PLA and PCL mixtures were found to be about −57 °C for blends with glycerol content of 36%. Increasing the concentration of TPS in PLA resulted in a corresponding increase in fracture strain with a lowered elastic modulus and maximum strength. Natural rubber had also been mixed with starch to make nanocomposites [50]. SEM images of the fracture surface of the nanocomposite (unmodified natural rubber latex), in Figure 3.4, showed that both thermoplastic starch matrix (grey areas) and dispersed rubber phase (white areas) are clearly separated indicating weak adhesion at the interface. The use of modified natural rubber, in Figure 6.5.1b, resulted in a finer dispersion and stronger interfacial adhesion.

## 3.7 Merits and Demerits of Green Materials

The rising attention in natural fibers is mostly due to their economical manufacturing with few demands for apparatus and low specific weight, which yields a higher specific strength and stiffness when compared to glass reinforced composites. The current handling and working conditions are safer compared to synthetic reinforcements. Natural fibers are a renewable resource with production consuming less energy.

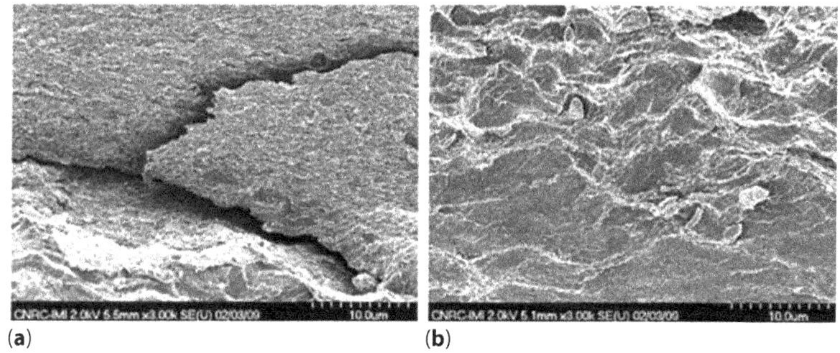

Figure 3.4 SEM Micrographs of Starch/Natural rubber/Clay Nanocomposites with (a) Unmodified Natural Rubber and (b) Modified Natural Rubber [51].

The processing environment is friendly with better working conditions and, therefore, there will be reduced dermal and breathing problems. Natural fibers have high electrical resistance and can undergo thermal recycling. The hollow cellular structure provides good sound insulating properties. The polar and hydrophilic nature of natural fibers and the non-polar characteristics of most thermoplastics results in major difficulties leading to non-uniform dispersion of fibers within the matrix which affects the efficiency of the composite. This is a massive disadvantage of natural fiber-reinforced composites [52].

One of the issues is that the processing temperature of composites is restricted to 200 °C, as vegetable fibers undergo degradation at higher temperatures; this reduces the choice of matrix material. The other drawback is the high moisture absorption of natural fibers causing swelling and presence of voids at the surface, which results in terrible mechanical properties and reduces dimensional consistency of composites. Restriction of biofibers for durable composite application is low microbial resistance and it being prone to rotting. These characteristics pose serious problems during shipping, storage, and composite processing. The non-uniformity and differences of dimensions and of their mechanical properties (even between individual plants in the same cultivation) poses another serious problem [53].

In Table 3.2, natural and synthetic fibers are compared. It also represents the natural fibers that have reasonable mechanical properties when compared to synthetic fibers such as glass, carbon, and kevlar. The usage of natural fibers is increasing in many industries such as automobiles, furniture, packing, and construction. This is due to their advantages related to synthetic fibers, that is, cheaper, light weight, less damage to processing equipment, better exterior finish of molded parts, good relative mechanical

**Table 3.2** Comparison between Natural Fibers and Synthetic Fibers [56].

| Aspects | Property | Natural fibers | Synthetic fibers |
|---|---|---|---|
| Technical | Mechanical Properties | Moderate | High |
| | Moisture Sensitivity | High | Low |
| | Thermal Sensitivity | High | Low |
| Environmental | Resource | Infinite | Limited |
| | Production | Low | High |
| | Recyclability | Good | Moderate |

properties, and abundant and renewable resources [54]. Natural fibers are used in various areas such as building materials, particle boards, insulation boards, human food and animal feed, cosmetics, medicine, and for other biopolymers and fine chemicals [55].

However, the most prominent disadvantages of natural fibers are as follows: (i) the poor compatibility between fiber and the matrix in composites and (ii) the relatively high moisture absorption.

## 3.8 Recent Progress in Improvement of Material Properties

The expansion of textile technologies such as weaving, knitting, and braiding has resulted in the development of composites that have higher mechanical properties, as unceasing location of fibers is not restricted at any point. In usages where more than one fiber alignment is required, a fabric utilizing 0° and 90° fiber orientations is beneficial. Woven fabrics are produced by the interlacing of horizontal (0°) fibers and vertical (90°) fibers in a consistent pattern or weave style. The fabric's integrity is maintained by the mechanical meshing of the fibers. Surface smoothness and stability of a fabric are measured mainly by the weave style [57].

In Figure 3.5, the common weave style found is shown. Each horizontal fiber passes under and over each vertical fiber. The fabric is symmetrical, with good consistency and sensible porosity. However, it is the hardest of the weaves to drape (the characteristic of a fabric to conform to a complex surface), and the high amount of fiber crimp imparts relatively low mechanical properties with respect to other weave styles. By large fibers, this weave style produces excessive crimp and, therefore, it is not commonly used for very heavy fabrics [58].

**Figure 3.5** Bi-Directional Fiber Glass.

### 3.8.1 Hybridization

Hybrid composite materials are created by combining two or more different types of fibers in a common matrix. They offer a vast range of properties compared to conventional composite materials, where a single kind of reinforcement is present [59]. Hybridization would ideally allow designers to tailor the composite properties to the exact needs of the structure under consideration. This can be achieved provided the properties of the hybrid composite are predictable from those of the originating composites.

Many reasons can be accounted for this difficulty, including the presence of internal cavities, limiting the consistency of plant fibers and the varying diameters of these fibers that may lead to more hindrances such as water absorption [60].

Plant fibers also have varying geometries that increase the number of defects in a single length. This problem can be dealt with the use of woven structures such as tissues or mats.

The manufacturing of hybrid composites has been implemented as a final step in reducing the environmental impact of fiberglass through the incomplete removal of glass fibers with plant fibers such as jute, hemp, flax, sisal, and kenaf [61]. The restrictions of this removal arise especially when the service performance is simulated through dynamic testing, such as impact and fatigue [62]. In this situation, the remaining performance obtained from these materials is also significant, as the occurrence of vegetable fibers can cause ineffective damage absorption and complex degradation modes [63].

## 3.9 Current Applications of Biocomposites and Biopolymers

In many uses, such as mirror covers on numerous vehicle types, door handles, engine covers, etc., have been utilizing nanocomposites. Other uses currently being considered include impellers and blades for vacuum cleaners, power tool housings, mower hoods, and covers for portable electronic equipment such as mobile phones and pagers [64]. Its incredible barrier properties, chemical resistance, and surface appearance make it a brilliant material for packaging applications such as in beer, carbonated drink bottles, and paperboard for fruit juice and dairy. Nanocomposites also have a future in aerospace applications owing to their light weight [65].

### 3.9.1 Green Fibers and their Potential in Diversified Applications

In uses like building materials, particleboards, insulation boards, human food and animal feed, cosmetics, medicine, and sources of other biopolymers and "fine chemicals," numerous portions of plants such as woody core, bast, leaf, cane, straw from cereals, grass, and seed can be utilized. They do not affect the ecosystem, they can grow in many climate zones, and they reduce the carbon dioxide in the atmosphere. These plants contribute to a better agricultural balance in Europe, and they will assist in the increasing request from an expanding population for cellulosic pulp in the future.

Some of these plants such as flax and hemp can be used for soil cleaning, polluted by heavy metals, by mining and removing cadmium, lead, copper and others [66]. The estimated world-wide production of fibrous raw materials from agricultural crops is provided in Table 3.3:

### 3.9.2 Textile Applications

Although difficult in handling, traditional long flax, ramie, and hemp are still produced and processed by slightly modernized methods [68]. Flax and ramie provide yarns of high strength, luster, and low linear mass.

These products include the following:

- 100 percent linen or ramie knitted apparels, perfectly hygienic and comfortable properties make them especially attractive for hot, humid climates [69].
- 100 percent linen and ramie fabrics—from almost transparent to terry type, having high "Wash and Wear" properties when finished with liquid ammonia [70].
- Hemp and jute are mainly used in sport wear and Friday-wear [71].
- Hemp and linen are hygienic; they are used in the making of comfortable boots and shoes.

**Table 3.3** Estimated Global Tonnage of Fibrous Raw Materials from Agricultural Crops [67].

| Crop | Plant component | Availability (kilo tonnes) |
|---|---|---|
| Barley | Straw | 219 |
| Oats | Straw | 51 |
| Rice | Straw | 465 |
| Rye | Straw | 42 |
| Wheat | Straw | 740 |
| Corn | Stalks | 727 |
| | Lint | 18 |
| Cottons | Linters | 2 |
| | Stalks | 36 |
| | Mote | 900 |
| Bast fibrous plants | Straw | 25 |
| Seed grass | Straw | 2 |
| Oil flax | Straw | 3 |
| Sorghum | Stalks | 105 |
| Sugarcane | Bagasse | 100 |
| **Total** | | 3435 |

### 3.9.3 Green Fibers for Pulp

In the cellulose industry, attention is rising on the manufacturing of pulp and paper from agro-based lignocellulosic raw materials (feedstock to the pulp and paper industry) such as bamboo, hemp, flax, and sisal. The amount of pulp developed from non-wood resources is still growing. The contribution of cellulose of agro-plant origin to the world production of pulp and paper is vastly expanding. Further expansions in the contribution of agro-plant cellulose is predicted, and it should acquire about 15% in 2010 (when the total production is estimated to be about 480 million tons) [72].

A sample of the use of jute for pulp (for fibers, but also suitable for non-textile applications) is given by the South India Viscose Company [73].

In the United States, when the Department of Agriculture conducted a research for an agricultural source of fiber for paper industries, kenaf was chosen to extend the life of North American forests [74].

Similar advances are in progress through the entire tropical zone and involve jute, kenaf, sisal, as well as crops such as bamboo and grass [75].

The Institute of Natural Fibers introduced an environmentally safe technology for producing bleached pulp from hemp grown on land polluted by metallurgic industry.

The productivity of hemp biomass is very high: 2–2.5 times higher than can be produced from the same area of forest. The concentration of alpha-cellulose in the raw fiber (bast), reaches the level of 90% while the concentration of cellulose in softwood and hardwood ranges from 50% to 54%. Promising solvent methods for the pulping of annual plants include the following [76]:

- Acetosolve (Germany)—using solution of 85–93% acetic acid and addition of 0.1% Hydrochloric Acid (HCL), and MILOX (Finland)—using formic acid.
- Alcell method, employing ethyl alcohol.
- Steam Explosion Pulping (SEP)—using chemically infused raw material cooked with saturated steam.
- Ethanol Pulping (Egypt)—using ethanol/water, ethanol/water/NAOH solutions.
- Peroxy compounds using Peroxymonosulfate Anion ($HSO_5$).

### 3.9.4 Green Fiber for Biocomposites, Based on Lignocelluloses

Biocomposites comprise of a wide variety of products for different applications ranging from construction or insulation panels made of wood pieces, particles, and fibers, through special textiles, to plastic products based on polymers infused with lignocellulosic particles. Fiber plants are promising lignocellulosic raw materials for different applications [77].

These lignocellulosic are recyclable, and when joined with natural resin, they are as solid as steel yet of a lower density. Such composites may be used in motor vehicles, building materials, furniture, machine construction, insulating materials, gardening, agricultural equipment, and other uses. Instances of usage of composite materials containing lignocellulosic mechanisms include glue-lam wood, plywood, particle-board, fiber board, Medium Density Fiber (MDF), and Oriented Strand Board (OSB) [78].

### 3.9.5 Applications of Composites

Composite materials are used in (transportation industry) aircraft, railway, and automotive industries. Uses of natural fibers in the automotive industry include different types of fillings, reinforcing fiber and in some cases replacement of glass fiber, one of the components of hybrid composites, degradable composite of natural fibers and natural polymers. The great thing about composites reinforced with fibers is that when the fibers are orientated parallel to the direction of applied forces (unidirectional laminates), the possibility of utilizing anisotropic properties of material for structure arises (crushed fibers, embroidered and 3-D weaved structures).

The properties of this new material can be compared with regular glass fiber or reinforced polymers in normal conditions, for an example, in the automotive industry, road construction, furniture industry, and in sport. New nano-binders have been proposed for jute and hemp composites resulting in flame retardant construction elements with a compressive strength of 3.5 MPa by a specific weight of only 0.13 g/cm$^2$ [79].

### 3.9.6 Particleboards

Composition boards, including particle board (extruded and plate-pressed), and fiber boards, especially Medium Density Fiber (MDF) board, are highly common in construction, furniture and interior paneling. The most common raw material used is wood, but many countries productively use other agriculturally based raw materials such as flax and hemp shives, jute stalks, reed stalks, cotton stalks, palm stalks, and other raw materials [80].

Medium Density Fiber (MDF) boards consist 82% of fiber from wood or annual plants. There is a growing interest in using annual plants to make boards because annual plants are renewed each year and produce three times more cellulose per year than ring-growth in trees. Residues of annual plants such as rape or canola straw, oil flax straw, small grain straw, reed, and reed wastes are necessary for production of insulating boards for building industry and important for houses in earth-globe areas. Hemp also can be used to make excellent insulating boards [81].

## 3.10 Futuristic Applications of Biocomposites and Biopolymers

In addition to traditional uses, new applications are arising in the automotive and building industries. New composite materials based on plant fibers and polymers are being progressively used in the building industry in the United States and in car manufacturing in Europe. Plant fibers, such as flax or hemp, can be used as a strengthening agent for Polymerizing Vinyl Chloride (PVC), Polyethylene (PE), or Polypropylene (PP)-type polymers in place of synthetic fibers (glass, Kevlar, carbon, etc.). This replacement offers many perks [82]:

- Economic: lower costs because significantly reduced cycle times, energy savings during processing and low density.
- Technical: mechanical properties identical to those of traditional reinforcements, reduced tool wear, high geometric stability of the manufactured parts, good insulation characteristics.
- Environmental: renewable resource, easy to recycle, no material toxicity, reduced fossil fuels content, CO2-neutral materials. In the automotive sector, the processing technologies currently used are hot compression molding, injection, and extrusion.

Hot compression molding technology normally needs a nonwoven mat, containing either partially or entirely of plant fibers that may or may not be fused with synthetic fibers. While processing, the synthetic fibers melt at a temperature of approximately 180 °C and hence develop a matrix in which the plant fibers are embedded. These nonwovens are used as interior linings for car doors, parcel shelves, and boots or pillars. Most car manufacturers have at least one model equipped with these kinds of linings in their range [83].

The processing of plant fibers by means of injection and extrusion is a growing trend. These technologies use plant fibers in combination with polypropylene, polyethylene, and Acrylonitrile Butadiene Styrene (ABS). This kind of material is marketed in the form of pellets that can be used in injection or extrusion processes. The primary applications for injection molding include structural parts for the car industry. Extrusion is used to produce profiles, hollow components, etc. [84].

In the building sector, the use of plant fiber/polymer composites as construction materials is still very limited, but, it is in favor of increasing due to the high concerns regarding the toxicity of treatment agents for wood preservation. Products developed include window profiles, fences, shutters, gates, decking (terrace floors and outdoor profiles), and railings [85].

### 3.10.1 Development Prospects for Plant Fiber/Polymer Composites:

The advancement of plant fiber/polymer composites is at an emerging stage since they currently represent. This market has escalated quickly owing to the increasing cost of oil and the crisis affecting the automotive sector. Certainly, some specialists believe that the widespread application of these materials could bring innovation into the sector and provide it with renewed momentum.

The main aspects of research include the enhancement of fiber properties through a better understanding of their ingredients and their physical/chemical characteristics, mastering their reproducibility and their adaptability to specific technical applications [86]. Researchers are also focusing on the use of processing technologies, plant fiber extraction techniques, the processing of fibers to facilitate their adhesion to the polymer or biopolymer matrix, the optimization of current processes to adapt them to plant fibers (temperature and pressure resistance, stability, etc.), and understanding the development over time of the properties of fibers used in products (ageing, etc.).

As far as the structuring requirements of the plant fiber/polymer composites industry are concerned, Agence de l'Environnement et de la Maîtrise de l'Énergie (ADEME) a French Environment and Energy Management Agency is providing global support to the plant fibers industry, by bringing together the various researchers (scientists, technologists, plant fiber producers, processors, and users) in achieving general studies and by organizing technical or brainstorming sessions on this topic. Such studies include the market study of the new uses for plant and other studies carried out that aims to assess the current market for industrial bio-products and biofuels and soon the forecast developments [87].

## 3.11 Conclusion

The global push toward sustainable materials has brought to light natural, recyclable, and biodegradable raw materials, amongst them natural fibers. Recent accomplishments and new uses of these green fibers allow us to conclude the following:

- Fast rising population, eco and health awareness has created a large space for future growth of other than cottons natural cellulosic fibers.

- All the textile goods based on green fibers can boast high comfort/health properties and are ecological beneficial.
- Green fibers are increasingly used with non-wood pulps.
- Green fibers will also be used in growing amounts in a wide spectrum of biocomposites materials. Being lignocellulosic, they can be combined with man-made or natural polymers to provide a wide range of useful composites in textiles (including geotextiles and non-woven).

Reinforced natural fiber composites that are using biodegradable polymers as a matrix are said to be the most "green" materials. The establishing of novel polymers with inherent environmentally favorable characteristics such as renewability and degradability, a sequence of interesting polymers has been apprehended through multiple investigation actions ranging from thermoplastic starch and its blends, PLA and its alterations, cellulose, etc. Natural fiber being favored over synthetic fibers for environmental reasons has also been synthesized from agricultural sources such as kenaf, hemp, banana, sisal, and coconut coir. There is thus a wide range of possible applications of nanocomposites. However, the many problems of poor adhesion of matrix and fiber, difficulty of fiber orientation, the achievement of nanoscale sizes, and the evolution of truly green polymers, which will be environmentally friendly and renewable, are still to be addressed fully.

## References

1. Plackett, D., In: *Natural Fibres, Biopolymers and Biocomposites,* (Eds.), A.K. Mohanty, M. Misra, L.T. Drzal, S.E. Selke, B.R. Harte and G. Hinrichsen, CRC Press, Boca Raton, p. 569, 2005.
2. Njuguna, J., Pielichowski, K. and Desai, S., Nanofiller-reinforced polymer nanocomposites. *Polym. Adv. Technol.*, 19, 947–959, 2008.
3. Leja, K. and Lewandowicz, G., Polymer biodegradation and biodegradable polymers—a review. *Polish J. Environ. Stud.*, 19, 255–266, 2010.
4. Evans, W.J., Isaac, D.H., Suddell, B.C. and Crosky, A., In: *Natural Fibres and Their Composites: A Global Perspective, Proceedings of the 23rd Risr International Symposium on Materials Science: Sustainable Natural and Polymeric Composites- Science and Technology,* H. Lilholt *et al.* (Eds.), p. 1, Risr National Laboratory, Roskilde, Denmark, 2002.
5. Fuzhong Wang, Lawrence T. Drzal, Yan Qin, Zhixiong Huang. Multifunctional graphene nanoplatelets/cellulose nanocrystals composite paper. *Composites Part B: Engineering,* 79, 521–529, 2015.

6. Jamshidian, M., Tehrany, E.A., Imran M., Jacquot M. and Desobry S., Polylactic acid: Production, applications, nanocomposites, and release studies. *Compr. Rev. Food Sci. Food Saf.*, 9, 552–571, 2010.
7. Amass, W., Amass, A. and Tighe, B.A., Review of biodegradable polymers: Uses, current developments in the synthesis and characterization of biodegradable polyesters, blends of biodegradable polymers and recent advances in biodegradation studies. *Polym. Int.*, 47, 89–144, 1998.
8. Chandra, R. and Rustgi, R., Biodegradable polymers. *Prog. Polym. Sci.*, 23, 1273–1335, 1998.
9. Mohanty, A.K., Misra, M. and Hinrichsen, G., Biofibres, biodegradable polymers and biocomposites: An overview. *Macrmol. Mater. Eng.*, 276/277, 1–24, 2000.
10. Siracusa, V., Rocculi, P., Romani, S. and Rosa, M.D., Biodegradable polymers for food packaging: a review. *Trends Food Sci. Technol.* 2008, 19, 634–643.
11. Pandey, J.K., Chu, W.S., Lee, C.S. and Ahn, S.H., Preparation characterization and performance evaluation of nanocomposites from natural fibre reinforced biodegradable polymer matrix for automotive applications. *Presented at the International Symposium on Polymers and the Environment: Emerging Technology and Science, BioEnvironmental Polymer Society (BEPS)*, Vancouver, WA, USA, 17–20 October 2007.
12. Sinha, S.R. and Bousmina, M., Biodegradable polymer/layered silicate nanocomposites, in: *Polymer Nanocomposites*, Mai, Y., Yu, Z. (Eds.), pp. 57–129, Woodhead Publishing and Maney Publishing: Cambridge, England.
13. John, M.J. and Thomas, S. Biofibres and biocomposites. *Carbohyd. Polym.*, 71, 343–364, 2008.
14. Carvalho, A.J.F., Curvelo, A.A.S. and Agnelli, J.A.M.A., First insight on composites of thermoplastic starch and kaolin. *Carbohyd. Polym.*, 45, 189–194, 2001.
15. Pandey, J.K. and Singh, R.P., Green nanocomposites from renewable resources: Effect of plasticizer on the structure and material properties of clay-filled starch. *Starch/Stärke*, 57, 8–15, 2005.
16. Guan, J. and Hanna, M A., Selected morphological and functional properties of extruded acetylated starch-cellulose foams. *Bioresource Technol.*, 97, 1716–1726, 2006.
17. Kumar, A.P. and Singh, R.P., Biocomposites of cellulose reinforced starch: Improvement of properties by photo-induced crosslinking. *Bioresource Technol.*, 99, 8803–8809, 2008.
18. Lu, Y., Weng, L. and Cao, X., Morphological, thermal and mechanical properties of ramie crystallites—reinforced plasticized starch biocomposites. *Carbohyd. Polym.*, 63, 198–204, 2006.
19. Ma, X.F., Yu, J.G. and Wang, N., Fly ash-reinforced thermoplastic starch composites. *Carbohyd. Polym.*, 67, 32–39, 2007.
20. Svagan, A., Bio-inspired cellulose Nanocomposites and foams based on starch matrix. PhD thesis, Department of Fibre and Polymer Technology, KTH Chemical Science and Engineering, SE-100 44, Stockholm, Sweden, 2008.

21. Famá, L., Gerschenson, L. and Goyanes, S., Starch-vegetable fibre composites to protect food products. *Carbohyd. Polym.*, 75, 230–235, 2009.
22. Kaushik, A., Singh, M. and Verma, G., Green nanocomposites based on thermoplastic starch and steam exploded cellulose nanofibrils from wheat straw. *Carbohyd. Polym.*, 82, 337–345, 2010.
23. Liu, D., Zhong, T., Chang, P.R., Li, K. and Wu, Q., Starch composites reinforced by bamboo cellulosic crystals. *Bioresource Technol.*, 101, 2529–2536, 2010.
24. Guimarães, J.L., Wypych, F., Saul, C.K., Ramos, L.P. and Satyanarayana, K.G., Studies of the processing and characterization of corn starch and its composites with banana and sugarcane fibres from Brazil. *Carbohyd. Polym.*, 80, 130–138, 2010.
25. Kaith, B.S., Jindal, R., Jana, A.K. and Maiti, M., Development of corn starch based green composites reinforced with *Saccharum spontaneum* L fibre and graft copolymers—Evaluation of thermal, physico-chemical and mechanical properties. *Bioresource Technol.*, 101, 6843–6851,2010.
26. Ogata, N., Jimenez, G., Kawai H. and Ogihara T., Structure and thermal/mechanical properties of poly(L-lactide)-clay blend. *J. Polym. Sci. Part B: Polym. Phys.*, 35, 389–396, 1997.
27. Sinha, R.S., Okamoto, K., Yamada, K. and Okamoto, M., Novel porous ceramic material via burning of polylactide/layered silicate nanocomposite. *Nano Letts.*, 2, 423–426, 2002.
28. Sinha, R.S., Yamada K., Okamoto, M. and Ueda, K., New polylactide/layered silicate nanocomposite: A novel biodegradable material. *Nano Letts.*, 2, 1093–1096, 2002.
29. Sinha, R.S., Maiti P., Okamoto, M., Yamada, K. and Ueda, K., New polylactide/layered silicate nanocomposites. 1. Preparation, characterization and properties. *Macromolecule*, 35, 3104–3110, 2002.
30. Sinha, R.S., Yamada, K., Ogami A., Okamoto, M. and Ueda, K., New polylactide layered silicate nanocomposite: Nanoscale regulation of multiple properties. *Macromol. Rapid Commun.*, 23, 493–497, 2002.
31. Sinha, R.S., Okamoto, M., Yamada, K. and Ueda, K. New biodegradable polylactide/layered silicate nanocomposites: Preparation, characterization and materials properties. *Macromolecules*, 35, 659–660, 2002.
32. Sinha, R.S., Okamoto, M., Yamada, K. and Ueda, K., New polylactide/layered silicate nanocomposites: Concurrent improvement of materials properties and biodegradability. *Polymer*, 44, 857–866, 2003.
33. Yamada, K., Ueda, K., Sinha, R.S. and Okamoto, M., Preparation and properties of polylactide/layered silicate nanocomposites. *Kobunshi Robunshu*, 59, 760–765, 2002.
34. Maiti, P., Yamada, K., Okamoto, M, Ueda, K. and Okamoto, K., New polylactide/ layered silicate Nanocomposites: role of organoclay. *Chem. Mater.*, 14, 4654–4661, 2002.
35. Paul, M.A., Alexandre, M., Degee, P., Calberg, C., Jerome, R. and Dubois, P., Exfoliated polylactide/clay nanocomposites by *in-situ* coordination-insertion polymerization. *Macromol. Rapid Commun*, 24., 561–566, 2003.

36. Lee, J.H., Park, T.G., Park, H.S., Lee, D.S., Lee, Y.K., Yoon, S.C. and Nam, J.D., Thermal and mechanical characteristics of poly (L-lactic acid) nanocomposite scaffold. *Biomaterials*, 24, 2773–2778, 2002.
37. Chang, J., An, Y.U., Cho, D. and Giannelis E.P., Poly (lactic acid) nanocomposites: Comparison of their properties with montmorillonite and synthetic mica (II). *Polymer*, 44, 3715-3720, 2003.
38. Bondeson, D. and Oksman, K., Dispersion and characteristics of surfactant modified cellulose whiskers nanocomposites. *Compos. Interface.*, 14, 617–630, 2007.
39. Lee, S., Kang, I., Doh, G., Yoon, H., Park, B. and Wu, Q., Thermal and Mechanical Properties of Wood Flour/Talc-filled Polylactic Acid Composites: Effect of Filler Content and Coupling Treatment. *J. Thermoplast. Compos. Mater.*, 21, 209–223, 2008.
40. Qu, P., Gao, Y., Wu, G. and Zhang, L., Nanocomposites of poly (lactic acid) reinforced with cellulose nanofibrils. *BioResources*, 5, 1811–1823, 2010.
41. Misra, M., Park, H., Mohanty, A.K. and Drzal, L.T., Injection molded 'Green' nanocomposite materials from renewable resources. *Presented at the Global Plastics Environmental Conference*, Detroit, MI, USA, 18–19 February 2004.
42. Mahadeva, S.K., Yun, S. and Kim, J., Flexible humidity and temperature sensor based cellulose-polypyrrole nanocomposite. *Sensor. Actuator. A Phys.*,165, 194–199, 2011
43. Tunç, S.; Duman, O. Preparation of active antimicrobial methyl cellulose/carvacrol/montmorillonite nanocomposite films and investigation of carvacrol release. *Food Sci. Technol.*, 44, 465–472, 2011.
44. Zimmermann, K.A., LeBlanc, J.M., Sheets, K.T., Fox, R.W. and Gatenholm, P., Biomimetic design of a bacterial cellulose/hydroxyapatite nanocomposite for bone healing applications. *Mater. Sci. Eng.*, 31, 43–49, 2011.
45. Zadegan, S., Hosainalipour, M., Rezaie, H.R., Ghassai, H. and Shokrgozar, M.A., Synthesis and biocompatibility evaluation of cellulose/hydroxyapatite nanocomposite scaffold in 1-n-allyl-3-methylimidazolium chloride. *Mater. Sci. Eng.*, 31, 954–961, 2011.
46. Sithique, M.A. and Alagar, M., Preparation and properties of bio-based nanocomposites from epoxidized soy bean oil and layered silicate. *Malaysian Polym. J.*, 5, 151–161, 2010.
47. Azeredo, H.M.C., Mattoso, L.H.C., Wood, D., Williams, T.G., Avena-Bustillos, R.J. and Mchugh, T.H., Nanocomposite edible films from mango puree reinforced with cellulose nanofibres. *J. Food Sci.*, 74, 31–35, 2009.
48. Tate, J.S., Akinola, A.T. and Kabakov, D., Bio-based Nanocomposites: An Alternative to Traditional Composites. *J. Technol. Stud.*, 1, 25–32, 2010.
49. Ke, T.Y. and Sun, X.Z., Effects of moisture content and heat treatment on the physical properties of starch and poly (lactic acid) blends. *J. Appl. Polym. Sci.*, 81, 3069–3082, 2001.
50. Uesaka, T., Nakane, K., Maeda, S., Ogihara, T. and Ogata, N., Structure and physical properties of poly (butylene succinate)/cellulose acetate blends. *Polymer*, 41, 8449–8454, 2000.

51. Kesel, C.D., Wauven, C.V. and David, C., Biodegradation of polycaprolactone and its blends with poly(vinylalcohol) by micro-organisms from a compost of house-hold refuse. *Polym. Degrad. Stab.*, 55, 107–113, 1997.
52. Averous, L., Fauconnier, N. and Moro, L., Fringant Blends of thermoplastic starch and polyesteramide: Processing and properties. *J. Appl. Polym. Sci.*, 76, 1117–1128, 2000.
53. Willett, J.L. and Shogren, R.L., Processing and properties of extruded starch/polymer foams. *Polymer*, 43, 5935–5947, 2002.
54. Martin, O. and Averous, L., Poly (lactic acid): Plasticization and properties of biodegradable multiphase systems. *Polymer* 42, 6209–6219, 2001.
55. Sarazin, P., Li, G., Orts, W.J. and Favis, B.D., Binary and ternary blends of polylactide, polycaprolactone and thermoplastic starch. *Polymer*, 49, 599–609, 2008.
56. Majdzadeh-Ardakani, K., Sadeghi-Ardakani, Sh., Experimental investigation of mechanical properties of Starch/natural rubber/clay nanocomposites. *Digest J. Nanomater. Biostruct.*, 5, 307–316, 2010.
57. Maiti, P., Batt, C.A., and Giannelis, E.P., Renewable plastics: Synthesis and properties of PHB nanocomposites. *Polym. Mater. Sci. Eng.*, 88, 58–59, 2003.
58. Zheng, J.P., Li, P., Ma, Y.L. and Yao, K.D., Gelatine/montmorillonite hybrid nanocomposite. I. Preparation and properties. *J. Appl. Polym. Sci.*, 86, 1189–1194, 2002.
59. Takegawa, A., Murakami, M., Kaneko, Y. and Kadokawa, J., Preparation of chitin/cellulose composite gels and films with ionic liquids. *Carbohyd. Polym.*, 79, 85–90, 2010.
60. Nunes, M.R.S., Silva, R.C., Silva, J.G., Jr., Tonholo, J.,and Ribeiro, A.S., Preparation and morphological characterization of chitosan/clay nanocomposites. In *Proceedings of the 11th International Conference on Advanced Materials*, Rio de jenero, Brazil, 20–25 September pp. 20–25, 2009.
61. Pothan, L.A. and Thomas, S., Polarity parameters and dynamic mechanical behavior of chemically modified banana fibre reinforced polyester composites. *Compos. Sci. Technol.*, 63, 1231–1240, 2003.
62. Zemljic, L.F., Stenius, P., Stana-kleinschek, J. and Ribitsch, V., Characterization of cotton fibres modified by carboxymethyl cellulose. *Lenzinger Berichte*, 85, 68–76, 2006.
63. Ramesh, M., Palanikumar, K. and Hemachandra Reddy, K., Mechanical property evaluation of sisal–jute–glass fibre reinforced polyester composites, *Composites Part B: Engineering*, Vol. 48 1–9, 2013.
64. Kozlowski, R. and Manys, S., *Linen Knitted Apparel - Speciality for Hot Climates*. Proc. of IFKT 39$^{th}$ Intern. Congress. 4–8 October, Busto, Italy, 1998.
65. Kozlowski, R., and Manys, S., *Coexistence and Competition of Natural and Man-made Fibres*. Proc. of the 78$^{th}$ World Conference of the Textile Institute, 23–27 May, Thessaloniki, Greece, 1997.
66. Kozlowski, R. and Manys, S., *The Properties of Liquid Ammonia Treated Linen*. Proc. of 212$^{th}$ ACSNM, 25–29 August, Orlando, USA, 1996.

67. Manys, S. and Mazur, E., *Linen for aged/ill people.* Nord Flax Symposium, 10–12 August, Tampere, Finland, 1998.
68. Kaushik, R.C.D. and Agarwal, D.R., *Jute - The Golden Fibre - Rediscovered.* Proc. of World Textile Congress on Natural and Natural-Polymer Fibres, 9–11 July, Huddersfield, England, 1997.
69. Kessler, R.W. and Tubach, M., *Interdisciplinary approach for New Flax Products.* Proc. of the Fibre Flax World Congress, 22–25 May, New Haven, USA, 1994.
70. Weight, H., *Class Erntetechnik für Nachwachsende Rohstoffe.* Proc. of the Intern. Seminar, 9 July, Harsenwinkel, Germany,1992.
71. TEMAFA - FLACHS AUFBERITUNG, Proc. of the 3rd European Workshop on Flax, 15–17 June, Bonn, Germany, 1993.
72. Janosik, S., Maixner, V., Tessar, A., Manys, S., Manys, B. and Mazur,E., *New Era in Flax Processing.* Proc. of Hemp, Flax and Other Bast Fibrous Plants Production, Technology and Ecology Symposium, 24–25 September 1998, Poznan, Poland, 1999.
73. Keshavamutry, G.S., *Cellulose Fibre - Indian Scenario.* Proc. of the TI's 77th World Conference, 22–23 May, Tampere, Finland, 1996.
74. Jayachandran, K. and Rana, A.K., *Jute Fibre for Reinforced Composites and its Prospects.* Book of Abstracts. of the 5th Intern. Conference on Frontiers of Polymers and Advanced Materials, 21–25 June, Poznan, Poland, 1999.
75. Rowell, R.M. and Harrison, S.M., *Proc. of the Intern. Kenaf Conference*, March, Ca, USA, 1993.
76. Gusman Ferraz, J.M., *Sustainability and Land Reform Policy.* Proc. of SATHLA Intern. Conference, 9–13 March, 1998, Rio de Janeiro, Brazil, 1998.
77. Doppler, W., *Farmer's Business and Participation for Sustainable Agriculture.* Proc. of SATHLA Intern. Conference, 9–13 March, Rio de Janeiro, Brazil, 1998.
78. Geodicke, S., Kalleder, A., Reinhard, R. and Schmidt, S., *Nanobinders for Natural Composites.* Book of Abstracts of NATO Advanced Research Workshop on Polymers and Composites for Special Applications, 21–25 June, Poznan, Poland, 1999.
79. Kozlowski, R., and Przepiera, A., *Method of Continuous Production of Insulating Boards from Plant Wastes.* (Metoda ciaglej produkcji plyt izolacyjnych z odpadów roslin. Prace IKWN, Rocznik XXXI. (in Polish)), 1987.
80. Kozlowski, R., Mieleniak, B. and Przepiera, A., *Plant Residues as Raw Materials for Particleboards.* Proc. of the 28th Intern. Particleboards Composite Materials Symposium, W. S. U. Pullman, USA, 1994.
81. Kozlowski, R. and Helwig, M., 1996 *Critical Look on Cellulose Modification.* Proc. of the Intern. Symposium on Cellulose Modification, December, Honolulu, USA, 1996.
82. Kozlowski, R. and Helwig, M., *Lignocellulosic Polymer Composites.* Proc. of 4th Intern. Conference on Frontiers of Polymers and Advanced Materials, Cairo, 1997.

83. Kozlowska, J. and Biskupski, M., *Characteristic of Cannabinoid Profiles Occurring in Different Phenotypes of Cannabis.* Proc. of Hemp, Flax and Other Bast Fibrous Plants. Production, Technology and Ecology Symposium, 24–25 September, Poznan, Poland, 1998.
84. McIntyre, *The future of man-made fibres.* Proc. of the Intern. Symposium on Textile Science and Technology, 26–28 October, Yokohama, Japan, 1994.
85. Meadows, D.H., Meadows, D.L. and Randeres, J., *Beyond the limits.* Chelsea Green Publishing Company, 1992.
86. Rowell, R.M., and Harrison, S.M., *Proc. of the Intern. Kenaf Conference*, March, Ca, USA, 1993.
87. Rowell R.M., Young, R.A.J. and Rowell, K., *Paper and composites from agro-based resources.* Lewis Publishers, New York, London, Tokyo, 1997.

4

# Green and Sustainable Advanced Nanomaterials

**Alaa K. H. Al-Khalaf[1]* and Falah H. Hussein[2]**

*[1]College of Environmental Sciences, Al-Qasim Green University, Babylon, Iraq*
*[2]College of Pharmacy, Babylon University, Babylon, Iraq*

***Abstract***
Green chemistry and their principles with metrics of environmental efficiencies are considered the main approach to achieve sustainable development. The principles of green chemistry are summarized as prevention, atom economy, less hazardous chemical synthesis, designing safer chemicals, design for energy efficiency, and use of renewable feedstocks, catalysis, and design for degradation. Several studies have found the possibility of using alternative natural molecular dyes in the field of water purification. The natural dye of anthocyanin was used as a safe sensitizer for nanoparticles of titanium dioxide in the photodegradation process of organic pollutants in water.

Some of science policy reports have mentioned that there is great opportunity and benefit from providing green nano processes by producing nanomaterials and products without pollutants harming the environment or human health at stages of managing, design, manufacturing, and application. Thus, nanotechnology can help alleviate issues of the safe sustainable development such as environmental, human health, and safety risks and support for a sustainable environment with respect to energy, water, food production, raw materials, climate change, etc.

***Keywords:*** Green chemistry, nanomaterials, natural products, sustainable development

## 4.1 Introduction

During the past years, several topics of the search-strategic have appeared, one of them is the combination between techniques of green chemistry

**Corresponding author*: dralaa_al-khalaf@hotmail.co.uk

Shakeel Ahmed and Chaudhery Mustansar Hussain (eds.) Green and Sustainable Advanced Materials, Vol. 1 (93–106) © 2018 Scrivener Publishing LLC

and nanotechnology applications [1]. The nanomaterials are costly systems based on hazardous heavy metal such as CdS particles and Ru-compounds.

The new catalyst $TiO_2$/anthocyanin showed effectively catalyzed photodegradation under solar radiation than the earlier systems previously used, while avoiding their hazardous nature as well as the possibility of enhancing their efficiency and ease to recovery from catalytic reaction by supported onto activated carbon particles, but the process need to reactivate the recovered catalyst by re-treated with fresh dye of anthocyanin owing to their degradation [2, 3].

Nowadays, scientists have focused on creating a sustainable green technologies used in the manufacture of nanomaterials, as in: Nanocatalysis, Self-Assembly, used of solid state-solventless process, Non-toxic starting materials, Renewable starting materials, Safe photochemical synthesis, Aqueous processing, Bio-inspired nanoscale synthesis, Improved synthesis-fewer steps, Microwave techniques, Renewable in nanocomposites, Non-toxic starting materials [4, 5].

It has become a trend specialist scientists to prepare biopolymers from natural sources, as in the corn plant and the roots of sugar beet where the starch is extracted from these plants and then prepare developed materials as a green nanocatalysts have a direct interest in reducing the reaction time and temperature in a safe reactors compared to common commercial processes [6].

### 4.1.1 Green Chemistry and Nanoscale Science

In recent years, there has been a growing interest in nanomaterials for their multiple physical properties and applications in different areas of human life, especially when scientists were directed to synthesize nanoparticles of 0.1–100 nm by using the 12 principles of green chemistry science, for the purpose of achieving comprehensive sustainable development [7, 8].

### 4.1.2 Examples of Such Green Nanoparticles:

#### *4.1.2.1 Beta-Carotene Molecule*

In theory, for example, the length of the non-dissolved beta-carotene molecule was found to be 0.294 nm, consisting of (10) carbon–carbon single bonds, (11) double carbon–carbon atoms, Thus producing a long hydrocarbon chain consisting of (22) carbon atoms as in the following chemical structure in Figure 4.1 [9]:

In 1996 [10] a research study was published showing the preparation and diagnosis of water-soluble beta-carotene molecules as a nanoparticles

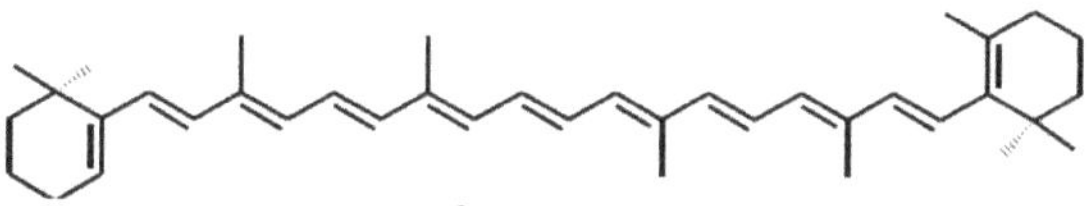


**Figure 4.1** Chemical composition of beta-carotene molecule.

**Figure 4.2** Control: Dispersion by water without dextrin. β-Carotene dispersion: Dispersion by water after complex formation with starch dextrin.

through the use of phase separation and polymerization techniques of the polymer surface of the packaging material used. The size of the prepared nanoparticles that coated with gelatin ranged from (15 to 250) nm with high thermal stability, giving new important applications to the pharmaceutical industry and the food industry.

Another study in 2013 [11] has shown that a nanocomposite of beta-carotene molecules can be synthesized through a complex by mixing with (30%) by weight of starch dextrin by using the water dispersion method for the non-mixed phases of hydrothermal starch and organic beta-carotene solutions and their solubility in water through the gradual evaporation of organic solvent during continuous mechanical stirring, resulting in aqueous solutions in which beta-carotene particles have been homogenously distributed with dextrin of corn starch, as in Figure 4.2, and thus reached a range of (16–30) nm and therefore gave great benefit to nanoparticles prepared in terms of high stability against oxidation and enzymatic digestion.

### 4.1.2.2 *Anthocyanin Molecule*

Natural anthocyanin is a water soluble organic compound, their color gradually changes from violet-blue-red, depending on the change in pH value. It has multiple functions as an antioxidant, scavenger of free radicals, and anti-cancer through their ability to absorb and protect cells from UV-radiation damage.

Several recent studies have included the use of alternative natural dyes which composite with nanomaterials that available in nature to provide them with feature of safe green sustainable reactions within the photochemical degradation processes of industrial organic pollutants.

The researchers explained the use of anthocyanin dye in combination with the optical catalyst of ($TiO_2$) nanoparticle as a photo sensor in photo-degradation of the industrial dye of orange methyl for the purpose of improving the properties of ($TiO_2$) by increasing its effectiveness and the possibility of linking the available biomass with the technique of nanotechnology and making it a sustainable green process.

The extracted anthocyanin dye from Hibiscus flowers was used as natural alternatives in a nanocomposite formula with $TiO_2$ instead of the use of high-cost toxic heavy metal compounds (cadmium sulphide and ruthenium compounds) through their use as photo catalyst ($TiO_2$/anthocyanin), as a safe photo sensor for $TiO_2$ particles during the photo-degradation reaction of organic pollutants in the water purification process [12].

Figure 4.3 shows the chemical structure of the anthocyanin molecule before and after the process of physical and chemical adsorption with ($TiO_2$) by bonding between the metal and the oxygen atoms of the molecule.

This dye is a promising alternative to expensive systems based on heavy, toxic, and non-essential elements. It was found to be very effective in the process of photo-degradation of organic pollutants under the influence of solar radiation with catalyst according to mechanism and the reactions

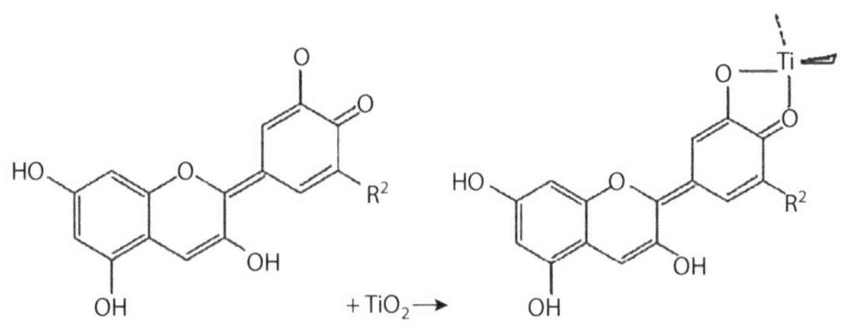


**Figure 4.3** Describe the structure of anthocyanin before and after adsorption on $TiO_2$ surface.

suggested in (Figures 4.4 and 4.5), through which the high efficiency and non-dissolving in water, even after the reinforcement by activated carbon, the system showed an increase in efficiency and ease of improvement of the amount of catalyst used in the reaction mixture compared to previous systems as well as avoid the toxic nature.

Figure 4.4, shows the role of anthocyanins as a photo sensor by its high ability to form pairs of (electron-gap) on its outer surface. Thus, the excited electrons will move to the catalyst for the reduction process, as well as their participation in the process of oxidation together by the effect of the solar radiations as an energy of electrons excitation in the equivalence band and transfer to the empty conductive band (the reduction process occurs), leaving behind an empty positive charge gap (the oxidation process occurs).

From the observation of the proposed photoreactions in Figure 4.5, the role of the composite dye in the excitation process by the solar radiation and the formation of electron-gap pairs on its outer surface and then transfer them to the organic pollutant using air oxygen in the formation of free radicals (Hydroxyl radical) after bonding with liberated protons and eventually forming the byproducts of the photo-degradation process of organic pollutant (water vapor and carbon dioxide).

In general, with regard to the process of photosensitization to a system containing semiconductors in organic dye solutions with a polychromatic

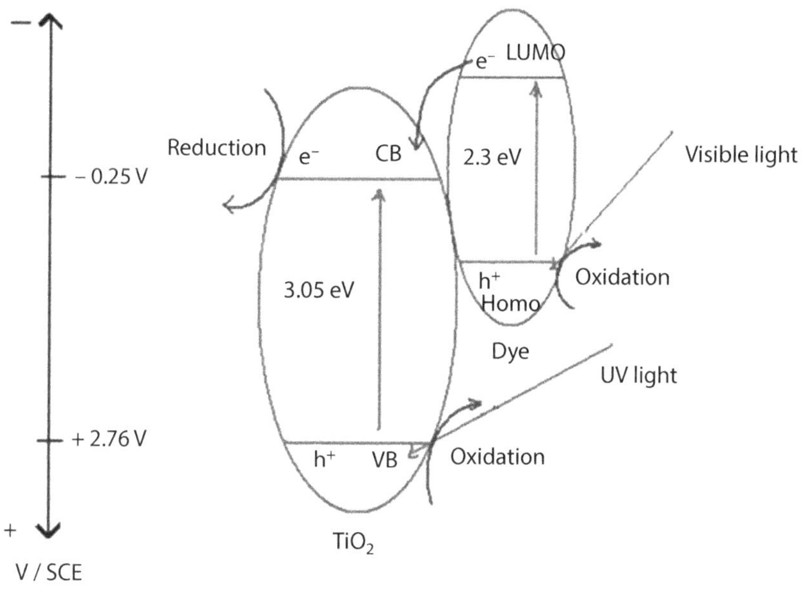


**Figure 4.4** Proposed mechanism of the photo-sensor for the $TiO_2$ catalyst.

$$TiO_2/A + h\nu \rightarrow TiO_2/A^*$$

$$TiO_2/A^* \rightarrow TiO_2/(A^+ + e^-_{CB})$$

$$TiO_2/(A^+ + e^-_{CB}) + MO \rightarrow TiO_2/(A + e^-_{CB}) + MO^{+*}$$

$$TiO_2/(A + e^-_{CB}) + O_2 \rightarrow TiO_2/A + {\cdot}O^-_2$$

$${\cdot}O^-_2 + H^+ \rightarrow {\cdot}HO_2$$

$${\cdot}O^-_2/{\cdot}HO_2 + MO/MO^{+*} \rightarrow \text{Products}$$

(hv = visible light; $TiO_2/A$ = $TiO_2$/anthocyanin; $e^-CB$ = is electron injected to the semiconductor conduction band; MO = Methyl orange)

**Figure 4.5** Suggested photoreactions of the photodegradation for the orange dye using catalyst.

light, this leads to the possibility of the presence of two mechanisms of the process which depends on the amount of energy from the incident light, whether equal and higher than the band gap of the semiconductor or less than it [13, 14].

In the first mechanism: the positive holes in the valence band will be created because the promoted electrons are moving to the conduction band due to the light energy that is equal or higher than the band gap of the illuminated semiconductors.

The formed photo holes and electrons will react with both water and oxygen molecules absorbed on the surface of semiconductors and thus produce hydroxyl free radicals (·OH) and perhydroxyl free radicals ($HO_2{\cdot}$), respectively, as in Scheme 4.1.

In the second mechanism: the adsorbed dyes molecules on the surface of the semiconductor have ability to absorb a radiation of visible spectra in addition to the other part of light with energy which is less than the band gap of the illuminated semiconductor. The excited dyes can inject an electron to the conduction band of the semiconductor. However, the dye molecules will be suffering from photocatalytic decolorization owing to photosensitization process, as in Scheme 4.2.

### *4.1.2.3 Hydro Gel*

In previous years, other studies have shown that organic hydrogel is a three-dimensional water-phobic network that is bound together by chemical or

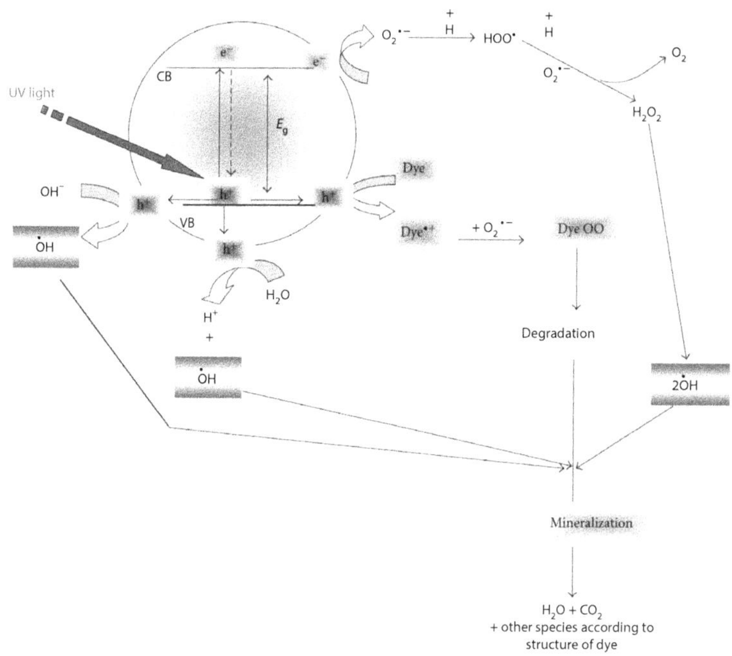


**Scheme 4.1** Dye/semiconductor/UV light system.

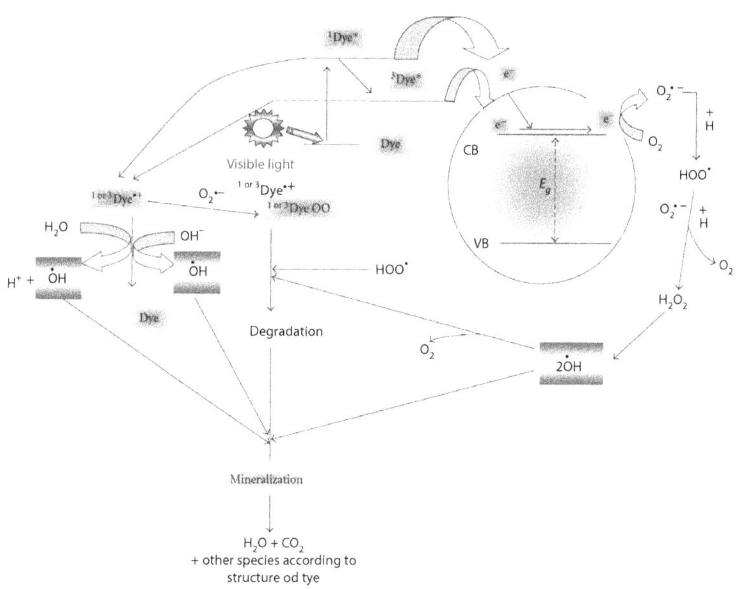


**Scheme 4.2** Nonregenerative dye/semiconductor/visible light system.

physical bonds, so that the absorbed water is not released under normal conditions because of the presence of water-phobic aggregates in the generation of hydrogel, has the ability to absorb and store the water molecules by filling the space that permeates inside the network, so that the water molecules become trapped and immobile. The hydrogel properties of water absorption and storage eventually led to significant and specific benefit in the treatment of wounds [15, 16].

Recently, hydrogels have been used from natural polymers, especially polysaccharides, from corn starch because they have unique advantages that apply to the principles of green chemistry and sustainable development, by being available in nature, non-toxic, biologically compatible, and biodegradable [17].

However, polysaccharides are easily dissolved in water, so they cannot form a stable hydrogel. The need and interest at both levels of the academic and industrial have been demonstrated by finding an effective way of making them in the form of hydrogel-based polymer networks by mixing and bonding the natural and synthetic polymers with a specific bonding material to form the required broad-use membranes [18] such as in the manufacture of contact lenses [19] as bone separation and bone transplantation [20] as a natural carrier related to blood, and applications related to enzymes, hormones, contraceptives, anticoagulants [21].

## 4.2 Applications of Natural NanoOrganic Materials

### 4.2.1 Application of Beta-Carotene

In 2013 [22] a study showed the possibility of using many of the techniques available for nanoscale packaging for organic compounds, such as beta-carotene. For example, self-assembly or aggregation, high homogeneous pressure, nanoscale emulsification, supercritical liquids, etc.

In 2015 [23] Japanese study showed the possibility of converting the lamellar microscale into the spherical nanoscale of the dissolved beta-carotene particles in solvent of ethyl acetate using the supercritical fluid method by increasing the pressure of the $CO_2$ flow rate (a supercritical fluid) at 40 °C as in Photo 4.1.

### 4.2.2 Application of Anthocyanin

On the other hand, in 2015 [24] study showed the possibility of treatment the problem of instability of anthocyanin dye in the food industry through the use of micro-encapsulation technology or packaging of

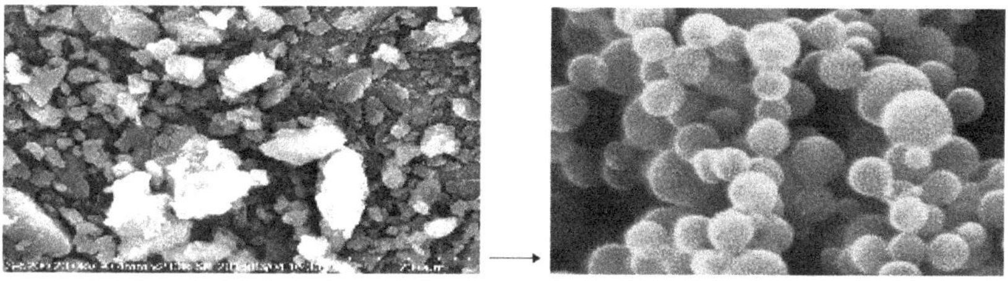

**Photo 4.1** SEM images for the transformation of the micro-size of raw material of beta-carotene crystals into nanoparticle size.

natural anthocyanin pigment particles extracted from red cabbage. In this study, the surface response method was used to improve the technology of fine-grained process for anthocyanin by using a spray drying technique for freeze-dried anthocyanin in an air free atmosphere as a core material (capsule center), in addition to maltodextrin (derived from starch used as food additive) with an Arabic gum (Gum Acacia), which are used as a materials for the capsule wall coated for anthocyanin dye, where the anthocyanin capsule showed better stability when exposed to different temperatures.

### 4.2.3 Application of Hydrogel

In 2006 [25] research study was conducted on the wide medical applications of the hydrogel extracted from corn starch, such as in artificial skin, which showed the use of hydrogel in protecting the infected skin and maintaining the surface of the wound in a wet to accelerate the healing process by absorption of secretions with maintaining the repair of damaged tissues, including the growth of new tissues.

As a result of the published literature, the hydrogel membrane was consisted from the corn starch bonded with both polyvinyl alcohol used in the manufacture of tissue and the bilateral binding unit (Glutaral solution used in sterilization and medicine), as demonstrated by the Infra-red spectra disappearance of the aldehyde groups and thus reduces the chance of poisoning the prepared membrane, in addition, the resulting membrane has sufficient strength to be used as an artificial skin, as in the compositions that show its components in Figure 4.6 [26–28].

In 2011, a research study showed that super-absorbent hydrogels have the potential to be used as a carrier systems of an oral medicines through the production of super-absorbent organic polymers derived from starch used as a carriers depending on the pH function in order to control on the process of the transport of loaded drugs and characterized by technique of scanning electron microscope (SEM).

The effect of the acidic function on the process was studied, in addition to the level of the released water dissoluble drug to its quantity loaded on

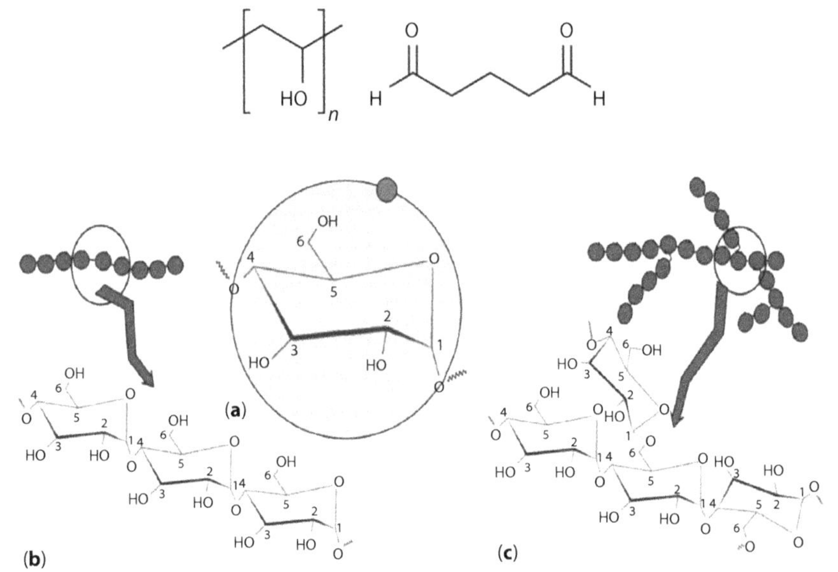

**Figure 4.6** Chemical structures of hydrogel membrane-making materials as artificial skin.

the hydrogels polymers under conditions of acidity of the stomach and intestines of the human body.

It was found that the best release of the drug in the acidic medium is much higher than the pH of the stomach and thus ensure the passage of the drug to the small intestine and the completion of the liberation process in the basic medium because of the presence of hydroxyl groups, which increases the electrostatic repulsion between carboxylic groups and thus increase the degree of hydrogel swelling, which increases the release of metronidazole [29].

In 2012, a super absorbent hydrogel was prepared with reverse behavior (water lost in the acidic medium and acquired at basic) to respond to changes in pH values, which can be used in effective drug transport systems and biomarkers.

The method of preparation included a co-polymerization reaction in aqueous medium with the precursor (ammonium persulfate) via the linkage (N,N'-Methylene bis-acrylamide) for a mixture of acrylic acid and a compound (2-Hydroxy ethyl methacrylate) (HEMA) on the carbonic structure of starch to form a (starch-polyacrylate) polymer as in the following structure in the Figure 4.7, demonstrated by the SEM spectra [30].

From the above diagram we observed the formation reaction of the ether bond between starch oxygen atoms by the precursor with the double bond in acrylic acid, the other end of which interacts with the linkage that it collects with the (HEMA) compound to form the hydrogel polymer.

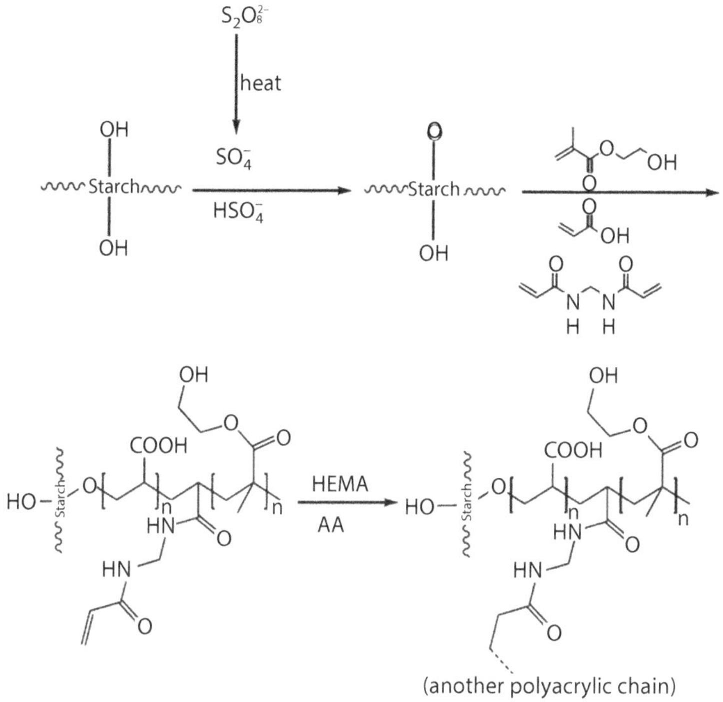


**Figure 4.7** Method of preparation of the hydrogel polymer.

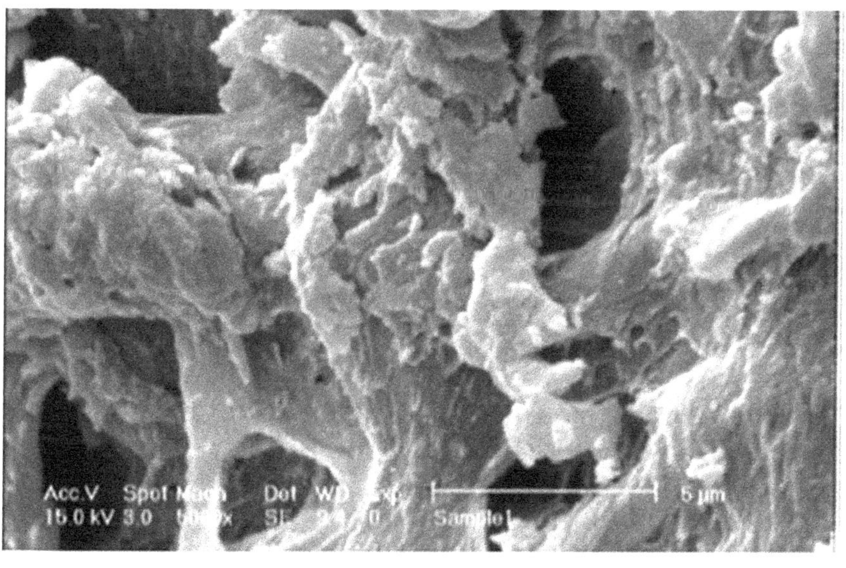


**Photo 4.2** SEM images for the porous surface of hydrogel polymer.

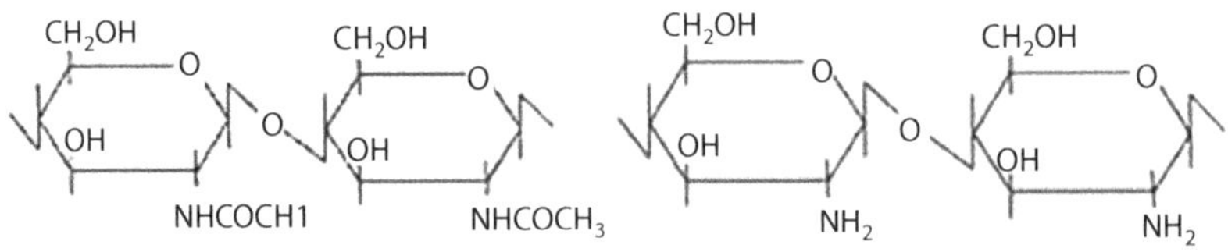


**Figure 4.8** Chemical structure of chitosan polymer (containing $NH_2$-groups) after removal of acetyl groups.

**Table 4.1** The amount of absorbed copper after add (50 mg) of chitosan at 30 °C and (pH = 5.7).

| Concentration of $Cu^{2+}$ adsorbents after addition of chitosan/unit (ppm) | Concentration of copper ions $Cu^{2+}$ in the original solution/unit (ppm) |
|---|---|
| 2.7837 | 3 |
| 4.6066 | 5 |
| 6.9451 | 8 |

In the Photo 4.2 we observed the porosity of the prepared product, which is a clear indication of its wide ability of water absorption in the basic medium and its released in the acidic medium, where the maximum water absorption reached (95 g/g) in the basic solutions (pH = 8). On the other hand, a study showed that the biopolymer (Chitosan) was prepared from the shellfish after four chemical processes to remove proteins, metals, colors and acetyl groups as in Figure 4.8:

This polymer has the ability to absorb heavy metals through physical or chemical adsorption between amine and hydroxyl groups with metal ions as in copper and thus can be used to remove them and other heavy metals from water effluents as well as being available in nature and thus economically beneficial for processing wastewater containing heavy metals, as in Table 4.1 [31]:

## 4.3 Conclusion

The applied chemical and physical properties of natural nano organic compounds that extracted or derived from biomass, such as the beta-carotene, porous hydrogel, and photo sensor of natural anthocyanin and their color changes gradually depending on the pH changing.

These compounds and their unique properties have brought the interest and product of researchers in medical and industrial applications to a large extent and put them firmly into the newly published scientific literatures, as a case of combining applications of nano biotechnology, green chemistry, and sustainable development and their unification into sustainable green nanoscale applications.

## References


1. Al-Khalaf, A.K.H., Effect of Solid Catalysts on Selective Formation of Naphthalene Derivatives, *PhD Thesis*, Cardiff University, Wales-UK, 2012.
2. Smith, S., Al-Khalaf, A.K.H., El-Hiti, G.A. and Pattisson,, S., Highly regioselective di-*tert*-amylation of naphthalene over reusable H-mordenite zeolite, *Green Chem.*, 14, 1103-1110, 2012.
3. Al-Khalaf, A.K.H., Green chemistry and sustainability of chemical safety and security, *Irq. Nat. J. Chem.*, 16, 71–81, 2016.
4. Mousa, A.O., Al-Khalaf, A.K.H. and Zghair, M.A., Preparation and study the structural properties and magnetic susceptibility of a liquid crystal, *J. Babylon Univ,, Pure Appl. Sci.*, 24, 718–726, 2016.
5. Al-Khalaf, A.K.H., Haddawi, S.M. and Yasser, O.M., New types of liquid crystalline semiconductors, *Irq. Nat. J. Chem.*, 24, 475-487, 2006.
6. Al-Khalaf, A.K.H., Green chemical applications: Highly efficient process on selective formation of industrial materials, *Int. J. Curr. Res. Aca. Rev.*, 2, 15–19, 2015.
7. Lancaster, M., *Green Chemistry: An Introductory Text*, pp. 2–6, RSC Paperbacks, University of York, UK, 2002.
8. Anastas, P.T., John C., and Warner, J., *Green Chemistry: Theory and Practice*, pp. 10–20, Oxford Press, Oxford, USA, 1998.
9. Atkins, P., and de Paula, J., *Physical Chemistry*, pp. 205–250, 8th edition, Oxford Press, Oxford, USA, 2006.
10. Horn, D., and Luddecke, E., Preparation and characterization of nono sized carotenoid hydrosols, *Fine Particles Science and Technology*, E. Pelizzeni (Ed.), pp. 761–775, Kluwer Academic Publishers, 1996.
11. Kim, J., Seo, T. and Lim, S., Preparation of aqueous dispersion of β-carotene nano-composites through complex formation with starch dextrin, *Food Hydrocolloids*, 33, 256–263, 2013.
12. Zyoud, A., Zaatar, N., Saadeddin, I., Helal, M.H., Campet, G., Hakim, M., Park, D. and Hilal, H., Alternative natural dyes in water purification: anthocyanin as TiO2-sensitizer in methyl orange photo-degradation, *J. Solid State Sci.*, 13, 1268–1275, 2011.
13. Hussein, F., Alkhateeb, A. and Ismail, J., Solar photolysis and photocatalytic decolorization of thymol blue, *E-J. Chem.*, 5, 243–250, 2008.
14. Hussein, F.H., Photochemical Treatments of Textile Industries Wastewater, Advances in Treating Textile Effluent, Hauser, P. (Ed.), pp. 5–10, InTech, 2011.

15. LaPorte, R., *Hydrophilic polymer coating for medical devices*, pp. 19–50, Lancaster, USA: Technomic Publishing Co, 1997.
16. Mack, E.J., Okano T. and Kim, S.W., *Hydrogels in medicine and pharmacy-polymers*, pp. 65, Vol. II, CRC Press, Boca Raton, USA, 1988.
17. Cascone, M., Barbani, N., Cristallini, C., Giusti, P., Ciardelli, G., and Lazzeri, L., Bioartificial polymeric materials based on polysaccharides, *J. Biomat. Sci. Polymer Edition*, 123, 267–281, 2001.
18. Brinkman, E., Van der, D., and Bantjes, A., Poly(vinyl alcohol)-heparin hydrogels as sensor catheter membranes, *J. Biomater.*, 12, 63–70, 1991.
19. Netti, P., Shelton, J., Revell, P., Pirie, C., Smith, S., Ambrosio, L., Nicolais, L. and Bonfield, W., Hydrogels as an interface between bone and an implant, *J. Biomater.*, 14, 1098–1104, 1993.
20. Taguchi, T., Kishida, A., Sakamoto, N., and Akashi, M., Preparation of a novel functional hydrogel consisting on sulphated glycoside-bearing polymer: activation of basic fibroblast growth factor, *J. Biomed Mater. Res.*, 41, 386–391, 1998.
21. Abusafieh, A., Siegler, S., and Kalidindi, S.R., Development of self-anchoring bone implants. I. Processing and material characterization, *J. Biomed Res*, 38, 314–327, 1997.
22. Hongmei, Z. and Meng, Z., Microencapsulation of anthocyanins from red cabbage, *Int. Food Res. J.*, 22, 2327–2332, 2015.
23. Gutierrez F.J., Albillos, S., Casas-Sanz, E. and Mussons, M., Methods for the nanoencapsulation of $\beta$-carotene in the food sector, *Trends Food Sci. Technol.*, 32, 73–83, 2013.
24. Hazuki, N., Extraction of Carotenoids from natural products and nanoparticle formation using supercritical fluid, *PhD Thesis*, Nagoya University, Japan, 2015.
25. Kunal P., Banthia, A.K., and Majumdar, D.K., Starch based hydrogel with potential biomedical application as artificial skin, *African J. Biomed. Res.*, 9, 23–29, 2006.
26. Young, C.D., Wu, J.R., and Tsou, T.L., Fabrication and characteristics of poly-HEMA artificial skin with improved tensile properties, *J. Membrane Sci.*, 146, 83–93, 1998.
27. Kunal Pal, Banthia, A., and Majumdar, D., Preparation of transparent starch based hydrogel membrane with potential application as wound dressing, *Trends Biomater. Artif. Organs*, 20, 59–67, 2006.
28. Sadeghi, M. and Soleimani, F., Synthesis and characterization superabsorbent hydrogels for oral drug delivery systems, *Int. J. Chem. Eng. Appl.*, 2, 314–316, 2011.
29. Soleimani, F. and Sadeghi, M., Synthesis of pH-sensitive hydrogel based on starch-polyacrylate superabsorbent, *J. Biomater. Nanobiotechnol.*, 3, 310–314, 2012.
30. Angham, G.H., Synthesis of chitosan and its use in metal removal, *J. Chem. Mater. Res.*, 3, 55–58, 2013.
31. Zúñiga, M., Estremadoyro, L., León, C., Huapaya, J. and Cieza, J., Validation of the salivary urea test as a method to diagnose chronic kidney disease, *J. Nephrol.*, 25, 431–436, 2012.

# 5

# Biogenic Approaches for $SiO_2$ Nanostructures: Exploring the Sustainable Platform of Nanofabrication

**M. Hariram[1], P. Vishnukumar[1, 2] and S. Vivekanandhan[1*]**

[1]*Sustainable Materials and Nanotechnology Lab (SMNL), Department of Physics, VHNSN College, Virudhunagar, Tamil Nadu, India*
[2]*Department of Physics, Alagappa University, Karaikudi, Tamil Nadu, India*

## *Abstract*

Silica is one of the most common metal oxides, which has been explored for many potential applications. When its size reduced to nano, unique physico-chemical properties were appeared, which received extensive scientific and technological importance in many fields including optics, catalysis, drug delivery, and biomedical imaging. As the demand for $SiO_2$ nanostructures expands in many areas, the synthetic process plays a vital role in order to meet the specific requirements. A wide range of physical and chemical processes have been established and reported; however, there is a need for the sustainable greener alternative. Research and development activities have been promoted for the synthesis of $SiO_2$ nanostructures employing biological resources including bacteria, fungus, plants, and other bio-substances. Apart from the synthesis, the biogenic processes have also been explored for the modification of $SiO_2$ nanostructures to improve/create their functional properties. This chapter describes the recent developments in the biogenic approaches for $SiO_2$ nanostructures as well as their potential applications.

***Keywords:*** Biosilica, plant extract, rice husk, bio templates, functionalized silica

**Corresponding author*: svivekan@vhnsnc.edu.in

Shakeel Ahmed and Chaudhery Mustansar Hussain (eds.) Green and Sustainable Advanced Materials, Vol. 1 (107–142) © 2018 Scrivener Publishing LLC

## 5.1 Introduction

In recent years, a fundamental global transformation is occurring toward the green and sustainable solutions to all of our basic needs, which includes energy and materials. The key factors behind this transformation are renewability, economic benefits/cost reduction, sustainability of the feedstock, and environment benign approaches. Thus, the researchers are motivated to design and develop their products or protocols, which can follow the well-known 12 principles of green chemistry [1]. This made a great impact in the field of nanomaterials, which led to the global exploration of new materials and processes by employing biobased materials as the feedstock/precursor. As a result, a wide range of biobased nanomaterials include cellulose/protein nanofibres, starch/gum/chitosan nanoparticles, carbon nanostructures (carbon nanotubes [2], graphenes [3], carbon dots [4], and nanofibres [5]), and their composites have been fabricated using biomaterial as the source [6–10]. Further, the bioresources were extensively explored as catalyst or aides for the synthesis of a wide range of metals and metal oxides [2, 6, 8, 11–13]. These research works have been integrated into the emerging field of green nanotechnology, ultimately aimed in developing efficient synthetic strategies to produce nanomaterials and to reduce or eliminate the hazards to human health and environment through product design and optimization [14].

Among the various nanostructured materials, metal oxides receive significant potential owing to their physicochemical, structural, and morphological features, which enable them as the suitable candidates for many applications including, catalysis, energy storage/conversion, biomedical, sensor, polymer composites, cosmetics, and electronics. ZnO, $TiO_2$, $CeO_2$, $SiO_2$, and $SnO_2$ are the few examples of metal oxide materials, which have been extensively investigated in last two decades in bulk and nanosized forms for their application potential and commercial value. Among them, $SiO_2$, which is generally termed as silica has been attracted by the researchers not only due to its abundance in nature but also due to its diverse application platforms. The nano transformation of $SiO_2$ into their respective particulate and thin film nanoarchitecture created a revolution in many fields includes biomedical, catalysis, energy, electronics, and drug delivery [15]. This is due to the fact that the $SiO_2$ nanoparticles have excellent electronic and mechanical properties, high surface to volume ratios, high porosity, improved multi-functionality and biocompatibility [16]. The potential of $SiO_2$ nanoparticles is further enhanced by the various stabilization and functionalization processes [17]. On functionalizing, the $SiO_2$ nanoparticles exhibit unique properties that are entirely different from their pristine counterpart and results in new application potentials. Silica nanoparticles have been widely synthesized by adopting two different approaches that are top-down and bottom-up [16]. In the top-down approach, bulk $SiO_2$ is employed as the source and nanosized $SiO_2$ has been produced by the reduction of its dimension by employing various size reduction techniques that are known as physical methods including ball milling, laser ablation and sputtering [18–20]. In bottom-up approaches $SiO_2$ is produced from silicon precursors by adopting various chemical processes that are known to be sol–gel, reverse microemulsion and flame processes [21–23].

Recent years, there is an appropriate increase in the green approaches for the synthesis, stabilization, and functionalization of $SiO_2$ nanostructures. Such a biosynthesized $SiO_2$ nanoparticles are non-toxic to the environment and have good physical and chemical properties similar to the particle that is synthesized from the traditional physical and chemical approaches. A wide range of biological approaches has been explored for the synthesis of $SiO_2$ nanoparticles that includes extraction from various biomasses, bacterial/fungal assisted synthesis, exploring the bio templates. Each and every process has its own uniqueness in producing $SiO_2$ nanoparticles with unique physicochemical and functional properties. The $SiO_2$ nanoparticles that are extracted from the various biomasses or synthesized using bio-resources are named as biogenic silica. Biogenic silica nanostructures are an excellent alternative to synthetic silica because of its variable structure, density, and composition. As the demand for biogenic silica expands for many potential applications, there is an increasing interest in synthesizing biogenic $SiO_2$ in various nano-architectures. Thus, this chapter covers the biogenic approaches for the synthesis of silica nanostructures and its surface modifications for the stabilization and functionalization by means of biological approaches.

## 5.2 Synthesis of $SiO_2$ Nanostructures

The application potential of the $SiO_2$ nanoparticles are highly depends on its ability to disperse in various mediums and also the stability/ homogeneity of the dispersion. The physicochemical characteristics and the electrokinetic properties of nano $SiO_2$ aqueous dispersions are strongly associated with the primary particle size and their aggregation. This is due to the fact that the formation of agglomerated $SiO_2$ significantly hinders its interactions with the surroundings. These issues have been overcome by exploring various synthesis processes in physical, chemical and biological platforms, which are briefly discussed in this section.

### 5.2.1 Physical Processes

Physical synthesis of $SiO_2$ nanoparticles involves in the size reduction of bulk $SiO_2$ into nanoparticles by employing various top-down approaches, which include ball milling, laser ablation and sputtering. Various types of milling process including high speed vertical rotating mill, and planetary ball mill were explored for the conversion of bulk $SiO_2$ into nanoparticles. Saleh *et al.* [24] reported the production of $SiO_2$ nanoparticles from Ardma silica sand by employing ball milling and heating process. They were able to produce $SiO_2$ nanoparticles with ~ 50 nm from the high purity silica sand by milling for 30–50 h. Milling parameters such as milling speed, time and ball to material weight ratio play a vital role in getting nanosized $SiO_2$ particles. Rizlan and Mamat reported the investigation on the impact of various parameters for $SiO_2$ nanoparticle production using low-speed ball milling employing Taguchi method [25]. They concluded that the ball to powder weight ratio play vital role in effective milling process for the formation of nanosized powders. In some cases, ball milling was integrated with chemical processes with the aim of reducing the particle size of $SiO_2$ nanomaterials [26]. Furthermore, ball milling process also explored for the effective extraction of $SiO_2$ nanoparticles from various bioresources [18].

In addition to that, pulsed laser ablation has been extensively explored for the fabrication of diverse range of materials including $SiO_2$ as a simple and versatile process [27]. Geohegan *et al.* [28] investigated the gas phase synthesis of the $SiO_2$ nanoparticle by laser employing ablation. The synthesized $SiO_2$ nanoparticles have been found with the particle size range between 1 and 10 nm. Li *et al.* [29] reported the fabrication of $SiO_2$ nanowires by employing pulsed vapor deposition process equipped with Nd:YAG laser using nickel/carbon catalyst followed by $N_2$ annealing. The annealing process of the grown $SiO_2$ nanowires leads to becoming denser, longer, and more uniform, which further increases with the increasing annealing duration. Grigoriu *et al.* [30] explored the laser ablation process for the fabrication of Si/$SiO_2$ nanoclusters with the size range of 2–10 nm. In addition to this, KrF excimer laser was explored for the fabrication of $TiO_2$–$SiO_2$ binary composite thin film by Kunti *et al.* [31] and the amorphous composite films were made by manipulating the oxygen pressure. They found that the increased oxygen partial pressure increases the energy band gap due to the decrement of oxygen vacancy in the composite film [31]. Along with the $SiO_2$ based composite structures, the pulsed laser ablation process has also been effectively explored for the fabrication of Al-doped $SiO_2$ films [32] and $SiO_2$ based core-shell nanoparticles [33] for various applications. In addition to laser ablation process, sputtering also been extensively explored for the fabrication of various types of $SiO_2$ based nanostructures. Liao *et al.* [34] demonstrated the preparation of Au: $SiO_2$ nanocomposite films by co-sputtering of Au and $SiO_2$ employing sputtering processes equipped with the multitarget magnetron. Further, Mishra *et al.* [20] reported the effective fabrication of Ag nanoparticles distributed silica matrix by employing atom beam sputtering. Figure 5.1 shows the schematic of the atom beam sputtering set-up and the obtained composite materials with various Ag loading. Hamza and researchers reported the utilization of laser pulses and sputtering technique for the fabrication of electrically conductive $SiO_2$ nanofiber/Ag nanoparticle composite material [35]. First, Nd:YAG pulsed laser system was utilized for the fabrication of micro and nanofibrous structures on silicon wafer followed by the Ag deposition using sputtering techniques. As the demand for $SiO_2$ based nanostructured materials enhanced drastically, along with these physical processes various other chemical processes also explored for the fabrication of $SiO_2$ nanoparticles with various structural and morphological features.

### 5.2.2 Chemical Processes

In metal oxide synthesis, chemical processes play the extensive role in manipulating their composition, morphological features, and functionality.

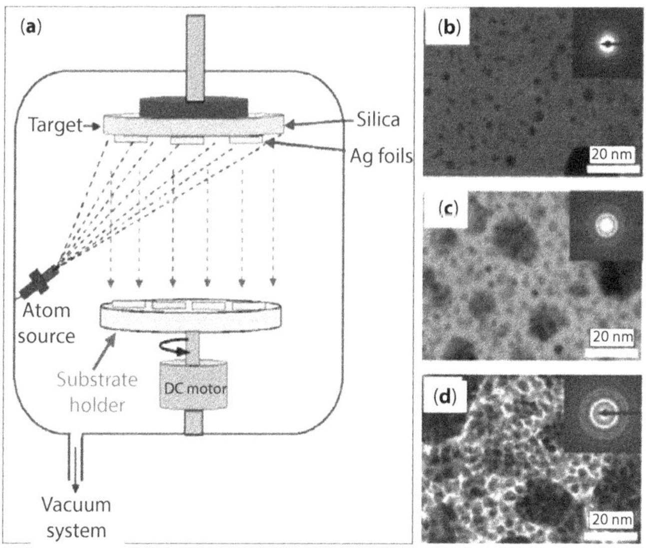


**Figure 5.1** (a) Schematic of the atom beam sputtering set-up and (b–d) the AgNP–$SiO_2$ composite materials with various Ag loading [20].

Particular with $SiO_2$ nanoparticles, the chemical process like sol–gel, reverse microemulsion and flame processes have been widely explored. Among them, the sol–gel process has been found to be most popular and versatile one, due to its ability in synthesizing $SiO_2$ nanoparticles with the precise control of size, distribution and the morphology by means of manipulating the reaction parameters [36]. The low-temperature sol–gel technique utilizes the inorganic or organic precursors for synthesizing $SiO_2$ based materials employing hydrolysis and condensation reactions in alcoholic solutions [37]. The processing conditions efficiently control the reaction kinetics that leads to the better control over their morphology and structure of the final $SiO_2$ nanoparticles [38]. In the sol–gel process, either acid or base catalyst can be employed for the synthesis of $SiO_2$ nanoparticles. Haque *et al.* [39] investigated the effect of acid concentration (acetic acid, hydrochloric acid, sulfuric acid, and oleic acid) in the synthesis of $SiO_2$ nanoparticles. They were able to synthesize the spherical shaped $SiO_2$ nanoparticles with the size ranging between 2 and 4 nm. In other research work, Qian *et al.* [40] investigated the effects of ammonia in fabricating the $SiO_2$ nanoparticles from tetraethyl orthosilicate (TEOS) employing polyethylene glycol (PEG) as the surfactant. They were able to synthesize the $SiO_2$ nanoparticles with 4–18 nm size without any agglomeration. In addition to $SiO_2$ nanoparticles, the sol–gel process has also been extensively utilized for the synthesis of various $SiO_2$ based nanocomposite systems. Ennas and team effectively synthesized the magnetic $Fe_2O_3$–$SiO_2$ nanocomposite employing sol–gel process using TEOS and iron nitrate as the metal ion sources [41]. The thermo stabilized $Fe_2O_3$-$SiO_2$ nanocomposite showed the formation of α- $Fe_2O_3$ amorphous material of 3–4 nm size, which leads to its superparamagnetic behavior [41]. Li *et al.* [42] reported the synthesis of $TiO_2$–$SiO_2$ composite nanoparticles by employing sol–gel-hydrothermal and sol–gel processes. Their research finding suggests that the $TiO_2$–$SiO_2$ composites prepared using by sol–gel-hydrothermal route showed excellent photocatalytic properties than the composite prepared from the sol–gel route. The review works reported by Rahman and Padavettan [16] as well as the Singh *et al.* [36] summarizes the various research activates on the synthesis of $SiO_2$ nanoparticles by employing sol–gel processes in detail.

Next to sol–gel process, the microemulsion process has also been widely used for the synthesis of $SiO_2$ nanoparticles with novel and unique properties. The microemulsions are colloidal nano-droplets which are thermodynamically stable and act as nanoreactors in synthesizing nanosized materials. There are two different types of micro-emulsions, (i) water in oil and (ii) oil in water that are commonly explored for the synthesis of $SiO_2$ nanoparticles. By varying the droplet texture, reactant/capping agent concentration and nature of surfactant, it is possible to tailor the size and shapes of the final products. Osseo-Asare and Arriagada reported the synthesis of $SiO_2$ nanoparticle with the size range of 50–70 nm in a non-ionic reverse micellar system [43]. They also reported the water-in-oil microemulsion process for the synthesis of $SiO_2$ nanoparticles with 30–70 nm diameter [44]. Their research investigation suggests that the water/surfactant molar ratio and the ammonia concentration in the reaction mixture play significant role in obtaining final products. This process has also been used for the synthesis of various $SiO_2$ based nanoparticles, which includes $SiO_2$ coated $Fe_2O_3$ nanoparticles [45], dye-doped $SiO_2$ nanoparticles [22] and $SiO_2$ coated Ag/Au nanoparticles [46]. Figure 5.2 shows the transmission electron microscopic (TEM) images of the silica-coated 12 nm gold nanoparticles with various particle sizes. Along with this, various other chemical process such as hydrothermal and flame pyrolysis processes are also experimented for the synthesis of various types of $SiO_2$ nanoparticles and their composite materials. In hydrothermal process, supercritical condition on the reaction mixture lead to the formation of $SiO_2$ nanoparticles with controlled morphological features relatively at lower temperatures. The challenges in chemical processes for the synthesis

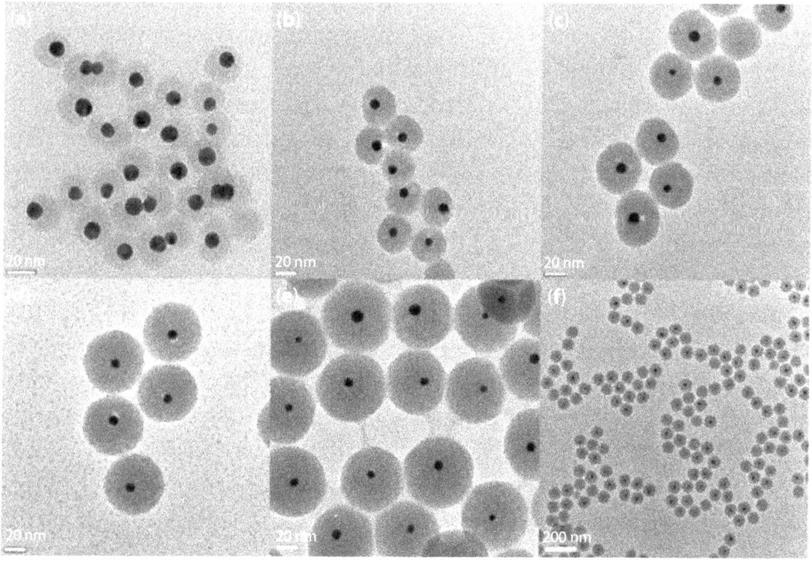


**Figure 5.2** TEM images of the $SiO_2$ coated 12 nm gold nanoparticles with various particle sizes [46].

of $SiO_2$ nanoparticles are their multistep protocol, expensive chemicals and the excessive utilization of solvents.

### 5.2.3 Template Assisted Process

Templating is the most common and frequently used method for the synthesis of nanostructures with the desired size, structural and morphological features. In these approaches, template acts as a structure-directing agent for the formation of $SiO_2$ nanoparticles and hence the selection of template is more important in the template-assisted process. The template should be thermally and chemically stable with the desired pore and size distribution, which is compatible with the wide range of processing conditions. Template process involves the effective replication of one structure (template) into another with structural inversion by employing following two stages. Initially, the stable template-nanomaterial hybrids are formed and the obtained hybrid materials undergo the selective removal of the template in order to detach the nanostructures. Hentze and researchers explored the nanosized silica hollow spheres with the diameter range of 60–120 nm by employing catanionic vesicles based templating process [47]. Further, Chen *et al.* [48] utilized the inorganic $CaCO_3$ nanoparticles as the template for the synthesis of porous hollow structured $SiO_2$ nanoparticles. The synthesized porous hollow $SiO_2$ nanoparticles with the diameter range between 60 and 70 nm were utilized for the drug delivery application [48]. Fuji *et al.* [49] also reported the synthesis of hollow silica nanoparticles by an inorganic particle template method using $CaCO_3$ as a template. They coated sol–gel silica over the surface of the template followed by the acid dissolution to remove the template. They obtained a variety of shapes including cubic, rough-surfaced spherical and rod-like particles, which precisely reflect the shape of the templates. Figure 5.3 shows the schematic representation of template process for the synthesis of hollow silica nanoparticles employing cubic template and the relevant TEM images. In addition to that, Yamashita *et al.* [50] used polyacrylic acid (PAA)-amine aggregates as a nanosized template for the synthesis of hollow silica nanoparticles. They investigated the effect of the amine as a template by comparatively studying different amines such as ethylenediamine (EDA), N,N,N',N'-tetramethylethylenediamine (TED), 3,3'-diaminodipropylamine (DDA), triethylenetetramine (TTA) in synthesizing template nanoparticles. They found that the PAA-DDA template effectively produced the spherical shaped hollow $SiO_2$ in shorter duration. Silica mesophases and nanoparticles were synthesized by Fowler

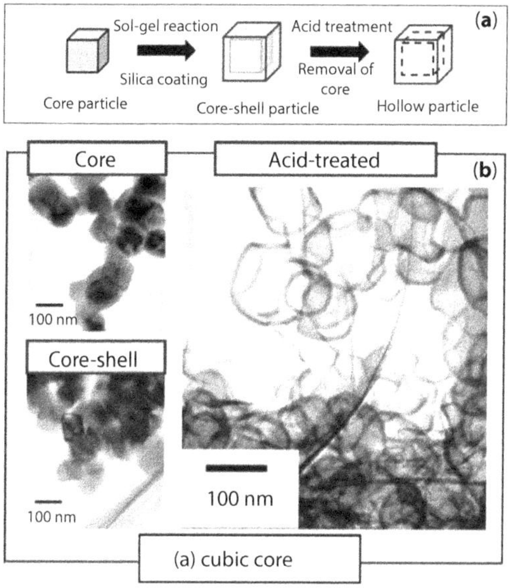


**Figure 5.3** Schematic representation of template process for the synthesis of hollow silica nanoparticles employing cubic template (a) and the relevant TEM images (b) [49].

*et al.* [51] using tobacco mosaic virus liquid crystals as bio-templates. Correa- Duarte *et al.* [52] investigated the utilization of carbon nanotubes as the templates for the fabrication of $SiO_2$ coated Au nanoparticles liner assemblies. This template process has also been explored for the synthesis of rattle-type $Fe_3O_4@SiO_2$ hollow mesoporous spheres employing colloidal carbon spheres as the template [53]. All the above-discussed physical, chemical, and template mediated processes are having their unique advantages as well as disadvantages. Since the demand for sustainable green materials is increasing, the synthesis of $SiO_2$ by biogenic process has received great attention in the last few years due to its cost-effectiveness and eco-friendliness than other traditional hazardous and sophisticated synthesizing processes.

## 5.3 Bio-Mediated Sustainable Processes for $SiO_2$ Nanostructures

Biological synthesis of nanosized $SiO_2$ particulates can be classified in to four main groups such as (i) the exploration of bacteria to produce nanoparticles, (ii) fungus mediated biogenic approaches for the synthesis of $SiO_2$ nanoparticles, (iii) the extraction of nanoparticles from bioresources such as rice husk, ash residuals and other biomasses, (iv) effective utilization of biobased products as templates or catalyst as discussed in this section.

### 5.3.1 Bacterial Assisted Synthesis Process

Earlier reports suggest that the exposure bacterium to various metal ions lead to the secretion of reductases and oxidising enzymes which lead to the formation of various metal and metal oxides nanoparticles. Similar mechanism also suggested for the synthesis of $SiO_2$ nanoparticles employing various bacterial strains. Westall *et al.* [54] reported that a number of silicates have been deposited on the surface of bacterial envelopes as thin (~100 nm) amorphous and granular coatings. Their research finding suggests that not all the microorganisms silicify; gram-negative bacteria are more recalcitrant, whereas gram-positive bacteria silicify readily [54]. Yee *et al.* [55] reported the effect of cyanobacteria *Calothrix sp.* on silica precipitation at neutral pH on the bacterial surface. Singh *et al.* [56] reported the synthesis of silicon/silica nanoparticle composites by employing *Actinobacter sp.* assisted biogenic process. The silicon/silica nanocomposite was synthesised by exposing $K_2SiF_6$ precursor to *Actinobacter sp.* medium under ambient conditions. They found that the particles embedded in the biomolecular matrix are found in 10 nm in size with the quasi-spherical morphology [56]. Jo *et al.* [57] reported the effective utilization of bacterial flagella isolated from *Salmonella typhimurium* as the bio-templates for the fabrication of silica nanotubes. The isolated *Salmonella typhimurium* flagella were repolymerized, treated with aminopropyl triethoxysilane (APTES), and exposed to tetraethoxysilane (TEOS), which led to the formation of silica-mineralized flagella nanotubes (SMFNs). Thicknesses of the silica layer can be precisely controlled by manipulating the TEOS concentration and the reaction time [57]. In another research work, they investigated the electrical property of the flagella-templated silica nanotube networks by employing its current-voltage characteristics [58]. They found that the integration of gold, palladium, and iron oxide nanoparticles enhances the electrical properties of the biosynthesized $SiO_2$ nanotube. Show and team demonstrated the synthesis of silica nanoparticles employing thermophilic bacteria (BKH1) assisted biogenic template process from silica-rich substrates [59]. $SiO_2$ nanoparticles were synthesized using magnesiumtri-silicate (inorganic), and tetraethyl orthosilicate (organic) as the source of Si ions. The extracellularly

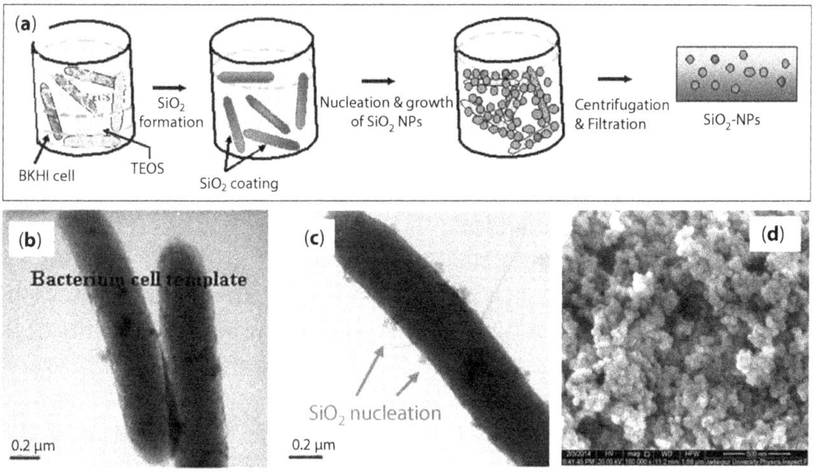


**Figure 5.4** (a) Schematic of bacterial assisted biogenic template process for the synthesis of $SiO_2$ nanoparticles, (b) TEM micrograph of bacterial template, (c) TEM micrograph shows the nucleation of $SiO_2$ over the bacterial cell, and (d) SEM images of the separated $SiO_2$ nanoparticles [59].

synthesised $SiO_2$ nanoparticles were found with the average particles size of 15 ± 5 nm. Figure 5.4 shows the schematic representation of bacteria assisted biogenic template process for the synthesis of $SiO_2$ nanoparticles. Apart from the nanoparticles, this bacterial assisted bio-template process has been explored for the synthesis of many types of hollow silica microparticles [60, 61]. This process also been used for the synthesis of $SiO_2$ based hybrid structures. Baumgartner *et al.* [62] reported the bacterial template based fabrication of hollow spirilla-shaped silica containing a magnetic chainn as hybrid silica/magnetic microparticles by employing magnetotactic bacteria as a precursor.

### 5.3.2 Fungal Mediated Biogenic Synthesis Process

The fungus has been widely utilized for the synthesis of various types of nanoparticles including, metals, bimetallic, metal oxides, and their functional nanostructures. Particular with the metal oxides, various types of nanoparticles such as $ZrO_2$, $Fe_3O_4$, $TiO_2$, and $SiO_2$ have been synthesized using fungus mediated biogenic processes [10, 63, 64]. The unique regenerative nature of fungal systems, capable of hydrolyzing metal complexes, shows great potential in the synthesis of metal oxide nanomaterials [10]. Fungal biomass has also been explored as the template for the synthesis of $SiO_2$ nanoparticles with the effective control of its size and shape. It has been suggested that the low molecular weight cationic proteins/ peptides present in the fungus were predominantly responsible for oxide synthesis. Bansal *et al.* [10] reported the synthesis of silica and titania nanoparticles by the fungus *Fusarium oxysporum* for the first time. When the fungus was explored to the anionic complexes of $SiF_6^{2-}$ and $TiF_6^{2-}$, the proteins and other biomolecules present in the fungus hydrolize them extracellularly that lead to the formation of silica and titania at room temperature. TEM analysis shows the formation of $SiO_2$ nanoparticles are in the range of from 5 to 15 nm and possess an average size of 9.8 ± 0.2 nm. In addition to this, $SiO_2$ nanoparticle has also been synthesized by exploring bioleaching process by various natural habitats such as fungus, bacteria, and algae. This process has been successfully explored for the synthesis of various metal and metal oxide nanoparticles. Bansal *et al.* [65] reported the bioleaching of sand by fungus *Fusarium oxysporum* in producing $SiO_2$ nanoparticles. Two key steps were suggested by the authors for the synthesis of the $SiO_2$ nanoparticle by this process, (i) leaching out silicon from the sand in the form of silicic acid by the fungus and (ii) the hydrolyzation of Si-based complex and their stabilization into $SiO_2$ nanoparticles. TEM investigation showed the formation of $SiO_2$ nanoparticles with the size range of 2–5 nm [65]. Figure 5.5 shows the TEM images of $SiO_2$ nanoparticles obtained from sand by using *Fusarium oxysporum* before and after calcination at 400 °C for 2 h with their respective Selective Area Electron Diffraction (SAED) patterns. Later, Kulkarni *et al.* [66] reported the synthesis of ultrafine silicate nanoparticles with 5 nm size by employing bioleaching process on glass using thermophilic, *Humicola sp* fungus.

Bansal *et al.* [67] demonstrated the *Fusarium oxysporum* mediated biotransformation of rice husk based amorphous $SiO_2$ into nanocrystalline $SiO_2$ The fungal mediated bio-transformed silica nanoparticles showed higher crystalline with quasi-spherical morphology and 2–6 nm size. The biosynthesized $SiO_2$ nanoparticles are extracellularly capped by stabilizing proteins of *Fusarium oxysporum.* Further, the calcination of synthesized $SiO_2$ nanoparticles led to the decomposition of occluded protein, which caused the formation of the porous microstructure. Pineda *et al.* [68] explored the similar bioconversion process using *Fusarium oxysporum* for the extracellular synthesis of $SiO_2$ particles from rice husk ash. Incubation of rice husk ash with *Fusarium oxysporum* at 28 °C and pH 6.8 dissolved 80% silica and caused their size reduction from 500 to 5 μm. Pineda-Vasquez *et al.* [69] reported the efficient way of $SiO_2$ nanoparticle synthesis by employing *Fusarium oxysporum* using rice husk ash as the source. *Fusarium oxysporum* assisted biotransformation of the rice husk ash was carried out using two different growth media: malt-glucose (MG) and

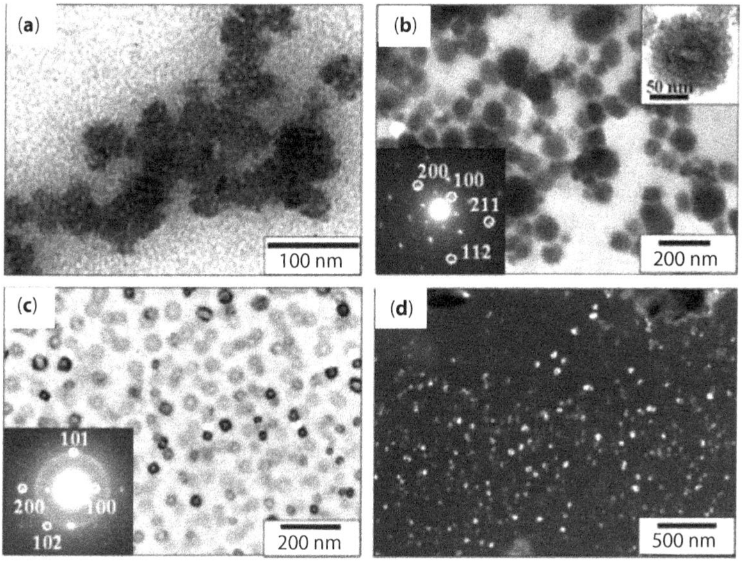


**Figure 5.5** TEM images of $SiO_2$ nanoparticles obtained from sand by using *Fusarium oxysporum* before (a, b) and after (c,d) calcination at 400 °C for 2 h with their respective SAED patterns [65].

malt-glucose-yeast-peptone (MGYP), that are commercially available. The process employing malt-glucose showed promising results in synthesizing semicrystalline, quasi-spherical $SiO_2$ nanoparticle with the range of 2–8 nm [69]. Kannan *et al.* [70] use the same fungus *Fusarium oxysporum*to synthesis silica NPs. They extract the *Fusarium oxysporum*from the wilted tomato plant. The synthesized $SiO_2$ NPs having the size between 20 and 50 nm and having good efficiency in DNA protection from enzymatic cleavages and dye decolorization of methyl orange. Kaur *et al.* [71] investigated the biotransformation of rice husks and wheat bran based $SiO_2$ into silicon/silica ($Si/SiO_2$) nanocomposites employing fungus mediated biogenic biosilification process. They have employed various eukaryotic microorganisms such as *Actinomycete, Aspergillus niger, Fusarium oxysporum, Penicillium sp., and Trichoderma sp.,* under ambient conditions [71]. All the microorganisms were able to transform amorphous $SiO_2$ to crystalline structures in 24 h with varying reaction kinetics. The fungal species, *Actinomycete* and *Trichoderma sp.* took more time about 48 h for the biotransformation of rice husk $SiO_2$, which is found to be only 24 h for wheat bran $SiO_2$ [71].

### 5.3.3 Plant Based Synthesis Process

Synthesis of $SiO_2$ nanoparticles from plant-based resources is a simple, most common and sustainable biogenic method among others. This is due to the fact that plant such as rice, grasses, wheat, sugarcane, pine cones and needles, horsetail, and genus Equisetum has high levels of biogenic silica in amorphous and crystalline nature. These biomasses are abundant, renewable, sustainable, eco-friendly and cost-effective. Silica nanoparticles are extracted from plants through pyrolysis of biomasses, by microwave hydrothermal processes, flame synthesis, sol–gel processes, microemulsion methods, and combustion processes. The obtained plant-based biogenic $SiO_2$ nanoparticles showed potential applications in many fields including biomedical, catalysis, energy storage/ conversion and drug delivery.

Among the various biomasses, rice husk has been extensively used for the synthesis of $SiO_2$ nanoparticles. Rice husk is a cost-effective, non-metallic and very common bio precursor for the synthesis of $SiO_2$ nanoparticles. It is a major agricultural waste product which contains 65–75% organic (lignin, proteins, vitamins, cellulose, and hemicellulose) and 15–20% inorganic substances ($SiO_2$, $Al_2O_3$, and CaO). The organic compounds in the rice husk can be removed employing the burning processes and the inorganic compounds other than silica can be removed by acid leaching and high-temperature annealing processes to obtain silica nanostructures.

Tang *et al.* [72] investigated the synthesis of $SiO_2$ aerogel from the rice hull ash by integrating sol-gel approach followed supercritical $CO_2$ drying. The rice hull ash based bio $SiO_2$ was first treated with NaOH solution to produce a sodium silicate, which was further treated with $H_2SO_4$ to form the $SiO_2$ gel. After performing water washing and the solvent exchange with $C_2H_5OH$, the aged gel was dried to produce aerogel using supercritical $CO_2$ drying. They reported the formation of $SiO_2$ aerogel with the specific surface area of 597.7 $m^2/g$, which consisted of ultrafine $SiO_2$ nanoparticle. Li *et al.* [73] prepared the aerogel made of nanostructured $SiO_2$ particles by using rice hull ash and employing the pyrolysis process at 600 °C. Then the ash was mixed with 30 mL, 1.0 mol/L, and NaOH aqueous solution and heated with constant stirring up to boiling for 1.5 h. Then they neutralized the solution with sulfuric acid to form silica hydrosol and an appropriate quantity of TEOS is added followed by gelation to form a gel. The obtained silica aerogel exhibited the bulk density of 0.33 $g/cm^2$ with high porosity, which is estimated to be ~87%.U. Many researchers including Nayak *et al.* [74], Tadjarodi *et al.* [75] and Vijayalakshmi *et al.* [76] also reported the similar process for the synthesis of $SiO_2$ aerogel from rice husk ash through NaOH extraction followed by drying at atmospheric pressure. More details can be found in the overview by Prasad and Pandey on the research accomplishments of rice husk ash as a renewable source for the production $SiO_2$ gel and its application [77].

Alshatwi *et al.* [78] extracted the $SiO_2$ from rice husk employing acid digestion process using HCl under pressurized conditions at 120 °C followed by a calcination process at different temperatures of 500 °C, 600 °C, and 700 °C. They found that the calcination temperature plays a vital role in the formation of $SiO_2$ nanoparticles. TEM analysis showed that the extracted particles were exist in the spherical shape with the diameter range between 10 and 30 nm. The obtained TEM images of the calcined nanoparticles at different temperatures confirmed that the increasing calcination temperature decreased the particle size. Figure 5.6 shows the schematic of the biogenic synthesis of $SiO_2$ nanoparticle and the TEM images of the obtained $SiO_2$ nanoparticles after calcining at different temperatures [78]. Athinarayanan *et al.* [79] also reported the similar work and found that with the increasing calcination temperature, the size of the particles decreased. Through this process, they were able to synthesis $SiO_2$ nanoparticles with irregular particles and the size range of 10–30 nm.

By adopting this process, Shankar *et al.* [80] able to produce mesoporous $SiO_2$ nanoparticles of 50 nm size with the surface area of 7.1548 $m^2/g$ by employing 700 °C as annealing temperature. Further, they investigated the effect of three types of rice husk ashes (sticky, red and brown) in the synthesis of mesoporous $SiO_2$ [81]. They found that the sticky rice husk ash derived silica have the cluster-type spherical shape while others have uniform spherical shape. The mean size of the $SiO_2$ nanoparticles derived from sticky, red and brown rice husk ashes are 50, 20 and 10 nm with the surface area of 7.5513, 201.45 and 247.18 $m^2g^{-1}$, respectively. Due to the lower particle size and higher surface area, they concluded that the brown rice husk ash derived mesoporous $SiO_2$ is most suitable for any applications [81]. Likewise, Carmona *et al.* [82] also compared the two different rice husks namely agulhinha and cateto for the extraction of $SiO_2$ nanoparticles. In this investigation, they found that the cateto rice husk the high percentage of silica content than agulhinha. Rice husk ash from two different sources, normal rice, and black rice was investigated for the synthesis of $SiO_2$ nanoparticles employing the precipitation method by Sinyoung and the team [83]. They found that the amorphous $SiO_2$ nanoparticles synthesized from black rice husk ash exhibited more crystalline impurities, which is suitable for producing belite cement. From these investigations, it is very

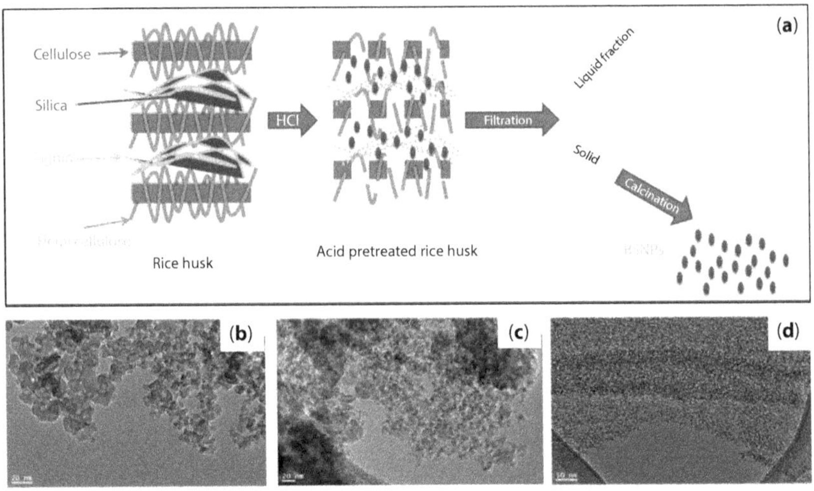


**Figure 5.6** (a) Schematic of the biogenic synthesis of $SiO_2$ nanoparticle and (b–d) TEM images of the biogenic $SiO_2$ nanoparticles calcined at different temperatures 500 °C (b), 600 °C (c), and 700 °C (d). [78]

clear that the source of rice husk plays the significant role in the structural, morphological and compositional features of the final $SiO_2$ nanoparticles.

Zulfiqar *et al.* [84] investigated the effect of white rice husk (obtained by treating rice husk by HCl) concentration in the sodium hydroxide to form sodium silicate and their transformation to $SiO_2$ nanoparticles. The size of the particles increased with increasing the concentration of rice husk but decreased with the addition of water. Unimodal silica particles in nanometer size were produced employing low concentrations of white rice husk (0.7wt.% and 1.4 wt.%) and using higher concentrations (2.8 wt.% and 5.6 wt.%) lead to the formation of bi- and tri-modal particles, which were converted to uni- and bi-modal by changing the reaction conditions. Rohani and team [85] produced pure $SiO_2$ in amorphous form from the hydrochloric acid and sulphuric acid leached rice husk followed by the combustion process at various temperatures. In addition to that Gomas *et al.* [86] also reported the synthesis of amorphous $SiO_2$ from rice-husk by their combustion on bubbling fluidized bed with different conditions. They investigated the effect of various purification methods and found they all result in same $SiO_2$ yield. [2016] Azmi *et al.* [87] and Noushad *et al.* [88] also reported the effective utilization of rice husk for the synthesis of $SiO_2$ nanoparticles.

Apart from rice husk, other biomasses are also used for the synthesis of $SiO_2$ nanoparticles. Bageru *et al.* [89] used the teff (Eragrostistef) straw, as a biological source for the synthesis of $SiO_2$ using thermal decomposition method. Biosilica with the purity 85, 91.2, and 91.82% were obtained by employing 500, 700 and 900 °C, respectively, as ash forming temperature. The obtained result showed that the increasing combustion temperature increases the purity of biosilica. Further, San *et al.* [90] extracted the $SiO_2$ from sugarbeet bagasse by adopting laser ablation and NaOH treatment. It is found that the silica produced by laser ablation method are smaller in size (38–190 nm) while NaOH treated silica are larger (531–825 nm). In addition to that, Rangaraj *et al.* [91] used bamboo leaf biomass as the source for the synthesis of $SiO_2$ nanoparticles. The synthesized nanoparticles showed the average size of 25 nm, with the high specific surface area of 428 $m^2g^{-1}$. Bamboo leaf mediated biosynthesized $SiO_2$ nanoparticles exhibited the very high purity of 99%. Mattos *et al.* [92] extracted $SiO_2$ nanoparticles from *Equisetum arvense* in aqueous media employing usual ball milling, high-energy planetary ball milling, and ultrasonication methods. Ball milling and ultrasonication result in the $SiO_2$ particles with the size of 10 nm. They also found that the extended milling time was not effective for particle size reduction by high-energy planetary ball milling. Athinarayanan *et al.* [93] used sugarcane bagasse as a precursor for synthesis of $SiO_2$ nanoparticles. The synthesis method includes the pre-treatment of biomass with acid followed by incineration at different temperatures. The obtained irregular amorphous $SiO_2$ nanoparticles showed the average particle size of 251, 210, and 191 nm, respectively, for the incineration temperatures of 550°C, 650°C, and 750°C.

### 5.3.4 Biomolecular Template Assisted Synthetic Process

Nature provides the perfect platform for synthesis of $SiO_2$ nanostructures by employing the various biomolecules as templates. In addition to the living organisms such as plants, viruses, diatoms, and bacteria (flagella), some self-assembling biomolecules including proteins, peptides, DNA, oligonucleotides, and collagen are also been explored as an effective template for the synthesis of various types of $SiO_2$ nanostructures. These templates are called biomolecular template in which the nucleation and self-assembly of silica take place. Electrostatic forces play a vital role in the formation of $SiO_2$ nanostructures through biomolecular templates. The presence of positive charges in the biomolecules causes the electrostatic force that interacts with the negatively charged templates. Such a biotemplated nanoparticles have several advantages includes environmental friendliness, high mechanical stability, fine tuneable surface area and so on. Here we discuss the biomolecular template-based synthesis of $SiO_2$ nanostructures.

Wang *et al.* [94] synthesis silica nanostructures by using *Bombyxmori* (*B. mori*) silk fibroin (SF) and sericin (SS) as a biotemplate. They use 3-Aminopropyl triethoxysilane (APTES), and TEOS as the precursor. The negative charges of SF and SS induces the self-assembly of silica from the positively charged silica precursors and causes the formation of Si/SF nanospheres and Si/SS nanofibers. They showed that the size and topology of silica were controlled by the concentration of biotemplates. Cao Feng *et al.* [95] explored the rape pollen grain as a bio-template for synthesising hollow surface structured $SiO_2$ microspheres by employing the facile sol–gel coating and calcination. They achieved its change in morphology by changing the ratio of the tetraethyl orthosilicate (TEOS) and water in sols. The pollen grain bio-template is removed by calcinating it around 500 °C. The obtained silica spheres are not perfectly spherical but some chemical modifications in pollen grain may improve the coating and morphology control and have potential applications in drug delivery. They have also reported the synthesis of $SiO_2$ microspheres by using ammonium oleate modified pollen grains as a bio-template [96]. It shows improved morphology control and spherical shape compared to the obtained silica in their previous work. SEM analysis showed that the structure and morphology of the silica microspheres can be adjusted by changing the amount of tetraethoxysilane (TEOS) added. Further, Guan *et al.* [97] also uses pollen grains such camellia pollen and rape pollen as a bio-template for the synthesis of $SiO_2$ particles by integrating sol–gel and calcination process. They obtained the variety of $SiO_2$ structures, which includes ellipsoid with meshes, sphere with taper-like protuberances and sphere with troughs. Leng *et al.* [98] synthesising biomimetic $SiO_2$ employing chitosan mediated template. They obtained the $SiO_2$ nanoparticles with the diameter range between 20 and 40 nm. They showed that the phosphate ions triggered the assembling of chitosan and this caused the formation of silica nanostructures. Bo Zhang *et al.* [99] prepared monodisperse $SiO_2$ micro particles (MSM) using spherical *synechocystis sp. PCC 6803* as bio-template. They found that the size, morphology and uniformity of the PCC 6803 cells, is reflected in the diameter (1.5 μm) of the synthesised MSM. Figure 5.7 shows the SEM micrograph of various states involved in the fabrication of $SiO_2$ microparticles. The biomolecular template mediated process found as a promising technique for the synthesis of silica nanostructures. But the

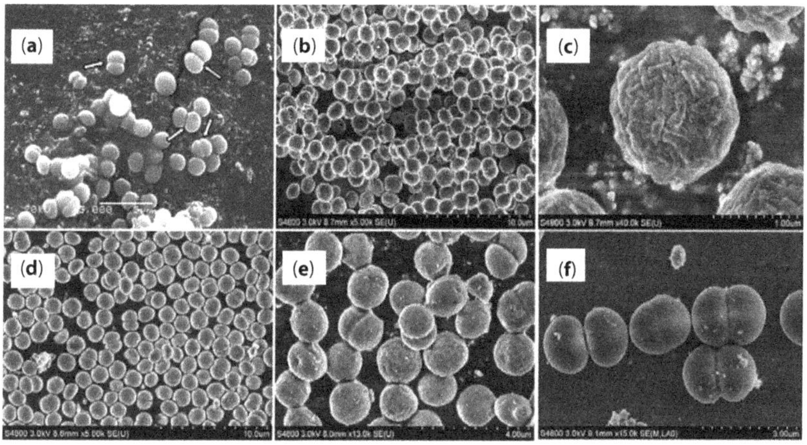


**Figure 5.7** (a) SEM image of PCC 6803, (b & c) FESEM images of PCC 6803/$SiO_2$ complex at different magnifications, (d, e, and f) FESEM images of solid $SiO_2$ after the removal of bio templates observed at different magnifications [99].

mechanisms happen during the synthesis process are still unclear even though there are many proposed mechanisms.

## 5.4 Biogenic $SiO_2$ Based Doped, Functionalized and Composite Nanostructures

On the way of green approaches, researchers not just terminate it for synthesizing nanoparticles; they also explored such green approaches for doping and functionalizing $SiO_2$ nanostructures. This biogenic process also utilized for the fabrication of $SiO_2$ nanoparticles combined with various materials such as composites and functional architectures.

### 5.4.1 Biogenic Synthesis of Doped and Functionalized $SiO_2$ Nanostructures

The doped and functionalized $SiO_2$ nanoparticles show some exciting properties. This is not just to stimulate the properties of the nanoparticles but also it stabilizes them by preventing agglomeration. Li *et al.* [100] reported the fabrication of cobalt-doped mesoporous $SiO_2$ by using Reeds (*Phragmitescommunis*) leaves as a bio template. The fabricated mesoporous structure showed the average pore diameter of 4.6 nm with the Brunauer, Emmett, and Teller (BET) surface area of 435 $m^2/g$. Liu *et al.* [101] reported the fabrication of carbon doped silica from the HCl treated rice husk by calcinations at 550 °C. The synthesized carbon-doped silica having the surface area of 203 $m^2/g$, pore diameter 6.4 nm, and pore volume of 0.33 $cm^3/g$ exhibited excellent high intense photoluminescence. Skolem reported the fabrication of Ti-doped $SiO_2$ based nanostructures formed by diatoms *Pinnularia sp.*and *Coscinodiscus wailesii* [102]. It had been found that *Pinnularia sp.* diatoms have high content of Ti on comparing with *Coscinodiscus* while the amount of Ti relative to silicon (Si) was higher in the biosilica located at the base of the largest pores compared to the biosilica located between the pores. Gale *et al.* [103] fabricated the Ge doped biosilica onto the two-dimensional photonic crystal slab by the by two-stage cell culture of the diatom *Pinnularia*. They further integrate it into an electroluminescent device by the spin coating method. The electroluminescence (EL) characteristics show that there was no photonic bandgap predicted; however, EL band gap between 500 and 640 nm was also observed between blue and red EL line emissions. The above characteristics are not absorbed previously shows unique characteristics of the diatom photonic crystal. Chandrasekaran *et al.* [104] fabricated the boron doped Silica by employing Aulacoseira sp. diatom and use it as a photocathode for water splitting. The experimental results showed that the fabricated p-type photocathode having, a photocurrent density of approximately −6 $\mu A\ cm^{-2}$ at a bias potential of −500 mV, which was able to produce 12.9 nmol of $H_2$ for 1 h, was produced for the boron doping concentration of about 0.953 ± 0.054 ppm. Chauton *et al.* [105] successfully incorporated Ti into silica frustules of the pennate diatom Pinnularia sp.

In addition to metal doping, the biogenic process also explored for the functionalization of $SiO_2$ nanostructures with metal, carbon and various chemical compounds for the enhanced physicochemical and functional properties. Wang *et al.* [106] reported the synthesis of carbon dot grafted $SiO_2$ nanoparticles by rice husk biomass. The synthesized nanostructure having 8 ± 2.5 nm, exhibits outstanding features includes surface modification, high water dispersibility, and biocompatibility. It also exhibits intense and versatile photoluminescence having outstanding potential in biomedical applications. Luo *et al.* [107] functionalized the mesoporous silica with amine and study its adsorption properties for Pb (II). They have synthesized the $SiO_2$ by using natural bamboo as a bio-feedstock by hydrothermal treatment and calcination at 650 °C. They functionalize silica by treating it with aminopropyltriethoxysilane (APTS) in dry toluene under the nitrogen atmosphere. Cui *et al.* [108] reported the preparation of amine-modified $SiO_2$ aerogel using rice husk ash and 3-(aminopropyl)triethoxysilane (APTES) as the modification agent. The amine-modified $SiO_2$ aerogel exhibited the high specific surface area of 654.24 $m^2/g$ with the pore volume of 2.72 $cm^3/g$, which lead to its excellent $CO_2$ adsorption–desorption performance. Jo *et al.* [109] reported the metallization of $SiO_2$ nanotube made using bacterial flagella as bio-template with gold, palladium, and iron oxide nanoparticles. Their functionalization process was performed under mild aqueous conditions. The precursors of Au and Pd nanoparticles were $HAuCl_4$ and $Na_2PdCl_4$, which are reduced with hexamethylenetetramine (HMT) and deposited on the $SiO_2$ nanotube using 3-Aminopropyl triethoxysilane (APTES) as the efficient binder. The metal functionalized silica nanotubes exhibited enhanced electrical, optical and catalytic properties. Cicco *et al.* [110] functionalized the *Thalassiosiraweissflogii* diatom-based biosilica with the cyclic nitroxide 2,6,6-tetramethylpiperidine-N-oxyl (TEMPO). Due to the functionalization, drug delivery and antioxidant properties of the functionalized biosilica enhanced significantly.

### 5.4.2 Biogenic $SiO_2$ Nanocomposites

Sujaridworakun *et al.* [111] reported the fabrication of $TiO_2$/ $SiO_2$ hybrid nanostructures from rice husk by employing hydrothermal process. The pre-extracted porous silica from rice husk ash was explored as the support for $TiO_2$ nanoparticles and the fabricated $TiO_2$/ $SiO_2$ hybrid nanostructures were used as a photocatalyst. Larichev *et al.* [112] reported the fabrication of carbon–$SiO_2$ nanocomposites using rice husk through carbonization in a fluidized-bed reactor using a copper-chromium catalyst. The $SiO_2$ nanoparticles with the narrow fractions (1–7nm) and the wider fraction (8–22 nm) are distributed in the carbon matrix. The increasing carbonization temperature leads to the increment of $SiO_2$ particle size from 5.9 (465 °C) to 7.6 nm (600 °C). Soltani *et al.* [113] reported the fabrication of macro-porous amorphous SiOC/ $SiO_2$ monoliths ceramic nanocomposites through freeze-cast method. The obtained amorphous silica from rice husk ash was freeze caste with commercially available methyl-phenyl-vinyl-hydrogen polysiloxane and tert-butyl alcohol, which led the cross-linking. The composite material was formed by pyrolyzing resulting freeze caste products at 1100 °C in nitrogen. Yang *et al.* [114] investigated the improved bacterial removal capacity of hierarchical porous $TiO_2$/$SiO_2$ synthesized by rice-husk templated biogenic process. Yener *et al.* [115] reported the synthesis of $TiO_2$/$SiO_2$ composites by employing thermal hydrolysis method and investigated the effect of synthesis temperature on the structural properties and photocatalytic activity of $TiO_2/SiO_2$ composites. They found that the surface areas of the $TiO_2/SiO_2$ composites structures were higher than that of the $TiO_2$ particles, which was contributed by the $SiO_2$ support with the high surface area. Hui *et al.* [116] reported the fabrication of nanoporous $TiO_2/SiO_2$ composite with high anatase content, by employing sol–gel process using rice husk (RH) as a template. They reported that the transformation of $TiO_2$ from anatase to rutile starts at 700–800 °C and completes at 1000 °C. In addition to this, Fatimah [117] also reported the fabrication of ZnO-$SiO_2$ nanostructures by a sol–gel process using rice husk as the $SiO_2$ source. Thus, rice husk mediated $SiO_2$ nanoparticles impregnated with MnOx and MnOx-CeOx was synthesized by the hydrothermal method stated by Liu *et al.* [118]. The reduced specific surface area and pore volume of the $SiO_2$ nanoparticles were observed due to the impregnation of MnOx and CeOx. Al-Amsyar *et al.* [119] fabricated the complex aluminium oxide-silica/carbon composites from rice, which was acted as bifunctional heterogeneous Lewis-Bronsted acid catalyst transformed the glucose to the hydroxymethylfurfural. Further, Khazaei *et al.* [120] reported the effective fabrication of nano-$Fe_3O_4$@$SiO_2$ supported Pd(0) ($Fe_3O_4$@$SiO_2$-Pd), utilizing the rice husk biomass-derived bio $SiO_2$ by adopting the simple and efficient protocol. Figure 5.8 shows the scheme for the fabrication of $Fe_3O_4$@$SiO_2$-Pd with their SEM and TEM images. Recently, Chen *et al.* [121] reported the fabrication ZnO nanorods on biogenic hierarchical rice husk $SiO_2$ by employing simple double-step annealing process. These synthesized nanocomposite materials were explored for the photocatalytic application. Roychoudhury *et al.* [122] reported the fabrication of Au-$SiO_2$ nanocomposite using a diatom, *Amphora copulata*, as reducing agent. They were using hydrogen tetra-auro chlorate ($HAuCl_4$) as a precursor for AuNPs and *Amphora* diatom as a $SiO_2$ source. The obtained nanomaterial contains triangular-, spherical-, and hexagonal-shaped Au-$SiO_2$ having the size in the range of 5–45 nm. Its DNA binding affinity is tested by agarose gel electrophoresis. TEM study after 30 min of interaction with isolated DNA from the green alga, *Pithophora* confirms its binding with DNA molecules without any surface modification shows its potential in medical science applications.

## 5.5 Applications of Bio-Fabricated $SiO_2$ Nanoparticles

Silica nanostructures derived from bioresources have been explored for several applications in diverse fields, which includes catalysis, biomedical, energy and environmental as summarized in this section.

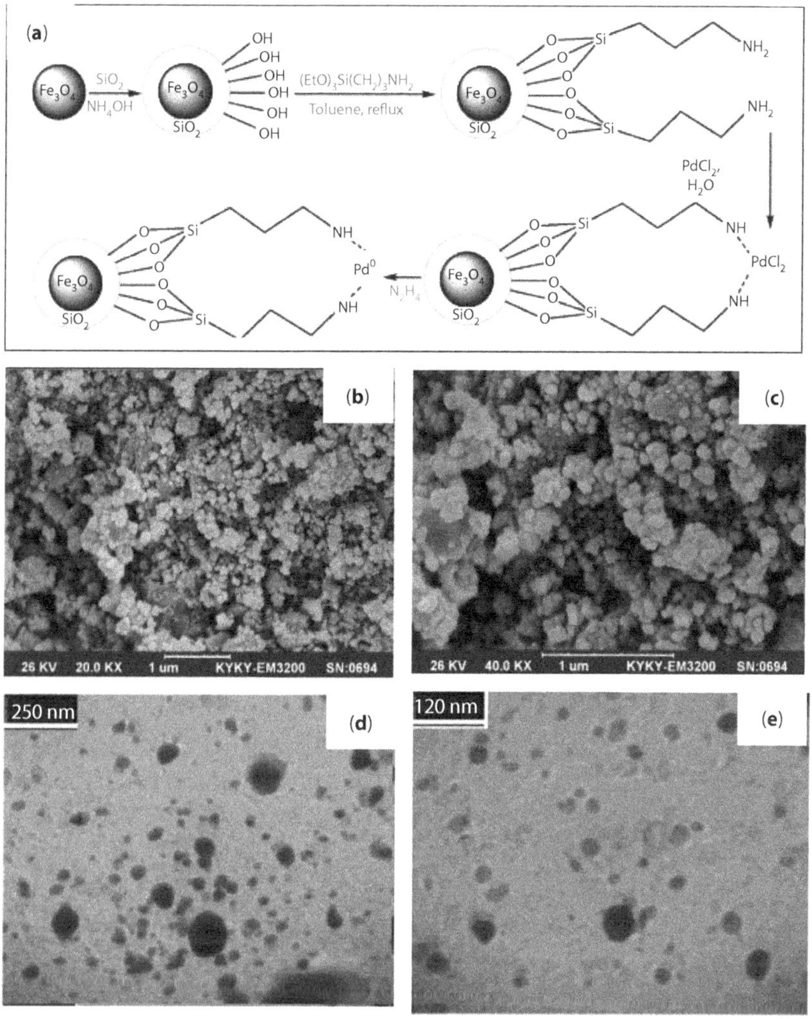


**Figure 5.8** (a) Scheme involved in synthesizing $Fe_3O_4$@$SiO_2$–Pd and their (b, c) SEM and (d,e) TEM images obtained with different magnification [120].

### 5.5.1 Catalysis

Silica nanostructures derived from bio-based resources exhibited excellent catalytic application. The biosynthesized $SiO_2$ nanoparticles have been utilized as the photocatalyst for the effective degradation of various organic dyes like methylene blue in combination with other metal oxide semiconductors such as $TiO_2$, and ZnO. This is due to the fact that semiconductor metal oxides exhibit good photocatalytic activity due to their suitable band gap, photo and chemical stability. Their efficiency can be increased further by improving the surface activities by supporting it on porous materials. Due to the high porosity, $SiO_2$ nanostructures have been widely explored as a effective supporting material. Sujaridworakun and his co-workers [111] investigated the hydrothermally synthesized $TiO_2$/$SiO_2$ hybrid for the degradation of methylene blue dye under UV irradiation. The $TiO_2$/$SiO_2$ composite shows good photo catalytic activity towards methylene blue degradation. Similarly, Hui *et al.* [116] also reported the photo catalytic activity of $TiO_2$/$SiO_2$ nanoporous composite, which showed the highest photo degradation efficiency of methyl blue, which is ~ 82% within 4 h. Fatimah also showed the photo catalytic activity ZnO-$SiO_2$ nanostructures for the degradation of MB dye [117]. Yener and his co-worker [115] investigated the photo catalytic activity for $SiO_2$-$TiO_2$ composites against liquid phase degradation of terephthalic acid (TPA) under UV-C illumination. The composite material exhibited excellent catalytic activity comparing to the pristine $TiO_2$. Apart from photocatalytic degradation, $SiO_2$ nanostructures act as an excellent heterogeneous catalyst. Al-Amsyar *et al.* [119] synthesizes the Al-$SiO_2$/C nanocomposite and used it as a bi-functional heterogeneous catalyst for the one-pot sequential reaction in the conversion of glucose into 5-hydroxymethylfurfural (HMF). It shows better catalytic activity by 52% yield and good recyclability. Li *et al.* [100] reported the catalytic activity of Co-doped mesoporous $SiO_2$ for the allylic oxidation of limonene to carvone. It exhibits high substrate conversion (100%) and good product (carvone) selectivity (40.2%) and good reusability.

### 5.5.2 Biomedicine

The $SiO_2$ nanostructures have created a great revolution in the field of biomedicine due to its unique structural and morphological features, which provide an opportunity for various applications. Silica nanostructures have high surface density silanol groups that enable their surface functionalization with a number of biologically active molecules including peptides, proteins, DNA and linking agents such as biotin and avidin, antigens, and antibodies, which results in its applications in biomedical fields which includes drug delivery, drug encapsulation, biomedical imaging, biosensing, bioassays and so on. Wang *et al.* [94] reported the anticancer drug delivery performance of the synthesized $SiO_2$ nanostructures employing bio template process. It shows high drug loading capability for hydrophilic doxorubicin (DOX) and hydrophobic paclitaxel (PTX), sustained release profile of an anticancer drug, DOX and higher cytotoxicity against human cervical carcinoma HeLa cells. Roychoudhury *et al.* [122] reported the better DNA binding affinity of the Au-$SiO_2$ composite to produce a particle-nucleic acid conjugate for biomedical applications. The biosynthesised silica nanoparticles also have good drug delivery which was showed by Alshatwi *et al.* [78] Athinarayanan *et al.* [79] and Rangaraj *et al.* [91] Apart from this, Noushad *et al.* [88] applied silica as a fillers nanohybrid dental composite due to its excellent mechanical properties. It was also shown by the researchers that biosynthesized silica nanostructures have good photoluminescence, which shows its potential applications in biomedical imaging [101, 103, 106].

### 5.5.3 Energy and Environment

It has been already mentioned that $SiO_2$ nanostructures act as a benign for the environment by acting as a good photocatalyst for the degradation of organic dyes. It also serves environment for the removal of toxic $Hg^0$, Pb (II) and NO by the process of adsorption. $SiO_2$ nanostructures act as a good adsorbent due to their unique large surface area and narrow pore size distribution. Liu *et al.* [118] reported the improved performance of MnOx- and MnOx-CeOx impregnated $SiO_2$ for the removal of $Hg^o$ and $NO_x$. Here $SiO_2$ composite acts as an adsorbent/ catalyst. Luo *et al.* [107] synthesized the amine-functionalized mesoporous silica and applied it for the removal of lead ions from aqueous solutions. Chandrasekaran *et al.* [104] utilized the boron doped $SiO_2$ diatom frustules as a photocathode for the hydrogen production from water splitting. Bo doped $SiO_2$ shows great potential for the future cost-effective green energy. Yang *et al.* [114] explored the as prepared $TiO_2/SiO_2$ nanostructures as the effective candidate for the chemical and physical removal of pathogenic bacteria.

## 5.6 Conclusions

Various types of chemical and physical approaches were extensively explored for the synthesis and surface modifications of the $SiO_2$ nanostructures. Though the chemical and physical processes exhibit their own unique advantages, they also showed challenges/ limitations such as the need for sophisticated equipment, chemical precursors, and multi-stage complicated protocols. Thus, the green approaches employing bioresources were explored for the synthesis of $SiO_2$ nanostructures as a versatile, sustainable and non-toxic alternative for the traditional physical and chemical processes. Natural resources such as bacteria, fungus, plants and other living organisms such as diatoms are used for the biogenic synthesis of silica nanoparticles. Apart from the pristine $SiO_2$ nanoparticles, the biogenic approaches were also used to synthesis $SiO_2$ based doped, functional and composites materials to attain certain new properties that its virgin material does not have. This approach elevates the potential of $SiO_2$ nanostructures towards various applications in the fields of biomedical, catalytic, energy and environment. Even though the biogenic approaches have several advantages, they have not explored as much as in physical and chemical approaches. Also, the undergoing mechanism for the various bio-based approaches still not clears. Further, the biogenic silica nanostructures are not explored for the many other potential applications than the field of catalysis and biomedical. However, the development of biogenic silica nanostructures in recent years shows its good progress and it might have several opportunities in upcoming years. It also creates a ray of hope on the global transformation towards green and sustainable nanofabrication.

## Acknowledgements

SV acknowledges University Grants Commission (UGC) for the financial support for this research activity through the Minor Research Project (MRP/UGC-SERO- Proposal No.: 1593).

## References

1. Anastas, P.T. and Warner, J.C., *Green chemistry: theory and practice*, Oxford University Press, New york, 2000.
2. Vivekanandhan, S., Schreiber, M., Muthuramkumar, S., Misra, M. and Mohanty A.K., Carbon nanotubes from renewable feedstocks: A move toward sustainable nanofabrication. *J. Appl. Polym. Sci.*, 4, 134, 2016.
3. Ray, A.K., Sahu, R.K., Rajinikanth, V., Bapari, H., Ghosh, M. and Paul, P., Preparation and characterization of graphene and Ni-decorated graphene using flower petals as the precursor material. *Carbon.*, 50, 4123, 2012.
4. De, B. and Karak, N., A green and facile approach for the synthesis of water soluble fluorescent carbon dots from banana juice. *RSC Adv.*, 3, 8286, 2013.
5. Lallave, M., Bedia, J., Ruiz-Rosas, R., Rodríguez-Mirasol, J., Cordero, T., Otero, J.C., Marquez, M., Barrero, A. and Loscertales, I.G., Filled and hollow

carbon nanofibers by coaxial electrospinning of alcell lignin without binder polymers. *Adv. Mater.*, 19, 4292, 2007.
6. Christensen, L., Vivekanandhan, S., Misra, M. and Mohanty, A.K., Biosynthesis of silver nanoparticles using murraya koenigii (curry leaf): an investigation on the effect of broth concentration in reduction mechanism and particle size. *Adv. Mat. Lett.*, 2, 429, 2011.
7. Chandran, S.P., Chaudhary, M., Pasricha, R., Ahmad, A. and Sastry, M., Synthesis of gold nanotriangles and silver nanoparticles using Aloevera plant extract. *Biotechnol. Progr.*, 22, 577, 2006.
8. Petla, R.K., Vivekanandhan, S., Misra, M., Mohanty, A.K. and Satyanarayana, N., Soybean (Glycine max) leaf extract based green synthesis of palladium nanoparticles. *J. Biomater. Nanobiotechnol*, 3, 14, 2012.
9. Sangeetha, G., Rajeshwari, S. and Venckatesh, R., Green synthesis of zinc oxide nanoparticles by aloe barbadensis miller leaf extract: Structure and optical properties. *Mater. Res. Bull.*, 46, 2560, 2011.
10. Bansal, V., Rautaray, D., Bharde, A., Ahire, K., Sanyal, A., Ahmad, A. and Sastry, M., Fungus-mediated biosynthesis of silica and titania particles. *J. Mater. Chem.*, 15, 2583, 2005.
11. Vishnukumar, P., Vivekanandhan, S. and Muthuramkumar, S., Plant-mediated biogenic synthesis of palladium nanoparticles: Recent trends and emerging opportunities. *ChemBioEng. Rev.*, 4, 18, 2016.
12. Vivekanandhan, S., Christensen, L., Misra, M. and Mohanty, A.K., Green process for impregnation of silver nanoparticles into microcrystalline cellulose and their antimicrobial bionanocomposite films. *J. Biomater. Nanobiotechnol.*, 3, 371, 2012.
13. Vivekanandhan, S., Tang, D., Misra, M. and Mohanty, A.K., Novel Glycine Max (Soybean) leaf extract based biological process for the functionalization of carbon nanotubes with silver nanoparticles. *Nanosci. Nanotechnol. Lett.*, 2, 240, 2010.
14. Hutchison, J.E., Greener nanoscience: a proactive approach to advancing applications and reducing implications of nanotechnology. *ACS Nano.*, 2, 395, 2008.
15. Li, Z., Barnes, J.C., Bosoy, A., Stoddart, J.F. and Zink, J.I., Mesoporous silica nanoparticles in biomedical applications. *Chem. Soc. Rev.*, 41, 2590, 2012.
16. Rahman, I.A. and Padavettan, V., Synthesis of silica nanoparticles by sol-gel: size-dependent properties, surface modification, and applications in silica-polymer nanocomposites—a review. *J. Nanomater.*, 2012, 1, 2012.
17. Liberman, A., Mendez, N., Trogler, W.C. and Kummel, A.C., Synthesis and surface functionalization of silica nanoparticles for nanomedicine. *Surf. Sci. Rep.*, 69, 132, 2014.
18. Salavati-Niasari, M., Javidi, J. and Dadkhah, M., Ball milling synthesis of silica nanoparticle from rice husk ash for drug delivery application. *Comb. Chem. High Throughput Screening.*, 16, 458, 2013.

19. Geohegan, D.B., Puretzky, A.A., Duscher, G. and Pennycook, S.J., Photoluminescence from gas-suspended SiO x nanoparticles synthesized by laser ablation. *Appl. Phys. Lett.*, 73, 438, 1998.
20. Mishra, Y., Mohapatra, S., Kabiraj, D., Mohanta, B., Lalla, N., Pivin, J. and Avasthi, D., Synthesis and characterization of Ag nanoparticles in silica matrix by atom beam sputtering. *Scr. Mater.*, 56, 629, 2007.
21. Trewyn, B.G., Slowing, I.I., Giri, S., Chen, H.T. and Lin, V.S.Y., Synthesis and functionalization of a mesoporous silica nanoparticle based on the sol–gel process and applications in controlled release. *Acc. Chem. Res.*, 40, 846, 2007.
22. Bagwe, R.P, Yang, C., Hilliard, L.R. and Tan, W., Optimization of dye-doped silica nanoparticles prepared using a reverse microemulsion method. *Langmuir*, 20, 8336, 2004.
23. Kammler, H.K., Beaucage, G., Mueller, R. and Pratsinis, S.E., Structure of flame-made silica nanoparticles by ultra-small-angle X-ray scattering. *Langmuir*, 20, 1915, 2004.
24. Saleh, N.J., Ibrahim, R.I. and Salman, A.D., Characterization of nano-silica prepared from local silica sand and its application in cement mortar using optimization technique. *Adv. Powder Technol.*, 26, 1123, 2015.
25. Rizlan, Z. and Mamat, O., Process parameters optimization of silica sand nanoparticles production using low speed ball milling method. *Chin. J. Eng.*, 2014, 1, 2014.
26. Laranjo, M.T., Ricardi, N.C., Arenas, L.T., Benvenutti, E.V., de Oliveira, M.C., Buchner, S., Santos, M.J. and Costa, T.M.H., Influence of ball milling on textural and morphological properties of $TiO_2$ and TiO2/$SiO_2$ xerogel powders applied in photoanodes for solar cells. *J. Solid State Electrochem.*, 20, 1731, 2016.
27. Ashfold, M.N., Claeyssens, F., Fuge, G.M. and Henley, S.J., Pulsed laser ablation and deposition of thin films. *Chem. Soc. Rev.*, 33, 23, 2004.
28. Geohegan, D.B., Puretzky, A.A., Duscher, G. and Pennycook, S.J., Time-resolved imaging of gas phase nanoparticle synthesis by laser ablation. *Appl. Phys. Lett.*, 72, 2987, 1998.
29. Li, H., Guan, L., Xu, Z., Zhao, Y., Sun, J., Wu, J. and Xu, N., Synthesis and characterization of amorphous $SiO_2$ nanowires via pulsed laser deposition accompanied by N 2 annealing. *Appl. Surf. Sci.*, 389, 705, 2016.
30. Grigoriu, C., Kuroki, Y., Nicolae, I., Zhu, X., Hirai, M., Suematsu, H., Takata, M. and Yatsui, K., Photo and cathodoluminescence of Si/SiO 2 nanoparticles produced by laser ablation. *J. Optoelectron. Adv. Mater.*, 7, 2979, 2005.
31. Kunti, A., Chowdhury, M., Sharma, S., Gupta, M. and Chaudhary, R., Influence of $O_2$ pressure on structural, morphological and optical properties of $TiO_2$–$SiO_2$ composite thin films prepared by pulsed laser deposition. *Thin Solid Films.*, 629, 79, 2017.
32. Uchida, N., Okami, T., Tagami, H., Fukata, N., Mitome, M., Bando, Y. and Murakami, K., Synthesis of silicon nanocrystals in aluminum-doped $SiO_2$ film by laser ablation method. *Physica E.*, 38, 31, 2007.

33. Seto, T., Koga, K., Takano, F., Akinaga, H., Orii, T., Hirasawa, M. and Murayama, M., Synthesis of magnetic CoPt/$SiO_2$ nano-composite by pulsed laser ablation. *J. Photochem. Photobiol., A.*, 182, 342, 2006.
34. Liao, H., Xiao, R., Fu, J., Yu, P., Wong, G.K.L. and Sheng, P., Large third-order optical nonlinearity in Au:$SiO_2$ composite films near the percolation threshold. *Appl. Phys. Lett.*, 70, 1, 1997.
35. Hamza, S., Ignaszak, A. and Kiani, A., Synthesis of electrical conductive silica nanofiber/gold nanoparticle composite by laser pulses and sputtering technique. *Nanoscale Res. Lett.*, 12, 432, 2017.
36. Singh, L.P., Bhattacharyya, S.K., Kumar, R., Mishra, G., Sharma, U., Singh, G. and Ahalawat, S., Sol-Gel processing of silica nanoparticles and their applications, *Adv. Colloid Interface Sci.*, 214, 17, 2014.
37. Wright, J.D. and Sommerdijk, N.A., *Sol-gel materials: chemistry and applications*, CRC press, London, 2000.
38. Sakka, S., Sol–gel technology as representative processing for nanomaterials: case studies on the starting solution, *J. Sol-Gel Sci. Technol.*, 46, 241, 2008.
39. Haque, F.Z., Nandanwar, R., Singh, P., Dharavath, K. and Syed, F.F., Effect of Different Acids and Solvents on Optical Properties of SiO2 Nanoparticles Prepared by the Sol-Gel Process, *Silicon.*, 10, 413, 2016.
40. Guo, Q., Huang, D., Kou, X., Cao, W., Li, L., Ge, L. and Li, J., Synthesis of disperse amorphous $SiO_2$ nanoparticles via sol–gel process. *Ceram. Int.*, 43, 192, 2017.
41. Ennas, G., Musinu, A., Piccaluga, G., Zedda, D., Gatteschi, D., Sangregorio, C., Stanger, J., Concas, G. and Spano, G., Characterization of iron oxide nanoparticles in an Fe2O3– SiO2 composite prepared by a sol– gel method. *Chem. Mater.*, 10, 495, 1998.
42. Li, Z., Hou, B., Xu, Y., Wu, D., Sun, Y., Hu, W. and Deng, F., Comparative study of sol–gel-hydrothermal and sol–gel synthesis of titania–silica composite nanoparticles. *J. Solid State Chem.*, 178, 1395, 2005.
43. Osseo-Asare, K. and Arriagada, F., Preparation of SiO2 nanoparticles in a non-ionic reverse micellar system. *Colloids surf.*, 50, 321, 1990.
44. Arriagada, F. and Osseo-Asare, K., Synthesis of nanosize silica in a nonionic water-in-oil microemulsion: effects of the water/surfactant molar ratio and ammonia concentration. *J. Colloid Interface Sci.*, 211, 210, 1999.
45. Santra, S., Tapec, R., Theodoropoulou, N., Dobson, J., Hebard, A. and Tan, W., Synthesis and characterization of silica-coated iron oxide nanoparticles in microemulsion: the effect of nonionic surfactants. *Langmuir*, 17, 2900, 2001.
46. Han, Y. Jiang, J., Lee, S.S. and Ying, J.Y., Reverse microemulsion-mediated synthesis of silica-coated gold and silver nanoparticles. *Langmuir.*, 24, 5842, 2008.
47. Hentze, H.P., Raghavan S.R., McKelvey, C.A. and Kaler, E.W., Silica hollow spheres by templating of catanionic vesicles. *Langmuir.*, 19, 1069, 2003.

48. Chen, J.F., Ding, H.M., Wang, J.X. and Shao, L., Preparation and characterization of porous hollow silica nanoparticles for drug delivery application. *Biomater.*, 25, 723, 2004.
49. Fuji, M., Shin, T., Watanabe, H., andTakei, T., Shape-controlled hollow silica nanoparticles synthesized by an inorganic particle template method. *Adv. Powder Technol.*, 23, 562, 2012.
50. Takai-Yamashita, C., Imabeppu, H. and Fuji, M., Synthesis of hollow silica nanoparticles using poly (acrylic acid)-3, 3'-diaminodipropylamine template. *Colloids Surf., A.*, 483, 81, 2015.
51. Fowler, C.E., Shenton, W., Stubbs, G. and Mann, S., Tobacco mosaic virus liquid crystals as templates for the interior design of silica mesophases and nanoparticles. *Adv. Mater.*, 13, 1266, 2001.
52. Correa-Duarte, M.A., Sobal, N., Liz-Marzán, L.M. and Giersig, M., Linear assemblies of silica-coated gold nanoparticles using carbon nanotubes as templates. *Adv. Mater.*, 16, 2179, 2004.
53. Zhu, Y., Kockrick, E., Ikoma, T., Hanagata, N. and Kaskel, S., An efficient route to rattle-type Fe3O4@ SiO2 hollow mesoporous spheres using colloidal carbon spheres templates. *Chem. Mater.*, 21, 2547, 2009.
54. Westall, F., Boni, L. and Guerzoni, E., The experimental silicification of microorganisms. *Palaeontol.*, 38, 495, 1995.
55. Yee, N., Phoenix, V.R., Konhauser, K.O., Benning, L.G. and Ferris, F.G., The effect of cyanobacteria on silica precipitation at neutral pH: implications for bacterial silicification in geothermal hot springs. *Chem. Geol.*, 199, 83, 2003.
56. Singh, S., Bhatta, U.M., Satyam, P., Dhawan, A., Sastry, M. and Prasad, B., Bacterial synthesis of silicon/silica nanocomposites. *J. Mater. Chem.*, 18, 2601, 2008.
57. Jo, W., Freedman, K.J., Yi, D.K. and Kim, M.J., Fabrication of tunable silica-mineralized nanotubes using flagella as bio-templates. *Nanotechnol.*, 23, 055601, 2012.
58. Jo, W., Darmawan, M., Kim, J., Ahn, C.W., Byun, D., Baik, S.H. and Kim, M.J., Electrical property measurements of metallized flagella-templated silica nanotube networks. *Nanotechnol.*, 24, 135704, 2013.
59. Show, S., Tamang, A., Chowdhury, T., Mandal, D. and Chattopadhyay, B., Bacterial (BKH1) assisted silica nanoparticles from silica rich substrates: a facile and green approach for biotechnological applications. *Colloids Surf., B.*, 126, 245, 2015.
60. Ebrahiminezhad, A., Najafipour, S., Kouhpayeh, A., Berenjian, A., Rasoul-Amini, S. and Ghasemi, Y., Facile fabrication of uniform hollow silica microspheres using a novel biological template. *Colloids Surf., B.*, 118, 249, 2014.
61. Nomura, T., Morimoto, Y., Ishikawa, M., Tokumoto, H. and Konishi, Y., Synthesis of hollow silica microparticles from bacterial templates. *Adv. Powder Technol.*, 21, 218, 2010.
62. Baumgartner, J., Lesevic, P., Kumari, M., Halbmair, K., Bennet, M., Körnig, A., Widdrat, M., Andert, J., Wollgarten, M. and Bertinetti, L., From

magnetotactic bacteria to hollow spirilla-shaped silica containing a magnetic chain. *RSC Adv.*, 2, 8007, 2012.
63. Bharde, A., Rautaray, D., Bansal, V., Ahmad, A., Sarkar, I., Yusuf, S.M., Sanyal, M. and Sastry, M., Extracellular biosynthesis of magnetite using fungi. *Small.*, 2, 135, 2006.
64. Bansal, V., Rautaray, D., Ahmad, A. and Sastry, M., Biosynthesis of zirconia nanoparticles using the fungus Fusarium oxysporum. *J. Mater. Chem.*, 14, 3303, 2004.
65. Bansal, V., Sanyal, A., Rautaray, D., Ahmad, A. and Sastry, M., Bioleaching of sand by the fungus Fusarium oxysporum as a means of producing extracellular silica nanoparticles. *Adv. Mater.*, 17, 889, 2005.
66. Kulkarni, S., Syed, A., Singh, S., Gaikwad, A., Patil, K., Vijayamohanan, K., Ahmad, A. and Ogale, S., Silicate nanoparticles by bioleaching of glass and modification of the glass surface. *J. Non-Cryst. Solids.*, 354, 3433, 2008.
67. Bansal, V., Ahmad, A. and Sastry, M., Fungus-mediated biotransformation of amorphous silica in rice husk to nanocrystalline silica. *J. Am. Chem. Soc.*, 128, 14059, 2006.
68. Pineda, T., Soares, C.H.L., Hotza, D., Casas-Botero, A.E., Ramírez-Carmona, M. and Torres-Taborda, M., Extracellular synthesis of silica oxide particles by Fusarium oxysporum from rice husk ash, in: *Materials Science Forum*, L. Salgado, F.A. Filho (Ed.), pp. 1153–1157, Trans Tech Publ., Switzerland, 2012.
69. Pineda-Vásquez, T.G., Casas-Botero, A.E., Ramírez-Carmona, M.E., Torres-Taborda, M.M., Soares, C.H. and Hotza, D., Biogeneration of Silica Nanoparticles from Rice Husk Ash Using Fusarium oxysporum in Two Different Growth Media. *Ind. Eng. Chem. Res.*, 53, 6959, 2014.
70. Marikani, K., Sangareswari, K.U., Suganya, P., Ganesan, R. and Rajarathinam, K., Biobased approach for the synthesis, characterization, optimization and application of silica nanoparticles by fungus Fusarium oxysporum. *Pharm. Biol. Eval.*, 2, 223, 2016.
71. Kaur, T., Singh, G.P., Kaur, G., Kaur, S. and Gill, P.K., Synthesis of biogenic silicon/silica (Si/$SiO_2$) nanocomposites from rice husks and wheat bran through various microorganisms. *Mater. Res. Express.*, 3, 085026, 2016.
72. Tang, Q. and Wang, T., Preparation of silica aerogel from rice hull ash by supercritical carbon dioxide drying. *J Supercrit Fluids.*, 35, 91, 2005.
73. Li, T. and Wang, T., Preparation of silica aerogel from rice hull ash by drying at atmospheric pressure. *Mater. Chem. Phys.*, 112, 398, 2008.
74. Nayak, J. and Bera, J., Preparation of silica aerogel by ambient pressure drying process using rice husk ash as raw material. *Trans. Indian Ceram. Soc.*, 68, 91, 2009.
75. Tadjarodi, A., Haghverdi, M. and Mohammadi, V., Preparation and characterization of nano-porous silica aerogel from rice husk ash by drying at atmospheric pressure. *Mater. Res. Bull.*, 47, 2584, 2012.

76. Vijayalakshmi, U., Vaibhav, V., Chellappa, M. and Anjaneyulu, U., Green synthesis of silica nanoparticles and its corrosion resistance behavior on mild steel. *J. Indian Chem. Soc.*, 92, 675, 2015.
77. Prasad, R. and Pandey, M., Rice husk ash as a renewable source for the production of value added silica gel and its application: an overview. *Bull. Chem. React. Eng. Catal.*, 7, 1, 2012.
78. Alshatwi, A.A., Athinarayanan, J. and Periasamy, V.S., Biocompatibility assessment of rice husk-derived biogenic silica nanoparticles for biomedical applications. *Mater. Sci. Eng., C.*, 47, 8, 2015.
79. Athinarayanan, J., Periasamy, V.S., Alhazmi, M., Alatiah, K.A. and Alshatwi, A.A., Synthesis of biogenic silica nanoparticles from rice husks for biomedical applications. *Ceram. Int.*, 41, 275, 2015.
80. Sankar, S., Sanjeev, S. and Sharma, D., Synthesis and characterization of mesoporous SiO2 nanoparticles synthesized from Biogenic Rice Husk Ash for optoelectronic applications. *Int J Eng Sci.*, 17, 353, 2016.
81. Sankar, S., Sharma, S.K., Kaur, N., Lee, B., Kim, D.Y., Lee, S. and Jung, H., Biogenerated silica nanoparticles synthesized from sticky, red, and brown rice husk ashes by a chemical method. *Ceram. Int.*, 42, 4875, 2016.
82. Carmona, V., Oliveira, R., Silva, W., Mattoso, L. and Marconcini, J., Nanosilica from rice husk: extraction and characterization. *Ind. Crops Prod.*, 43, 291, 2013.
83. Sinyoung, S., Kunchariyakun, K., Asavapisit, S. and MacKenzie, K.J., Synthesis of belite cement from nano-silica extracted from two rice husk ashes. *J. Environ. Manage.*, 190, 53, 2017.
84. Zulfiqar, U., Subhani, T. and Husain, S.W., Towards tunable size of silica particles from rice husk. *J. Non-Cryst. Solids.*, 429, 61, 2015.
85. Bakar, R.A., Yahya, R. and Gan, S.N., Production of high purity amorphous silica from rice husk. *Procedia Chem.*, 19, 189, 2016.
86. Gomes, G.M.F., Philipssen, C., Bard, E.K., Dalla Zen, L. and de Souza, G., Rice husk bubbling fluidized bed combustion for amorphous silica synthesis. *J. Environ. Chem. Eng.*, 4, 2278, 2016.
87. Azmi, M., Ismail, N., Rizamarhaiza, M. and Taib, H., Characterisation of silica derived from rice husk (Muar, Johor, Malaysia) decomposition at different temperatures, in: AIP Conference Proceedings, pp. 020005, AIP Publishing, United States, 2016.
88. Noushad, M., Ab Rahman, I., Husein, A. and Mohamad, D., Nanohybrid dental composite using silica from biomass waste. *Powder Technol.*, 299, 19, 2016.
89. Bageru, A.B. and Srivastava, V.C., Preparation and characterisation of biosilica from teff (eragrostis tef) straw by thermal method. *Mater. Lett.*, 206, 13, 2017.
90. San, N.O., Kurşungöz, C., Tümtaş, Y., Yaşa, Ö., Ortaç, B. and Tekinay, T., Novel one-step synthesis of silica nanoparticles from sugarbeet bagasse by laser ablation and their effects on the growth of freshwater algae culture. *Particuol.*, 17, 29, 2014.

91. Rangaraj, S. and Venkatachalam, R.,A lucrative chemical processing of bamboo leaf biomass to synthesize biocompatible amorphous silica nanoparticles of biomedical importance. *Appl. Nanosci.*, 7, 1, 2017.
92. Mattos, B.D., Rojas, O.J. and Magalhães, W.L., Biogenic $SiO_2$ in colloidal dispersions via ball milling and ultrasonication. *Powder Technol.*, 301, 58, 2016.
93. Athinarayanan, J., Periasamy, V.S., Alhazmi, M. and Alshatwi, A.A., Synthesis and biocompatibility assessment of sugarcane bagasse-derived biogenic silica nanoparticles for biomedical applications. *J. Biomed. Mater. Res. Part B.*,105, 340, 2017.
94. Wang, J., Yang, S., Li, C., Miao, Y., Zhu, L., Mao, C. and Yang, M., Nucleation and assembly of silica into protein-based nanocomposites as effective anticancer drug carriers using self-assembled silk protein nanostructures as biotemplates. *ACS Appl. Mater. Interfaces.*, 9, 22259, 2017.
95. Cao, F. and Li, D.X., Morphology-controlled synthesis of SiO2 hollow microspheres using pollen grain as a biotemplate. *Biomed. Mater.*, 4, 025009, 2009.
96. Cao, F., Li, D.X. and Wu, Z.Y., *In Situ* Synthesis of the Hollow Silica Microspheres with Porous Surface by Using Chemical Modified Pollen Grains as Biotemplate, in: *Key Engineering Materials*, F. Tang (Ed.), pp. 1273–1278, Trans Tech Publ, 2015.
97. Guan, Z.S., Zhang, Y., Lu, C.H. and Xu, Z.Z., Morphology-controlled synthesis of SiO2 hierarchical structures using pollen grains as templates. *Chin. J. Chem.*, 26, 467, 2008.
98. Leng, B., Chen, X., Shao, Z. and Ming, W., Biomimetic synthesis of silica with chitosan-mediated morphology. *Small.*, 4, 755, 2008.
99. Zhang, B., Ren, T.R., Song, B.A. and Wang, Q.X., Fabrication of monodisperse silica microspheres using synechocystis cell as a template. *Mater. Chem. Phys.*, 138, 762, 2013.
100. Li, J., Li, Z., Zi, G., Yao, Z., Luo, Z., Wang, Y., Xue, D., Wang, B. and Wang, J., Synthesis, characterizations and catalytic allylic oxidation of limonene to carvone of cobalt doped mesoporous silica templated by reed leaves. *Catal. Commun.*, 59, 233, 2015.
101. Liu, Y., Wang, Z., Zeng, H., Chen, C., Liu, J., Sun, L. and Wang, W., Photoluminescent mesoporous carbon-doped silica from rice husks. *Mater. Lett.*, 142, 280, 2015.
102. Skolem, L.M.B., Biosynthesis and characterization of Ti-doped silica-based Nanostructures formed by the Diatoms Pinnularia sp. and Coscinodiscus wailesii, Institutt for bioteknologi, 2011.
103. Gale, D.K., Jeffryes, C., Gutu, T., Jiao, J., Chang, C.H. and Rorrer, G.L., Thermal annealing activates amplified photoluminescence of germanium metabolically doped in diatom biosilica. *J. Mater. Chem.*, 21, 10658, 2011.

104. Chandrasekaran, S., Macdonald, T.J., Gerson, A.R., Nann, T. and Voelcker, N.H., Boron-doped silicon diatom frustules as a photocathode for water splitting. *ACS Appl. Mater. Interfaces.*, 7, 17381, 2015.
105. Chauton, M.S., Skolem, L.M., Olsen, L.M., Vullum, P.E., Walmsley, J. and Vadstein, O., Titanium uptake and incorporation into silica nanostructures by the diatom Pinnularia sp.(Bacillariophyceae). *J. Appl. Phycol.*, 27, 777, 2015).
106. Wang, Z., Liu, J., Wang, W., Wei, Z., Wang, F., Gong, P., Wang, J., Li, N., Liu, B. and Zhang, Z., Photoluminescent carbon quantum dot grafted silica nanoparticles directly synthesized from rice husk biomass. *J. Mater. Chem. B.*, 5, 4679, 2017.
107. Luo, L.J., Liu, Y., Li, Q.L. and Mo, J.C., Hydrothermal Biotemplated Synthesis of Amine-functionalized mesporous silica and Adsorption Properties for lead (II), in: *Advanced Materials Research*, L. Yu, J. Guo, G, Yi, Q. Yu (Ed.), pp. 1958-1962, Trans Tech Publ, Switzerland, 2013.
108. Cui, S., Yu, S., Lin, B., Shen, X., Zhang, X. and Gu, D., Preparation of amine-modified SiO2 aerogel from rice husk ash for CO2 adsorption. *J. Porous Mater.*, 24, 455, 2017.
109. Jo, W., Freedman, K.J. and Kim, M.J., Metallization of biologically inspired silica nanotubes. *Mater. Sci. Eng., C.*, 32, 2426, 2012.
110. Cicco, S.R., Vona, D., De Giglio, E., Cometa, S., Mattioli-Belmonte, M., Palumbo, F., Ragni, R. and Farinola, G.M., Chemically modified diatoms biosilica for bone cell growth with combined drug-delivery and antioxidant properties. *ChemPlusChem.*, 80, 1104, 2015.
111. Sujaridworakun, P., Jinawath, S., Panpa, W., Nakajima, A. and Yoshimura, M., Hydrothermal Synthesis of TiO2/SiO2 Hybrid Photocatalyst from Rice Husk Ash, in: *Key Engineering Materials*, K. Komeya, Y. Matsuo, T. Goto (Ed.), pp. 281–285, Trans Tech Publ, 2007.
112. Larichev, Y.V., Yeletsky, P. and Yakovlev, V., Study of silica templates in the rice husk and the carbon–silica nanocomposites produced from rice husk. *J. Phys. Chem. Solids.*, 87, 58, 2015.
113. Soltani, N., Simon, U., Bahrami, A., Wang, X., Selve, S., Epping, J.D., Pech-Canul, M.I., Bekheet, M.F. and Gurlo, A., Macroporous polymer-derived SiO2/SiOC monoliths freeze-cast from polysiloxane and amorphous silica derived from rice husk. *J. Eur. Ceram. Soc.*, 37, 4809, 2017.
114. Yang, D., Du, B., Yan, Y., Li, H., Zhang, D. and Fan, T., Rice-husk-templated hierarchical porous TiO2/SiO2 for enhanced bacterial removal. *ACS Appl. Mater. Interfaces.*, 6, 2377, 2014.
115. Yener, H.B. and Helvacı, Ş.Ş., Effect of synthesis temperature on the structural properties and photocatalytic activity of $TiO_2/SiO_2$ composites synthesized using rice husk ash as a $SiO_2$ source. *Sep. Purif. Technol.*, 140, 84, 2015.
116. Hui, C., Lei, Z., Xitang, W., Shujing, L. and Zhongxing, L., Preparation of nanoporous TiO2/SiO2 composite with rice husk as template and its photocatalytic property. *Rare Metal Mat. Eng.*, 44, 1607, 2015.

117. I. Fatimah, Comparative study on the impact of synthesis route to the photocatalytic activity of ZnO-SiO2 from rice husk ash, 2016.
118. Liu, K.H., Chen, M.Y., Tsai, Y.C., Lin, H.P. and Hsi, H.C., Control of Hg 0 and NO from coal-combustion flue gases using MnO x-CeO x/mesoporous $SiO_2$ from waste rice husk. *Catal. Today.*, 297, 104, 2017.
119. S.M. Al-Amsyar, F. Adam and E.-P. Ng, Aluminium oxide-silica/carbon composites from rice husk as bifunctional heterogeneous catalysts for one-pot sequential reaction in glucose conversion. *Surf. Interfaces.*, 9, 1, 2017.
120. Khazaei, A., Khazaei, M., Nasrollahzadeh, M., Nano-Fe3O4@ $SiO_2$ supported Pd (0) as a magnetically recoverable nanocatalyst for Suzuki coupling reaction in the presence of waste eggshell as low-cost natural base. *Tetrahedron*, 73, 5624, 2017.
121. Chen, H., Zhao, L., Wang, G., He, X., Wang, X., Fang, W., Du, X., Direct growth of ZnO nanorods on biogenic hierarchical rice husk $SiO_2$ and their application to dye degradation. *Clean Technol. Envir.*, 19, 1335, 2017.
122. Roychoudhury, P., Nandi, C., Pal, R., Diatom-based biosynthesis of gold-silica nanocomposite and their DNA binding affinity. *J. Appl. Phycol.*, 28, 2857, 2016.

# 6

# Green and Sustainable Advanced Composite Materials

**Yahya F. Al-Khafaji[1] and Falah H. Hussein[2*]**

[1]*College of Science, Babylon University, Iraq*
[2]*College of Pharmacy, Babylon University, Iraq*

***Abstract***
This chapter investigates the main features of biodegradable polymer including, applications of polymers, the problems of synthetic polymers, why biodegradable polymers, biodegradable polymers, copolymers. Also, the chapter gives an example of biodegradable polymers which is aliphatic polyesters include: polylactide (PLA), polycaprolactone (PCL) and polyvalerolactone (PVL). Moreover, different approaches used to synthesis polymers are presented. Depending on the manner of prepare of the polymer there are different mechanisms are discussed in this chapter.

***Keywords:*** Green and sustainable, advanced materials, polymer

## 6.1 Introduction

During the past years, several topics of the search-strategic have been appeared, one of them the combination between techniques of green chemistry and nanotechnology applications. The nanomaterials are costly systems based on hazardous heavy metal such as CdS particles and Ru-compounds [1].

The green chemistry and their principles with metrics of environmental efficiencies are consider as a main approach to achieve the sustainable development that depends on three main categories: environmental, society, and economy. The principles of green chemistry are included:

**Corresponding author*: abohasan_hilla@yahoo.com

Shakeel Ahmed and Chaudhery Mustansar Hussain (eds.) Green and Sustainable Advanced Materials, Vol. 1 (143–154) © 2018 Scrivener Publishing LLC

prevention, atom economy, less hazardous chemical synthesis, designing safer chemicals, safer solvents and auxiliaries, design for energy efficiency, use of renewable feedstocks, reduce derivatives, catalysis, design for degradation, real-time analysis for pollution prevention, and inherently safer chemistry for accident prevention [2, 3].

There is a good possibility of using alternative natural molecular dyes in the field of water purification. The natural dye of anthocyanin used as safe sensitizer for nanoparticles of titanium dioxide in photo-degradation process of organic pollutants in water.

The new catalyst $TiO_2$/anthocyanin showed effectively catalyzed the photo-degradation under solar radiation than the earlier systems previously used, while avoiding their hazardous nature as well as the possibility of enhancing their efficiency and ease to recovery from catalytic reaction by supported onto activated carbon particles, but the process need to reactivate the recovered catalyst by re-treated with fresh dye of anthocyanin due to their degradation [4].

However, some of science policy reports have mentioned that there is great opportunity and benefit by providing green nano processes by producing nanomaterials and products without pollutants harming the environment or human health at stages of managing, design, manufacturing, and application. Thus, the nanotechnology can help alleviate issues of the safe sustainable development such as environmental, human health, and safety risks and support for a sustainable environment in related with terms of energy, water, food production, raw materials, climate change, etc. [5, 6].

When the green chemistry principles are applied to nanoscience the result surely would be practicing green nanoscience that depends over each principle and methods of designing nanomaterials.

Nowadays, scientists have focused on creating a sustainable green technologies used in the manufacture of nanomaterials, as in: Nanocatalysis, Self-Assembly, used of solid state-solventless process, Non-toxic starting materials, Renewable starting materials, Safe photochemical synthesis, Aqueous processing, Bio-inspired nanoscale synthesis, Improved synthesis-fewer steps, Microwave techniques, Renewable in nanocomposites, Non-toxic starting materials, etc.[7].

It has become a trend specialist scientists to prepare biopolymers from natural sources, as in the corn plant and the roots of sugar beet where the starch is extracted from these plants and then prepare developed materials as a green nanocatalysts have a direct interest in reducing the reaction time and temperature in a safe reactors compared to common commercial processes [8].

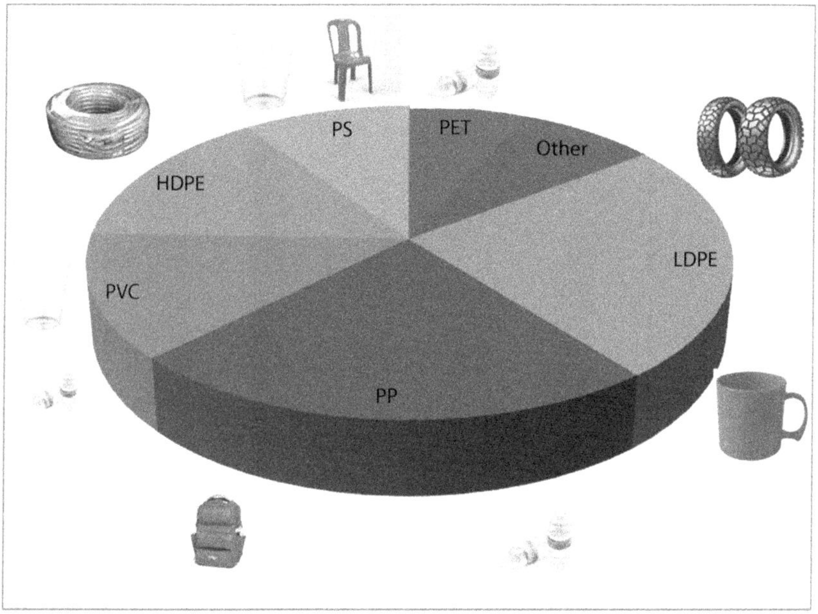


**Figure 6.1** Applications of polymers.

## 6.2 Applications of Polymers

In spite of the huge progresses in science and technology over the last few decades, humanity is facing gradually complicated challenges in life. Most the needs of current society, for example, energy, and materials are extremely reliant on diminishing resources. The progress technology has assisted humanity to consume the main world's resources. Polymer is an example of essential material based on depleting resources [9].

Polymers are used in every part of modern life due to their applications in all aspects of human life, for example, as packaging material, agriculture, electric and other every day purposes (Figure 6.1) [10].

Subsequently, the word plastic production increase rapidly, Figure 6.2 shows that the production of plastics during the period 2004 until 2014 and this will continue to increase in future because of the spread of human population [11].

## 6.3 The Problems of Synthetic Polymers

The majority of polymers are based on petrochemical resources such as polyethylene and polypropylene (Figure 6.3) [12]. Such materials brought

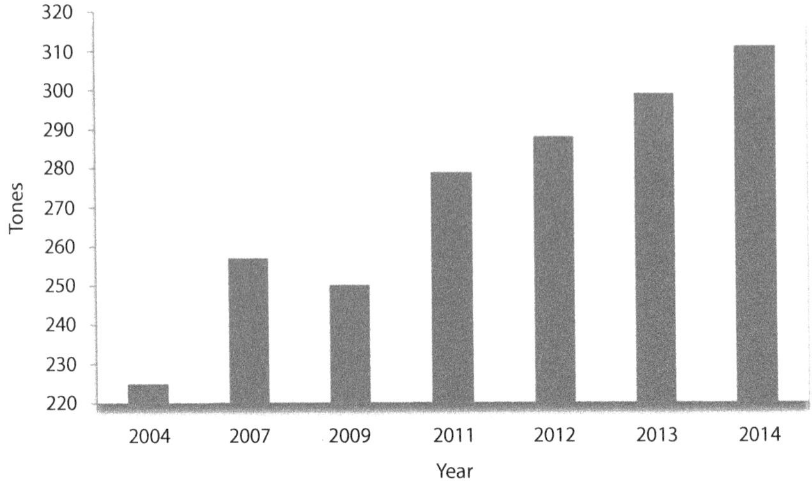


**Figure 6.2** World plastic production 2004–2014 in million tones units.

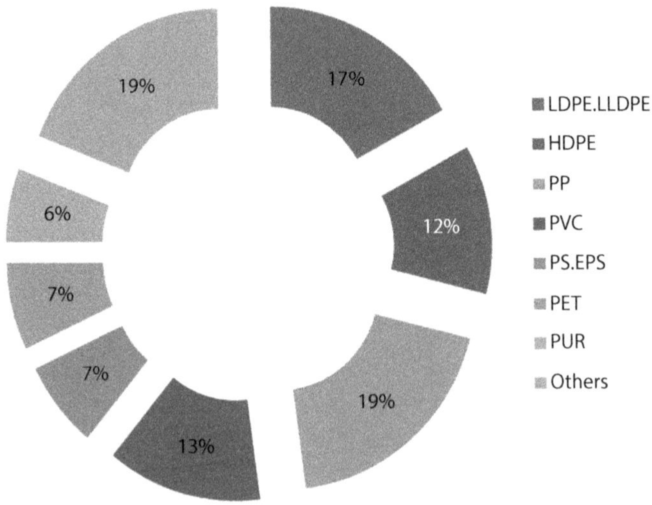


**Figure 6.3** Poly (olefin)s polymers used in the worldwide.

two issues (i) they pile up in the environment because their resistance toward bacterial degradation[13], For example, (100–400) years required time to degrade plastic bags in landfill sites as compare with degradable bags can take much less time [14] (ii) they are derived from diminished resources. In addition to petroleum price fluctuation can effect on production of plastic; bearing these facts in our mind we find that the only way to maintain producing plastic, by developing new approaches to synthesize polymers derived, natural, biodegradable and renewable resources.

## 6.4 Why Biodegradable Polymers

At present, the challenge is to be able to produce sustainable and biodegradable resources because of environmental issues, increasing human population and human modern lifestyle. Produce biodegradable polymers are investigated by a numerical study due to their biodegradability, biocompatibility and they derived from renewable resources, moreover, there is an expectation oil will finish in the future. In addition to the usage of petroleum resources is also causal to the rise of emission of carbon dioxide into the atmosphere [15].

## 6.5 Biodegradable Polymers

Biodegradable polymers were date back in 1980s. Biodegradable plastics can be derived from synthetic or natural sources. Renewable resources are the main source for biodegradable polymers compared with non-degradable polymers are derived from non-renewable resources.

The environmental conditions and, the chemical structure are effect on biodegradability of polymers also it is changed by copolymerization. For example, copolymerization between cyclic esters and chitosan by using the ring opening polymerization by using different catalysts [16, 17].

## 6.6 Copolymers

Copolymerization of monomers such as cyclicesters copolymerization (lactide and ε-caprolactone) provides a valuable way to get new properties for polymers. For example, flexibility and thermal properties of polylactide homopolymer are less compared to that of polycaprolactone homopolymer while new copolymers between lactide and lactone has new flexibility and thermal properties can be used in different application [18]. Co-polymer can be produced by two different approaches by a block or random co-polymer. Feng *et al.*, published work they revealed that copolymerization between polylactide with polycaprolactone, capable of producing a material with more rapid polylactide degradation and polycaprolactone [19].

# 6.7 Examples of Biodegradable Polymers is Polyesters

## 6.7.1 Aliphatic Polyesters Polylactide PLA, Polycaprolactone PCL and Polyvalerolactone PVL

Aliphatic polyesters (polylactide, polylactones) and their co-polymers have many applications such as; medical applications as drug delivery systems (DVL) and scaffolds in tissue engineering [20]. Industry applications in production of plastic bags due to their biodegradability [21]. Regarding of availability and toxicity their monomers are derived from renewable and natural resources (Scheme 6.1), for example, lactide can be prepared from corn starch, and according to literatures review, there is no toxicity related with these polymers and therefore they are used for food packaging [22]. (Scheme 6.2) represented polymerization of lactide, ε-caprolactone and d-valerolactone.

## 6.7.2 Preparation of Polyesters

There are two different commercial ways are used to produce polyester. The first way is polycondensation of hydroxyl-carboxylic acid and, second way by the ring opening polymerization of their monomers

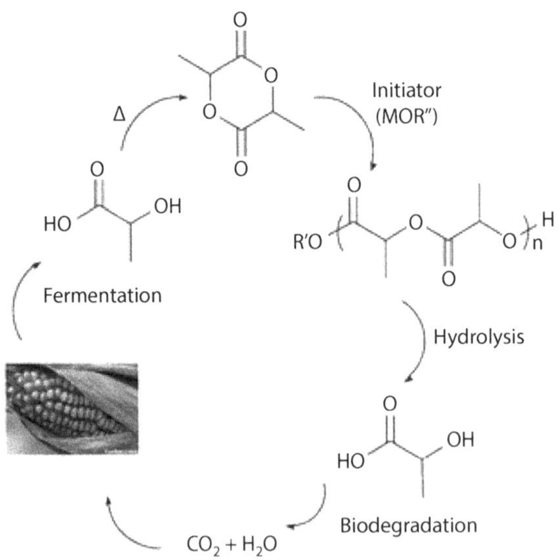


**Scheme 6.1** Polylactide source, polymerization and biodegradation.

### 6.7.2.1 *Polycondensation*

In this technique hydroxyl acid or diacidic react with diol to produce a polyester; this process includes losing water (Scheme 6.3) [23].

Recently, Braud *et al.* have reported prepare polycaprolactone oligomers without a catalyst in (6 h) at a temperature from 80 to 150°C [24].

The polycondensation is less expensive; however, not easy to get high molecular weight polymers.

### 6.7.2.2 *Ring Opening Polymerization (ROP)*

Ring opening polymerization is the typical technique with which to produce polyester from their monomers owing to the fact that control of the end group, got novel structures, and high molecular weight (Scheme 6.4) [25].

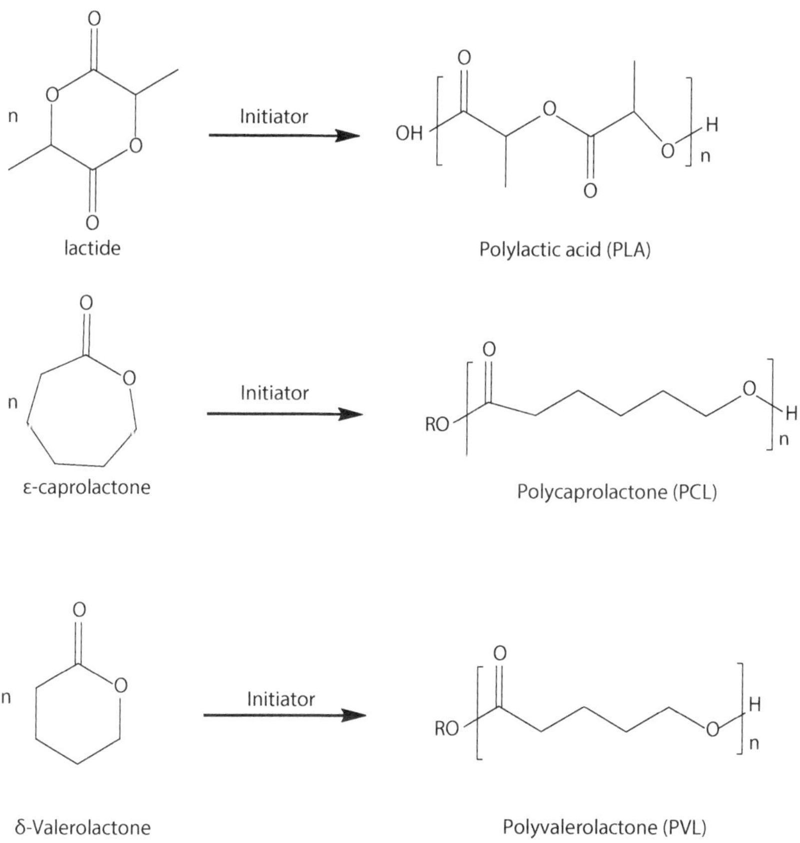


**Scheme 6.2** Cyclicesters (lactide, e-caprolactone and d-valerolactone) polymerization.

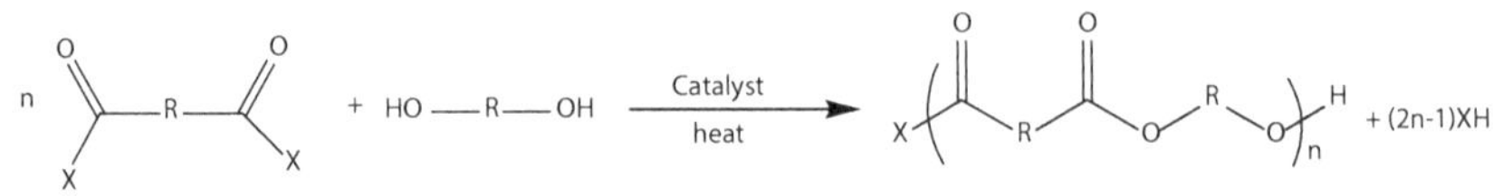


**Scheme 6.3** Condensation polyme.

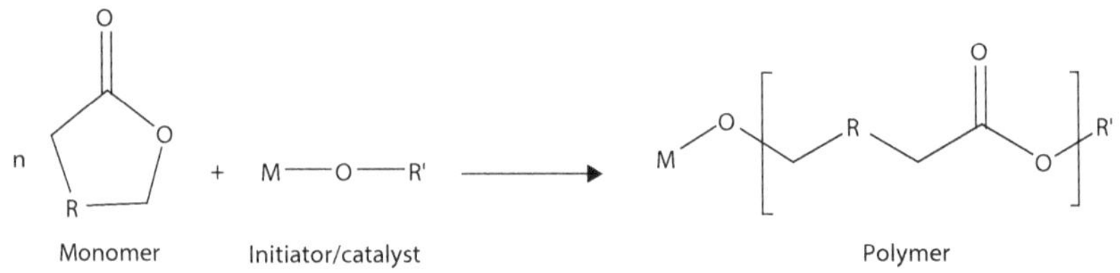


**Scheme 6.4** Polymerization of cyclic esters by ROP.

### 6.7.3 Mechanism of ROP

The ring opening polymerization of cyclicesters have been investigated by several study including metal alkoxides [26], carboxylates and Schiff base complexes [27] so as to achieve effective polymer synthesis.

Generally, the ring opening polymerization of cyclicesters can occur by three different mechanisms; anionic, cationic or coordination- insertion.

#### 6.7.3.1 *Cationic Ring Opening Polymerization (CROP)*

Cyclicesters react with cationic catalysts to form polyesters by creating of positive charge species. Electrophilic attack is the property of the (CROP) which is then attacked by a cyclicesters (Scheme 6.5) [28].

In this mechanism there are two mechanisms can happen; *active chain end* or *activated monomer*. But, using CROP to polymerize cyclicesters remains uncommon, because of side reaction may happen, difficult to control the structures and produce low molecular weight polymers [29].

#### 6.7.3.2 *Anionic Rring Opening Polymerization (AROP)*

In this mechanism, the carbon atom next to the acyl-oxygen or the carbonyl carbon of the cyclicester monomer broke by the nucleophilic attack. In this case, acyl-oxygen or alkyl-oxygen cleavage will happen (Scheme 6.6) [30].

#### 6.7.3.3 *Coordination-Insertion Polymerization*

Two steps can be occurred in the coordination-insertion reaction, first: the monomer coordinate to the metal and the alkoxide will bind with the acyl

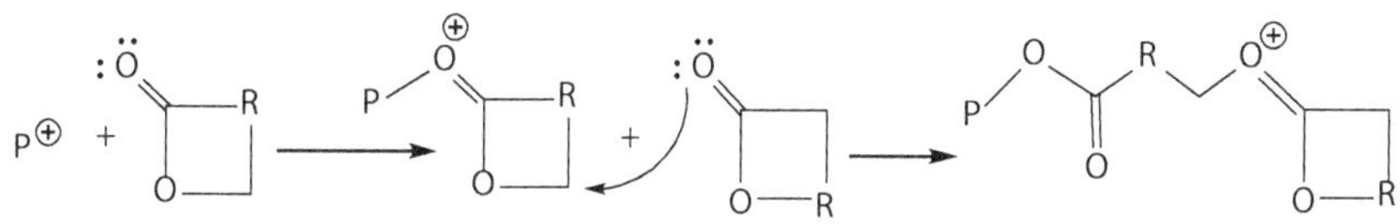

**Scheme 6.5** The CROP of a cyclicester.

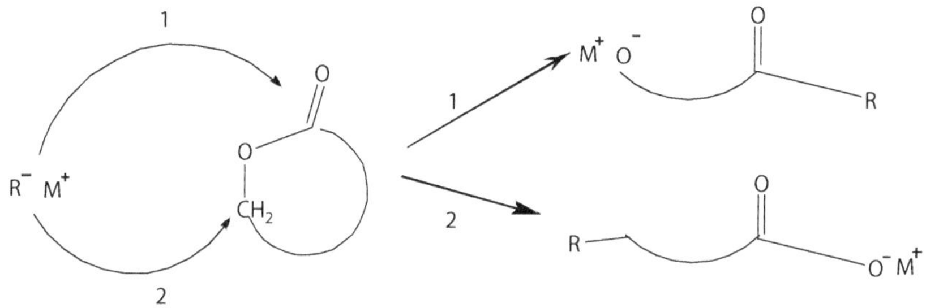

**Scheme 6.6** The AROP of a cyclicester.

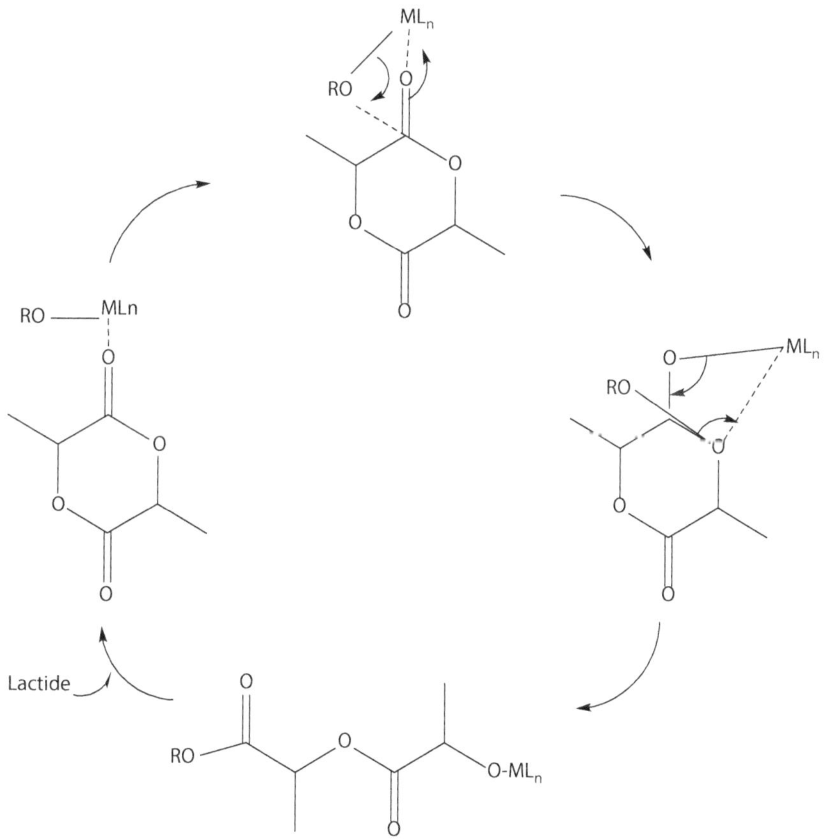


**Scheme 6.7** Lactide polymerization by coordination-insertion mechanism.

carbon atom. Second: attachment of the monomer with the alkoxide group to form new metal alkoxide group then extra insertion reactions happen (Scheme 6.7) [31].

## 6.8 Conclusion

Sustainability of biodegradable polymer is widely studied and employed in different applications. Consequently researching of production of biodegradable polymers via a number of methods can be of much importance. This article given a review on the produce of biodegradable polymers by different ways including polycondensation and ring opening polymerization. Mechanism of ring opening polymerization including: cationic ROP, anionic ROP and coordination-insertion polymerization are discussed.

## References

1. Zyoud, A., Zaatar, N., Saadeddin, I., Helal, M.H., Campet, G., Hakim, M., Park, D. and Hilal, H.S., Alternative natural dyes in water purification: Anthocyanin as TiO2-sensitizer in methyl orange photo-degradation, *Solid State Sci.*, 13, 1268, 2011.
2. Roco, M.C., Mirkin, C.A. and Hersam M.C., *J Nanopart Res*, 2011: National Science Foundation/World Technology Evaluation Center report, Springer, 2010, Boston (available on http://www.nsf.gov/nano/).
3. Dahl, J.A., Maddux, B.L.S. and Hutchison J.E., Toward greener nanosynthesis, *Chem. Rev.*, 107, 2228, 2007
4. Patet, J.M., Peng, X., Koenigsmann, C., Xu, Y., Karn, B. and Wong, S.S., Viable methodologies for the synthesis of high-quality nanostructures, *Green Chem.*, 13 (3), 482, 2011.
5. Smith, K., Al-Khalaf, A.K.H. and El-Hiti, G.A., Pattisson S. Highly regioselective di-tert-amylation of naphthalene over reusable H-mordenite zeolite, *Green Chem.*, 14, 1103, 2012.
6. Ravindra, R. and Bhat, B.R., An environmental friendly route to synthesis carbon nanofibers in bulk scale, *J. Nanoengin. Nanomanuf.*, 2, 31, 2012.
7. Reed, W. and Hutchison, W., Improved synthesis of small ($d_{CORE}$ =1.5 nm) phosphine-stabilized nanoparticles, *J. Am. Chem. Soc.*, 122, 12890, 2000.
8. Lu, Y. and Ozcan, S., Green nanomaterials: On track for a sustainable future, *Nano Today*, 10, 417, 2015.
9. Plastic Europe, The compelling facts about plastics, 2009. online on: http://www.plasticseurope.org/pl/document/the-compelling-facts-about-plastics-2009.aspx.

10. Baytekin, B., Baytekin, H.T. and Grzybowski, B.A., Retrieving and converting energy from polymers: deployable technologies and emerging concepts, *Energy Environ. Sci.*, 6, 3467, 2013.
11. PlasticsEurope in *the Facts. PlasticsEurope*, 2015. Online on: http://www.plasticseurope.org/en
12. P. Europe in *The Compelling Facts About Plastics*, 2008. http://www.pvc.org/en/p/the-compelling-facts-about-plastics-2008.
13. Vert, M., Polymères de fermentation. Les polyacides lactiques et leurs précurseurs, les acides lactiques, *Actual Chim.*, 11–12, 79, 2002.
14. Arbaoui, A. and Redshaw C., Metal catalysts for e-caprolactone polymerisation, *Polym. Chem.*, 1, 801, 2010.
15. Furkan, H., Isikgor, C. and Becer, B., Lignocellulosic biomass: a sustainable platform for the production of bio-based chemicals and polymers, *Polym. Chem.*, 6, 4497, 2015.
16. Willett, J.L., Mechanical properties of LDPE/granular starch composites, *J. Appl. Polym. Sci.*, 54, 1685, 1994.
17. Woo, C.J., Chun K.B.C. and Park, J.S., Ultraviolet reflective and mechanical properties of polyethylene mulching films, *Eur. Polym. J.*, 37, 1227, 2001.
18. Hadjichristidis, N., Pispas, S. and Floudas, G., Block copolymers: synthetic strategies, physical properties, and applications, *JWS.*, 1, 2003.
19. Song, C.X., Sun, H.F. and Feng, X.D., Microspheres of biodegradable blok copolymer for long-acting controled delivery of contraceptives, *Polym J.*, 19, 485, 1987.
20. Sinha, V.R., Bansal K., Kaushik R., Kumria, R., Trehan, A. Poly-epsilon-caprolactone microspheres and nanospheres: an overview, *Int. J. Pharm.*, 1, 278, 2004.
21. Kimura, Y., Molecular, structural, and material design of bio-based polymers, *J. Polym.* 41, 797, 2009.
22. Dechy-Cabaret, O., Martin-Vaca, B. and Bourissou D., Controlled ring-opening polymerization of lactide and glycolide, *Chem. Rev.*, 104, 6147, 2004.
23. Stridsberg, K.M. and Ryner, M., Albertsson, A. Controlled ring-opening polymerization: polymers withdesigned macromolecular architecture, *Springer Berlin Heidelberg, Berlin*, Heidelberg, 41, 2002.
24. Braud, C., Devarieux, R., Atlan, A., Ducos, C. and Michel, V., Capillary zone electrophoresis in normal or reverse polarity separation modes for the analysis of hydroxy acid oligomers in neutral phosphate buffer, *J. Chromatogr. B Biomed. Sci. Appl.* 706, 73, 1998.
25. Williams, C.K., Syntheses of functionalized aliphatic polyesters, *Chem. Soc. Rev.*, 36, 1573, 2007.
26. Ge, F., Dan, Y., Al-Khafaji, Y., Prior, T.J., Jiang, L., Elsegood, M.R.J. and Redshaw, R., Vanadium(V) phenolate complexes for ring opening homo- and co-polymerisation of e-caprolactone, Llactide and rac-lactide, *RSC Adv.*, 6, 4792, 2016.

27. Wang, X., Zhao K., Al-Khafaji Y., Mo S., Prior T.J., Elsegood M.R.J, and Redshaw C., Organoaluminium complexes derived from anilines or Schiff bases for the ring-opening polymerization of ε-caprolactone, δ-Valerolactone and *rac*-Lactide, *Eur. J. Inorg. Chem.* 1951, 2017.
28. Rozenberg, B.A., Cationic polymerization of e-caprolactone in the presence of diols via activated monomer mechanism, *Makromol. Chem. Macromol. Symp.*, 60, 177, 1992.
29. Penczek, S., Cationic ring-opening polymerization (CROP) major mechanistic phenomena, *J. Polym. Sci. A Polym. Chem.*, 38, 1919, 2000.
30. Nuyken, O. and Pask, S., Ring-opening polymerization—an introductory review, *Polymers* 5, 361, 2013.
31. Santos, V.S. and Herres-Pawlis, S., Lactide polymerisation with complexes of neutral n-donors – new strategies for robust catalysts, *Eur. J. Inorg. Chem.*,765, 2012.

7

# Design and Processing Aspects of Polymer and Composite Materials

**Hafiz M. N. Iqbal[1]*, Tahir Rasheed[2] and Muhammad Bilal[3]***

[1]*Tecnologico de Monterrey, School of Engineering and Sciences, Campus Monterrey, Monterrey, N.L., Mexico*
[2]*The School of Chemistry & Chemical Engineering, State Key Laboratory of Metal Matrix Composites, Shanghai Jiao Tong University, Shanghai, China*
[3]*School of Life Science and Food Engineering, Huaiyin Institute of Technology, Huaian, China*

## *Abstract*

In this chapter, an effort has been made to highlight the novel design and processing aspects of various polymers and composite materials. With ever-increasing scientific research, knowledge, and socioeconomic awareness, research and industrial communities are now more focused and interested in various types of polymers and composite materials with multi-functionalities. The polymeric materials' exploitation in a sophisticated manner can provide extensive opportunities for experimentation in the field of interdisciplinary and multidisciplinary scientific research with special reference to biotechnological and biomedical applications. Owing to the unique features of this trendy area, research scientists have already been directed or redirecting their interest in polymeric materials for targeted applications in different sectors of the modern world including cosmeceuticals, nutraceuticals, pharmaceuticals, and other biotechnological or biomedical. The present chapter highlights the unique opportunities to engineer composite materials using various types of natural and synthetic polymers. The information has also been given on the potential research activities with a close-up look at the future recommendations.

***Keywords:*** Polymeric materials, alginate, cellulose, chitosan, dextran, guar gum, xanthan, Poly (acrylic acid), Poly (acrylamide), Poly (vinyl alcohol), Poly (ethylene glycol), Poly (vinyl pyrrolidone), Poly (lactic acid), processing strategies

**Corresponding authors*: hafiz.iqbal@itesm.mx; bilaluaf@hotmail.com

Shakeel Ahmed and Chaudhery Mustansar Hussain (eds.) Green and Sustainable Advanced Materials, Vol. 1 (155–190) © 2018 Scrivener Publishing LLC

## 7.1 Introduction

During the past years, polymeric materials have gained special research interest as novel candidates and alternatives to the traditional petroleum-based synthetic counterparts [1, 2]. In this context, many research efforts have been made to engineer new types of high-performance materials–based innovative products [3–14]. This trendy area is moving towards the development of 'greener' technologies and in turn, the principle of 'going green' has directed this search towards eco-friendly polymeric materials. Words like renewability, recyclability, and sustainability are emphasized in growing environmental awareness. The sustainability concept is shown in Figure 7.1 [2]. The fact is that environmental legislation is the driving force behind the development of these polymeric materials [2]. A summary of biopolymers or bio-based polymers is shown in Figure 7.2 [2]. With ever-increasing scientific research, knowledge, and socioeconomic awareness, industrial communities are now more concerned about the environmental impact of persistent plastic-based wastes. Moreover, the disposal methods are limited and no longer acceptable. On the other hand, biopolymers have some advantages over petroleum-based polymers, such as being renewable, abundant, and biodegradable while also providing competitive mechanical properties. So, the focus has been shifted to polymers originating from bio-based renewable sources, which are often biocompatible and biodegradable [15]. Therefore, in this context, composite materials are being engineered for target applications in different sectors of the modern world.

There is an urgent need for the development of green polymeric materials/composites that would not involve the use of the toxic or noxious component in their manufacture and potentially be resistive against the wider community of various microbes to avoid some serious wound

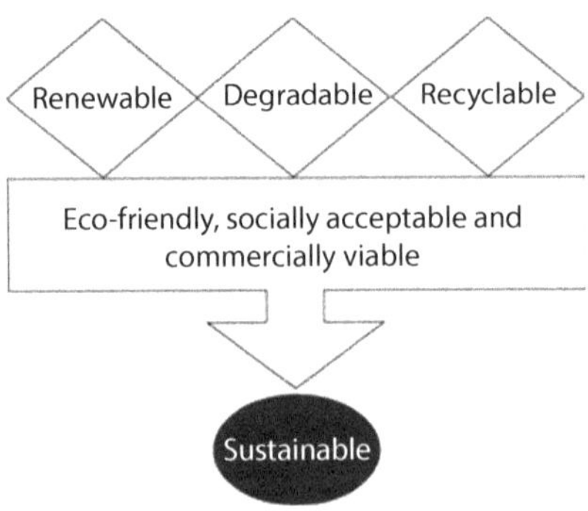


**Figure 7.1** Concept of 'sustainability'. (Reproduced with permission from Ref. [2, 16]).

contaminations [8]. One area that has received limited attention so far, but that will gain in importance as naturally conferring antimicrobial agents' use becomes further established, is the incorporation of such novel agents into the materials to provide an antibacterial effect on contact of that material with the target bacterium [17–20]. Such antimicrobial active materials might have great potential to respond to a new infection before the clinical signs are evident, with the potential to significantly improve patient prognosis. Antimicrobial agents–impregnated materials could be used as medical implants and in applications relevant to hospital hygiene. There are also clear industrial and biotechnological requests for materials loaded with natural agents that can quickly prevent deleterious microbial action following contamination events. It is intended that a technology platform for future exploitation, e.g. *in vivo* and *ex vivo* designs to find out other suitable potential applications such as biomedical implants of these newly developed novel materials, could also be established.

Among the potential causes, below are some possible explanations for an increased Antimicrobial resistance (AMR):

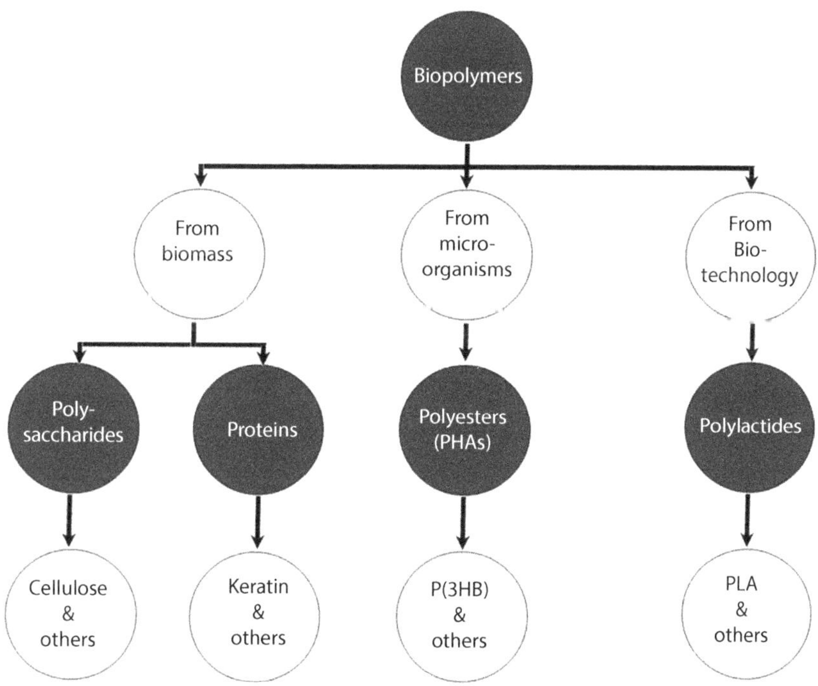


**Figure 7.2** Classification of biopolymers. (Reproduced with permission from Ref. [2]).

1. The genetic transformation from strain to strain.
2. Biofilm matrix forming potential of several strains.
3. Efflux pumps and other outer membrane structural variations.
4. Enzyme-mediated resistance against, in practice, antimicrobials.
5. Enhanced level of metabolic activity within the biofilm structure.
6. Lower/no perfusion of antimicrobial agents through the biofilm matrix.
7. Adaptability and interaction between antimicrobial agents and biofilm matrix.
8. Excessive/useless consumption of in practice antimicrobials in a random order.
9. Genetic variation and adaptability against the excessive antimicrobials' exposure.

## 7.2 Design and Processing

Research is underway around the globe to highlight the unique opportunities to develop the novel design and processing strategies using various types of natural and synthetic polymers. Considering the points discussed above, several successful efforts have already been made to transform polymeric materials into useful products. For this reason, several processing methodologies have been introduced in the literature. Depending on the processing technique, a wider spectrum of novel composites has been developed using various materials including pristine or modified cellulose, bacterial cellulose (BC), keratin, Polyhydroxyalkanoates (PHAs), and chitosan (CS), along with other synthetic counterparts though using different methodologies [2–9]. Depending on the preparation techniques, different properties of composite materials are summarized in Table 7.1. In this chapter, natural and synthetic polymers are comprehensively discussed, accompanied by their potential to be employed as tissue engineering candidates and drug delivery systems.

## 7.3 Natural Polymers and Their Applied Potentialities

### 7.3.1 Alginate – Physiochemical and Structural Aspects

Alginate is an anionic polysaccharide composed of covalently linked $\alpha$-L-guluronic acid (G) and $\beta$-D-mannuronic acid (M) residues [28]. It may

**Table 7.1** Processing strategies and properties of biocomposite materials.

| Composite material | Composite form | Preparation technique | Properties studied | Reference |
|---|---|---|---|---|
| Chitosan/starch | Films | Casting | Better mechanical properties | [21] |
| Keratin/poly(vinyl alcohol) | Films | Casting | Improved compactness, the compatibility, and the stability in water | [22] |
| Gelatin–alginate | Films | Casting | | [23] |
| Gelatin/pectin | Films | Air-drying casting | High transparency | [24] |
| Gelatin–alginate | Films | Extrusion | Oxygen barrier/turbidity | [25] |
| Chitosan/gelatin | Coatings | Solvent-casting | Antimicrobial/dissolvable | [26] |
| Starch/cellulose | Films | Air-drying casting | High tensile strength/poor optical properties | [27] |

occur in the form of G blocks (GGGGGG) and M blocks (MMMMMM) or alternating G and M blocks (GMGMGM). The typical mechanical properties of alginate-based gels and materials are improved by increasing the molecular weight and the length of G block. Commercially available alginates are widely extracted from brown algae such as *Ascophyllum nodosum*, *Laminaria digitata*, *Laminaria hyperborea*, *Laminaria japonica*, and *Macrocystis pyrifera* by treatment with aqueous alkali solutions. It can also be biosynthesized by bacteria like *Azotobacter* and *Pseudomonas*. Notably, bacterial-derived alginate has more defined chemical structures and physicochemical properties as compared with seaweed-originated alginate. The commercial grade sodium alginate with a wider molecular weight ranges from 32,000 and 400,000 g/mol are available in adequate quantity for research purposes. Such availability with a wider molecular weight ranges offers numerous possibilities to upgrade/modify the physiochemical features of alginate-based engineered constructs including hydrogels and microcapsules, etc. Figure 7.3 portrays the chemical structures of alginate monomers, $\alpha$-L-guluronic acid, D-mannuronic acid, alginate, and the

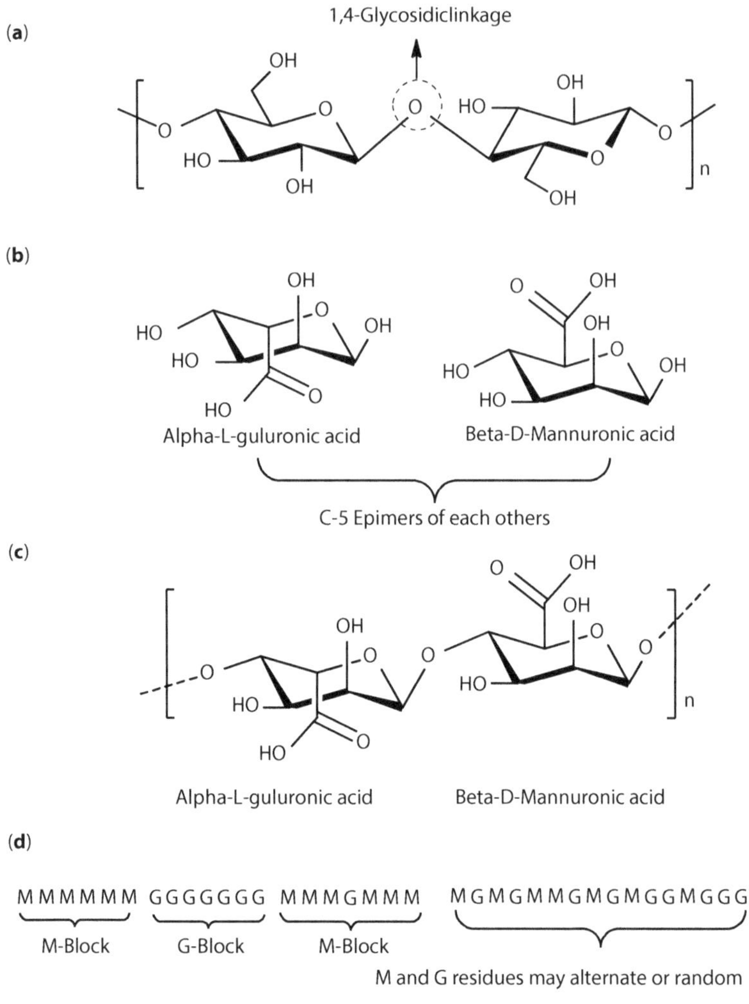


**Figure 7.3** Structure of (a) cellulose, (b) alpha-L-guluronic acid and beta-D-mannuronic acid (the epimers), (c) alginic acid (so-called alginate), (d) arrangements of M and G residues as M and G blocks. (Reproduced from Ref. [12], an open access article distributed under the terms and conditions of the Creative Commons Attribution (CC BY) license (http://creativecommons.org/licenses/by/4.0/)).

arrangement of G and M blocks. The copolymer is soluble in water, and solubility depends on certain factors such as medium ionic strength, pH, and divalent cations ($Ca^{+2}$, $Sr^{+2}$). Medium ionic strength influences change in viscosity, chain elongation, conformation, and eventually the solubility. At pH below 3.5, carboxylic groups remain protonated and increase the viscosity as a result of intramolecular hydrogen bonding. Therefore, solvents with pH greater than 3.0–3.5 are required for increased solubility. Divalent cations reduce the solubility, as the divalent cation–free solvent is required for the complete solubility of the alginate biopolymer. The tetrabutylammonium salt of alginate is completely soluble both in an organic solvent as ethylene glycol and water [29]. The characteristic alginate functionalities such as low toxicity, biocompatibility, biodegradability, mild gelation conditions, and economical availability renders it a particularly suitable candidate for use in numerous biomedical applications [30]. The alginate-based hydrogels developed using various crosslinking methods allow wide applications in wound healing and dressing and delivery of bioactive agents (small proteins and drugs), and cell/tissue transplantation. For example, Chen *et al.* [31] reported the pH-sensitive protein drug (bovine serum albumin (BSA)) release and swelling behavior by combining alginate, *N,O*-carboxymethyl CS and genipin (crosslinked) in physiological fluids. The composite hydrogel displayed a small amount of swelling at acidic milieu (pH 1.2), and released only 20% drug due to non-dissociation of the acidic groups and tight packing because of intramolecular H-bonding. At pH 7.4, the same composite hydrogel released 80% drug with extensive swelling as acidic group dissociation resulting in strong repulsion.

### 7.3.2 Carrageenan – Physiochemical and Structural Aspects

The word carrageenan (CG) is originated from the Irish name for the red seaweed, carrageen (meaning little rock). CG is extracted from the edible red seaweeds, as it is the major constituents of the cell wall of different species of Rhodophyta (red seaweeds). There are six different types of CG including Kappa-CG (κ), Iota-CG (ι), Theta-CG (θ), Lambda-CG (λ), Mu-CG (μ), and Nu-CG (ν). As a polysaccharide, CG consists of repeating *β* D-galactose and *α*-D-galactose or 3,6-anhydro-*α*-D-galactose units which are linked through *α*-1,3 and *β*-1,4-glycosidic linkages. Among the six types, Kappa, Iota, and Lambda are the most important and commercialized forms of CG. These CG forms differ from each other based on the number of sulfate groups that can be one to three on repeating parts of the disaccharides (Figure 7.4). Mu-CG and Nu-CG are biologically derived from Kappa and Iota-CG, respectively, whereas theta-CG is formed from Lambda-CG. All forms of CG are water soluble and insoluble in organic liquids. The viscosity, extent of water solubility, and gel-forming ability of CG depend upon the amount of sulfate and corresponding cations equilibrium, e.g. $Na^+$ and $K^+$ [32, 33].

CG is anionic in nature and is suitable for use as pH-sensitive hydrogels for restrained drug release. Piyakulawat group synthesized hydrogel beads

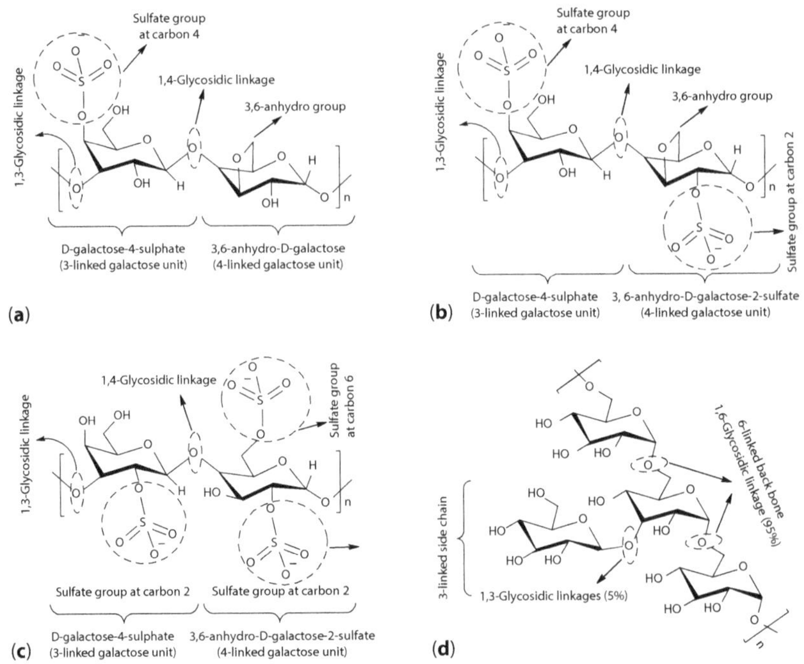


**Figure 7.4** Structure of (a) Kappa-carrageenan, (b) Iota-Carrageenan, (c) Lambda-carrageenan, (d) dextran. (Reproduced from Ref. [12], an open access article distributed under the terms and conditions of the Creative Commons Attribution (CC BY) license (http://creativecommons.org/licenses/by/4.0/)).

by crosslinking CS/CG with glutaraldehyde and examined the controlled release of diclofenac sodium drug over simulated body fluids. They investigated and optimized the ratio of CS/CG contents as 2:1 and diclofenac loading as 5%. Results revealed that the drug release could be delayed with no crosslinking of CS/CG up to 8 h, whereas drug release was delayed effectively up to 24 h with crosslinking of CS/CG with glutaraldehyde. Also, anionic nature of CG supports swelling in somewhat basic solvents [34].

### 7.3.3 Cellulose – Physiochemical and Structural Aspects

Cellulose is the natural polysaccharide consisting of $\beta$-D-glucose connected through 1,4-glycosidic bonds in a linear fashion (Figure 7.5). It is the most abundant organic polymer that is naturally found in plants, known as plant cellulose (PC), or is synthesized biologically by a certain type of microorganisms (e.g. *Acetobactor xylinum*), known as BC. BC possesses a number of benefits as compared to PC such as (1) BC is more

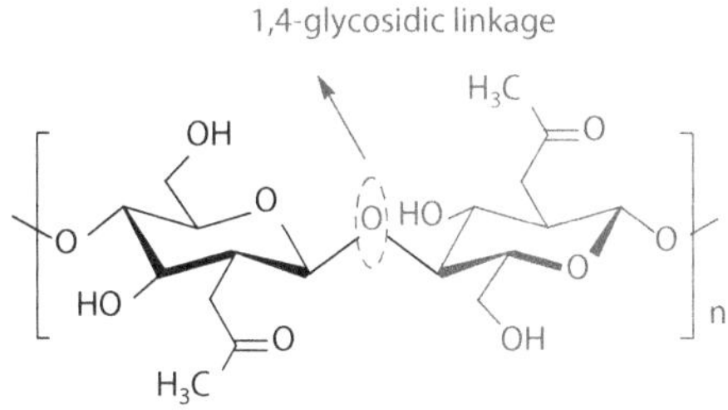


**Figure 7.5** Chemical structure of cellulose. (Reproduced from Ref. [12], an open access article distributed under the terms and conditions of the Creative Commons Attribution (CC BY) license (http://creativecommons.org/licenses/by/4.0/)).

crystalline than PC, (2) BC is completely pure while PC is mixed with some other plant materials like lignin and pectin etc., and (3) identical chemical properties but different physical characteristics as BC is nano-sized, two orders of magnitude smaller and ultra-fine as compared to PC accompanied by greater tensile strength and large water holding tendency. Cellulose and their derivatives have shown applications in the biomedical strategies owing to their excellent properties of biocompatibility, low cost, less toxicity, and biodegradability. It is generally a water-insoluble biopolymer; however, their derivatives such as carboxymethyl cellulose (CMC) and its sodium salt are water soluble [35]. In addition, various composite hydrogels have been developed and applied effectively for drug delivery in living systems. Lim *et al.* [36] developed outstanding cellulose nanocrystals/poly(acrylamide) (PAM) hydrogels that showed enhanced swelling in a neutral pH environment. In another study, Zhou *et al.* [37] prepared temperature/pH receptive cellulose-based hydrogels for controlled delivery of oral drugs such as methylene orange and blue drugs. The hydrogels presented excellent characteristics of pH sensitivity and prominent swelling at a slightly alkaline pH (pH 7.4).

### 7.3.4 CS – Physiochemical and Structural Aspects

CS is one of the most abundant natural polymers obtained from the alkaline deacetylation of chitin consisting of repeating units of *N*-acetylglucosamine and D-glucosamine coupled through 1,4-glycosidic linkage. The proportion of these two components is reflected as the degree of deacetylation (DD). CS is a non-toxic, bioactive, biocompatible, and bio-degradable polymer generated as a result of partial deacetylation of chitin [38, 39]. CS, the catch-all idiom refers to describe a series of polymers with different degrees of deacetylation. The deacetylation degrees and the molecular weight of commercially available CS typically range between 70%-95% and 10–1000 kDa, respectively. Inarguably, the physical properties, biodegradability, and biological perspective of CS are highly dependent on the relative amounts of D-glucosamine and *N*-acetyl-D-glucosamine residues [40].

After cellulose, chitin is the second most abundant natural polysaccharide and can be found in the exoskeleton of crustaceans such as crabs, lobsters, shrimps, etc. [41, 42]. Yeast and fungi are also the sources of chitin and CS can be directly recovered from the cell walls of some fungi such as *Mucor*, *Absidia*, and *Rhizopus* [43]. The chemical structures of chitin and CS are shown in Figure 7.6(a and b). Considering its functional properties like non-toxicity, biocompatibility, biodegradability, and abundant natural availability, CHS and its derivatives have fascinated much research interest and thereby potential applications in the field of biomedicine,

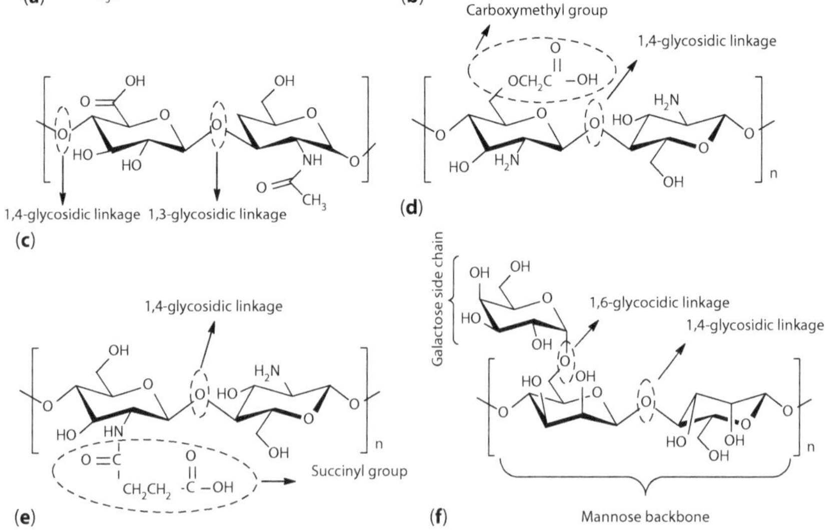


**Figure 7.6** Structure of (a) chitin, (b) chitosan, (c) glycosaminoglycan, (d) carboxymethyl chitosan, (e) *N*-succinyl chitosan, and (f) guar gum. (Reproduced from Ref. [12], an open access article distributed under the terms and conditions of the Creative Commons Attribution (CC BY) license (http://creativecommons.org/licenses/by/4.0/)).

biotechnology, cosmetics, textile, and gene/drug delivery [44]. CHS also plays a noteworthy role as an antibacterial agent by disrupting the cell wall of the negatively charged bacteria [45].

CHS has a structure similar to glycosaminoglycan – a chief component of extracellular matrix (Figure 7.6(c)). Glycosaminoglycan facilitates the cell propagation and attachment and thus improving the bio-feasibility and biocompatibility of the tissues and biomaterials [46]. CS is a polysaccharide insoluble in water as well as in most organic solvents. Many notable factors including the extent of deacetylation, the spatial configuration of acetyl group on CS, pH, ionization strength, isolation conditions, and acid properties control the solubility of the CS. CS with a more than 50% DD starts dissolving in acidic medium; nevertheless, in the presence of glycerol-2-phosphate, it can be dissolved even at neutral pH [47]. Restricted solubility and mechanical instability of CHS impede its wide-ranging applications in the field of biomedicine, as it undergoes distortion at high stresses resulting in excessive swelling in acidic medium.

Therapeutically, CHS has shown significant properties such as cholesterol reduction, antihypertension, immune and phytoalexin elicitation in addition to pain alleviation, development of epidermal tissue, and inhibition of microorganisms growth [42]. Due to its cationic nature, CHS bears a good mucoadhesive property that enhances the residence time in the intestine and the bioavailability of drugs in the gastrointestinal tract (GIT) [48]. In an acidic pH environment, the amino group on CHS is protonated and deprotonated upon neutral pH as $pK_a$ value for $-NH_2$ group is around 6.5 which is close to $pK_a$ value of the small intestine (pH 6.5–7.0). This explicates the reason why the ability of CS to open the tight junction is confined to the duodenum. In order to further broaden the scope of CHS in the GIT, researchers have made tremendous efforts to modify CS chemically or biologically over the past several years [49–51]. Chemical and physical properties of CS such as hydrophilicity, solubility, ionic nature, and mechanical properties can be improved by chemical modifications for broadening the CHS perspective. CHS bears three functional sites ($-NH_2$ at C-2, –OH at C-3, and –OH at C-6) that are more amenable to chemical modifications. Notably, the amino group at C-2 is more susceptible and likely to be the choice of reactions due to the presence of more reactive sites as compared to C-3 which is sterically hindered and less reactive. The replacement of alkyl, carboxymethyl, and succinyl groups in CHS has been reported as effective attempts to increase its solubility. *O*-carboxymethyl CHS is water soluble and displays the high capacity to bind $Ca^{2+}$ due to the occurrence of the carboxymethyl group. This led to an increased paracellular permeability of the epithelium [52]. The chemical structures of these carboxymethyl and succinyl are depicted in Figure 7.6(d and e).

### 7.3.5 Dextran – Physiochemical and Structural Aspects

Dextran is a biocompatible and bio-decomposable natural polysaccharide produced biologically by the *Leuconostoc mesenteroides* bacterium. It is composed of recurrent units of α-D-glucose joined through 1,6-glycosidic linkages (95%) with occasional 1,3-linkages (about 5.0%) as well [53]. Dextran degradability by the dextranase enzyme is considered to be one of the key features. Dextranase enzyme is secreted by the bacterium of genus bacteroids family present in the intestinal colon. This feature renders dextran an outstanding candidate for colon-targeted drug release since the dextranase catalyzes the degradation of 1,6-glycosidic linkage of dextran and disorganizes the dextran-based carrier matrix leading to effective drug delivery in the colon [54, 55]. Chiu *et al.* [56] synthesized a dextran-based hydrogel following activated dextran conjugated with 4-aminobutyric acid in the presence of 1,10-diaminodecane as a crosslinker. At higher pH (7.4), the addition of the carboxyl group facilitates the swelling of the hydrogel and drug release in a controlled manner in the colon. BSA release rate with and without dextranase enzyme has been reported between pH ranges from 2.0 to 7.4. Kim and Oh [57] developed a pH-sensitive crosslinked hydrogel composite of glycidyl methacrylate dextran (GMD) and poly(acrylic acid) (PAA) by ultraviolet (UV) irradiation for colon-targeted delivery of drugs. The hydrogel swelling tendency of GMD with PAA ratio and corresponding drug release behavior of 5-aminosalicylic acid (5-ASA) was studied at acidic (pH 2.0) and slightly alkaline pH (pH 7.4) environments with and without the presence of dextranase. The swelling ratio of GMD/PAA was found to be higher at pH 7.4 due to ionization of the acidic groups and swelling ratio increased significantly in the presence of dextranase indicating the GMD/PAA conjugated hydrogels as useful devices for controlled release of the colon-specific small protein molecules in particular drug delivery.

### 7.3.6 Guar Gum – Physiochemical and Structural Aspects

Guar gum (GG), also called guaran, is a non-ionic polysaccharide that contains sugar galactose and mannose units (Figure 7.7). The backbone chain consists of β-D-mannopyranose (β-D-mannose) residues linked through 1,4-glycosidic linkages having adjacent short branches of β-D-galactopyranose (β-D-galactose). GG is used as suspending, binding,

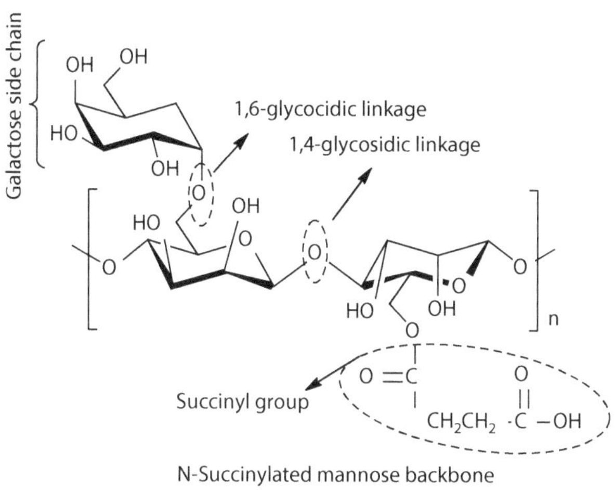


**Figure 7.7** Structure of guar gum succinate. (Reproduced from Ref. [12], an open access article distributed under the terms and conditions of the Creative Commons Attribution (CC BY) license (http://creativecommons.org/licenses/by/4.0/)).

stabilizing, and thickening agent in various pharmaceutical formulations [58]. Several important characteristics such as bio-degradability, non-toxicity, hydrophilicity, and adequate availability makes GG particularly alluring for pharmaceutical applications. The most distinguished feature of GG is its amenability to microbial depolymerization or degradation in the large intestine exclusively when it has acidic pendant groups which facilitate its swelling at neutral pH [59]. Wong *et al.* [60] examined the drug delivery performance by formulating a 60.5% (w/w) GG polysaccharide b tablets with budesonide and dexamethasone in simulated fluids. They noticed a minor drug discharge in simulated intestinal and gastric fluids; however, a substantial upsurge in drug release was observed in simulated colon fluids. GG succinate, one of the derivatives of GG, is employed for colon-specified drug delivery owing to its serviceable attributes such as hydrophilicity, pH sensitivity, and the liability toward biological/microbial degradation. The micro-particles of GG succinate developed by Seeli and Prabaharan [61] were found to be pH-sensitive colon drug carriers. The micro-particles showed better swelling rate and significant drug delivery at pH 7.4 in comparison with acidic pH of 1.2.

### 7.3.7 Xanthan – Physiochemical and Structural Aspects

Xanthan is an elevated molecular weight polysaccharide consisting of D-mannose, D-glucose, and D-glucuronic acid at 2:2:1 ratio, a backbone of $\beta$-D-glucose (connected through 1,4-glycosidic bond) having a trisaccharide branch associated by 1,3-glycosidic linkages on every alternative glucose unit. The trisaccharide branch consists of $\alpha$-D-mannose, $\beta$-D-glucuronic acid, $\beta$-D-mannose, and is connected via $\beta$-1,4, $\beta$-1,2, and $\alpha$-1,3 glycosidic linkages. Also, one acetyl group is attached to oxygen at C-6 of $\alpha$-D-mannose whereas one acetyl group which is joined to oxygen atom at C-6 of $\alpha$-D-mannose, and a pyruvic acid unit is joined to oxygens of C-4 and C-6 of $\beta$-D-mannose through a ketal unit [62] as shown in Figure 7.8. It is a highly branched polymer produced by the bacterium, *Xanthomonas compestris* from sugar cane under an anaerobic environment.

Being an anionic polymer, with a carboxylic group on one of the side chain of glucuronic acid, xanthan displays elevated swelling ratio at basic pH milieu presumably due to complete neutralization of the carboxylic group. The carboxylic group has $pK_a$ value of 4.6 and remains neutral below this value, but ionized and bears a negative charge accountable for swelling. Therefore, xanthan might be exploited as a controlled drug delivery candidate in the intestinal areas of pH 7.4. Bueno *et al.* [63] examined the BSA release from xanthan hydrogel matrices with and without

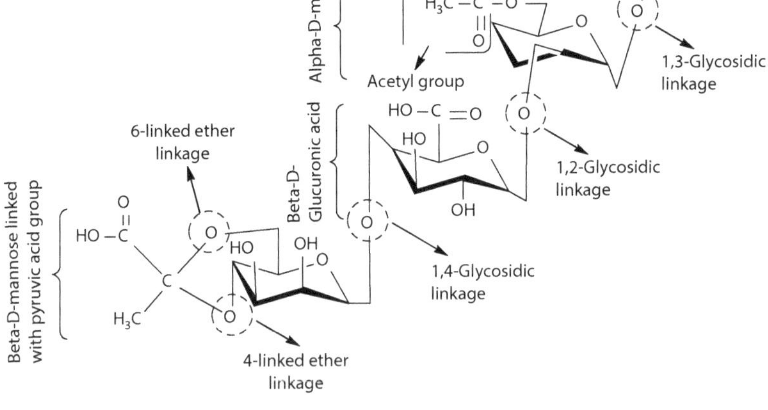


**Figure 7.8** Chemical structure of xanthan. (Reproduced from Ref. [12], an open access article distributed under the terms and conditions of the Creative Commons Attribution (CC BY) license (http://creativecommons.org/licenses/by/4.0/)).

the presence of citric acid. The results showed maximum BSA release at neutral and basic pH values because of carboxylic acid group ionization which causes swelling.

## 7.4 Synthetic Polymers and Their Applied Potentialities

### 7.4.1 PAA – Physiochemical and Structural Aspects

PAA (or Carbomer) is a high molecular weight synthetic polymer containing anionic groups on its polymeric chain (Figure 7.9(a)). PAA or its derivatives are suitable biomaterials for potential biomedical applications in particular for drug delivery [36]. PAA possesses high water solubility and may dissolve prior to drug release indicating its fragile nature [64]. This property explains the rationale why acrylic acid (AA) monomers are copolymerized with several other monomers comprising particular functional groups such as *N*-isopropyl acrylamide, *N,N*-methylene bisacrylamide, and 2-hydroxymethacrylate. The AA monomers can be crosslinked

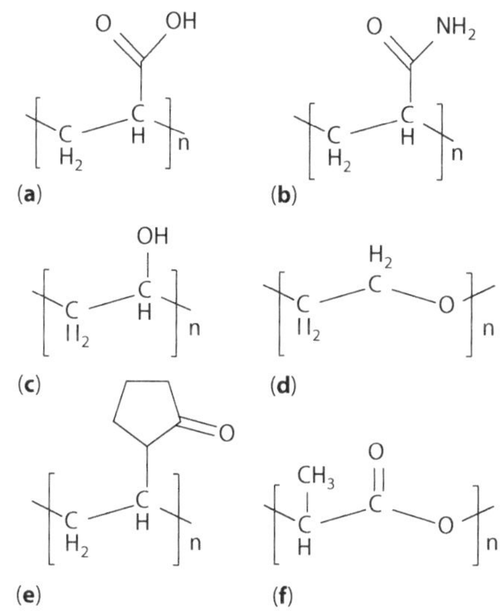


**Figure 7.9** Structure of (a) poly(acrylic acid), (b) poly(acrylamide), (c) poly(vinyl alcohol), (d) poly(ethylene glycol), (e) poly(vinyl pyrrolidone), and (f) poly(lactic acid). (Reproduced from Ref. [12], an open access article distributed under the terms and conditions of the Creative Commons Attribution (CC BY) license (http://creativecommons.org/licenses/by/4.0/)).

to poly(vinyl alcohol) (PVA) and poly(ethylene glycol) (PEG) via free radical mechanism or grafted to polysaccharide chains [65]. The occurrence of an acidic moiety on the PAA chains constitutes its suitability for drug delivery or other small protein molecules.

The dissociation constant ($pK_a$) of PAA lies in the range of 4.5 to 5.0. For pH > 5, acidic groups are ionized and negative charge is developed on the polymer. This creates strong repulsion in polymeric chains and results in significant swelling whereas it tends to least swelling under pH 4 which resembles stomach conditions (pH 4.0). For this characteristic, PAA can be applied effectively as pH-dependent intestinal drug carrier in combinations of PAA blends, pristine PAA, or natural polymers grafted on PAA [66, 67]. Sheikh *et al.* [68] synthesized PAA pH-dependent hydrogels using an electron beam and from PAA solution. It showed enhanced swelling at pH larger than $pK_a$ manifesting pH dependence behavior and applicability as a controlled drug carrier. Chang [67] prepared a superabsorbent, by employing grafted starch–AA (monomer), *N*,*N*-methylene bisacrylamide (crosslinker), and ammonium persulfate (reaction initiating agent). This superabsorbent was then mixed with Na-alginate and crosslinked to $Ca^{2+}$ to form pH-dependent hydrogel matrices. Results demonstrated that the hydrogels exhibited controlled drug discharge upon loading with diclofenac sodium at pH 7.4 corresponding to controlled intestinal drug release.

### 7.4.2 PAM – Physiochemical and Structural Aspects

The synthetic polymer, PAM, has its origin as old as 1959 when it was for the first time used as a support matrix for electrophoresis [69]. PAM is a polymer made up of acrylamide subunits with exceptional biocompatibility, and mucoadhesive properties that make it an important candidate for widespread applications in bio- and non-bio sectors ranging from protein fractionation to drug and protein delivery systems. Figure 7.9(b) exhibits the chemical structure of the PAM. In recent years, a great variety of natural biopolymers have been grafted by acrylamide/methacrylamide monomer to modify with unique properties such as pH sensitivity, mechanical stability, superabsorption, and mucoadhesive properties [70]. Mukhopadhyay *et al.* [71] reported a PAM-based hydrogel grafted with *N*-succinyl CS as an effective oral drug carrier for insulin. A 38% of loading efficacy along with 76% encapsulation efficacy was observed. At acidic stomach pH (1.2) and intestinal pH (7.4), the insulin release was observed to be 26% and 98%, respectively. Moreover, they found 4.43% bioavailability with no cytotoxicity of the reported hydrogel [71].

### 7.4.3 PVA – Physiochemical and Structural Aspects

PVA is the first ever formaldehyde-crosslinked hydrogel matrix fabricated in 1949 for medical implantations. Since that time, PVA-based hydrogel supports and biomaterials have been employed for many different types of biomedical and pharmaceutical uses such as vascular prosthesis, regeneration of bone tissues, and articular cartilage and skin replacement etc. PVA is a synthetic, non-toxic, water-soluble, biocompatible, hydrophilic, non-carcinogenic, and biodegradable polymer with good mechanical properties [72–74]. Due to the aforementioned distinctive features, PVA has received increasing research interest for controlled delivery of drugs and small bioactive molecules as well as scaffolds in tissue engineering [75, 76]. The chemical structure of PVA is shown in Figure 7.9(c). The PVA-based biomedical materials are being widely used in drug delivery systems, hemodialysis, implantable medical devices, and artificial pancreas. It is also used as a blending agent with a wide variety of natural biopolymers to increase their mechanical properties such as tensile strength and flexibility of the blended biomaterials.

Based on the crosslinking, Islam *et al.* [77] have engineered a PVA and CS (PVA/CS) conjugated gel biocomposite using tetra-ethoxy orthosilicate as a crosslinker to determine the loading and release profile of drug (dexamethasone). PVA exhibits influence on the degree of swelling (DS) of a hydrogel matrix and DS reduces with increasing the proportion of PVA in the hydrogels. The pH sensitivity of hydrogels was noticed to be the most important aspect since the gel matrix showed little swelling in acidic or alkaline media and maximized at neutral pH. A 9.37% of the loaded dexamethasone drug was released in simulated gastric fluid (SFG, pH 1.2) after 2 h, while rest of 90.67% was released in simulated intestinal fluid (SIF, pH 7.4).

### 7.4.4 PEG – Physiochemical and Structural Aspects

PEG is a water-soluble, and biocompatible but non-biodegradable polymer with noteworthy characteristics such as immunogenicity, non-carcinogenicity, and no or negligible toxicity [78]. The chemical structure of PEG is shown in Figure 7.9(d). The most imperative feature of PEG is its ability to conjugate proteins, peptides, and non-peptide drugs, which are less immunogenic, less antigenic, and display recalcitrance toward protein-degrading enzymes. PEG can be grafted or copolymerized with some other aliphatic esters like poly(lactic acid) (PLA) to boost the biocompatible behavior of the several polymers to be exploited as tissue regeneration/engineering and drug delivery technology. Atta *et al.* [39] engineered a PEG and CS (PEG/CS) copolymer as a successful candidate for the delivery of a model drug namely cefixime. The swelling in water was found to be increased with increasing the molecular weight of the PEG. The CS/PEG amalgamated polymer presented mild swelling at pH 7.0 (at neutral medium), while the highest swelling was noted in both acidic and basic media.

### 7.4.5 Poly(vinyl pyrrolidone) – Physiochemical and Structural Aspects

Poly(vinyl pyrrolidone) (PVP) is a high water-soluble polymer composed of *N*-vinyl pyrrolidone (Figure 7.9(e)) with biocompatible and absorbency features but meager mechanical properties. However, PVP can be engineered/grafted or its monomer could be conjugated with some other vinylic monomers such as AA, methacrylic acid, methacrylamide, etc. to resolve the shortcomings of poor mechanical properties. Sohail *et al.* [79] developed poly(vinyl pyrrolidone)/acrylic acid (PVP/AA) blended composites and assessed the swelling ratio and drug delivering tendency of a model drug tramadol at varying pHs. Both the factors such as swelling ratio and drug release were diminished by enhancing crosslinker concentration, while increased medium pH and AA monomer content led to a considerable enhancement in swelling and drug release. Moreover, the drug release was reduced at acidic pH 1.2 and maximized at basic pH (pH 7.5) reflecting PVP/AA as an efficient vehicle carrier for colon-specified drug delivery.

### 7.4.6 PLA – Physiochemical and Structural Aspects

PLA is an aliphatic synthetic polyester derived from naturally occurring lactic acid. The polymer exhibits fascinating traits like non-toxicity, biocompatibility, biodegradability, thermoplasticity, and eco-friendliness accompanied by excellent mechanical properties. Figure 7.9(f) displays the chemical structure of PLA. PLA and its copolymers or complexes are decomposable as approved by the *Food and Drug Administration* (FDA) and undergo degradation under biological conditions to innocuous and harmless products that can be easily eliminated through the kidneys. The above-mentioned characteristics of PLA witnessed it a promising tool/candidate for several biotechnological uses especially as cells/tissue replacement, bioactive molecules delivery, and medical transplants. In recent years, medical scientists have demonstrated the employment of PLA and its constructs poly(lactic-*co*-glycolic) acid, as a target-oriented drug delivering vehicles for targeted drug delivery based on their easy processability, biocompatibility, and biodegradability attributes [80].

In summary, Table 7.2 summarizes most of the above discussed biopolymers, their physicochemical characteristics, and applied potentialities.

## 7.5 Materials-Based Biocomposites

As mentioned above, there has been increasing research interest in the development of materials-based composites with multi-characteristics, i.e. (1) stronger, (2) stiffer, and (3) lighter along with other multi-functional properties for a variety of industrial and biotechnological applications [2]. A composite is defined as a 'material that consists of two or more distinct materials/polymers in order to obtain tailor-made characteristics or to improve or impart ideal properties'. More importantly, tailor-made characteristics include but are not limited to the specific strength, thermal properties, surface properties, biocompatibility, and biodegradability features that the individual material fails to demonstrate on its own [9, 11]. On the other hand, a biocomposite can be defined as 'composite materials derived from a biological origin and comprise on one or more phases are termed as bio-composites' [118]. A broad definition of a bio-composite is a composite material made up of natural or bio-derived polymers, e.g. BC/microbial cellulose (MC), PHAs, and PLA [11]. So far, a range of methodologies have been successfully adopted for the production of BC and BC-based composites [2]. Furthermore, potential applications of BC and BC-based composites are also provided in Table 7.3. In recent years, cellulose-based materials have been widely employed in the area of infection-free wound healing, tissue engineering/implants applications. Naturally occurring phenol grafted P(3HB)-EC bio-composites, on the other hand, are expected to exhibit an excellent HaCaT compatibility and potentially favorable for cell proliferation.

Recently, Iqbal *et al.* [6] developed a series of novel bio-composites with natural phenols as functional entities and P(3HB)-EC as a base material using laccase as a grafting tool. *In vitro* biocompatibility of CA-*g*-P(3HB)-EC composites, i.e. 0CA-*g*-P(3HB)-EC (control composite), 5CA-*g*-P(3HB)-EC, 10CA-*g*-P(3HB)-EC, 15CA-*g*-P(3HB)-EC, and 20CA-*g*-P(3HB)-EC was achieved with the human keratinocytes–like HaCaT cells [2, 7]. Additionally, the morphologies of cell cultured from all of the test composites displayed healthy shape at 5 days; nevertheless, the amount of HaCaT cells seeded on the surface of 15CA-*g*-P(3HB)-EC composite was higher than those of 20CA-*g*-P(3HB)-EC composite (Figure 7.10), which is again consistent with the results from viability/cytotoxicity analysis (Figure 7.11).

**Table 7.2** Biopolymers, their physicochemical characteristics, and applied potentialities.

| Name of biopolymer | Type | Physicochemical characteristics and applications | Reference |
|---|---|---|---|
| Alginate | Natural | Water soluble, biocompatible, biodegradable; low toxicity, relatively low cost, crosslinking/gelation under mild conditions; mechanical feebleness; difficulties in handling, storage in solution, and sterilization, safe to use and simple modifications to prepare alginate derivatives with novel properties | [81–84] |
| Carrageenan | Natural | Biocompatible, biodegradable, non-toxic, cheap and gel-forming, displays antibacterial activities against several strains, excellent stabilizing ability as well as gelling, emulsifying, and thickening agent, antioxidant activity | [85–88] |
| Chitosan | Natural | Excellent biocompatibility and good host response; unique biodegradability; broad-spectrum antimicrobial and antifungal activity; hydrophilic surface provides easy cell adhesion, proliferation, and differentiation; mechanical and thermal stability; water-soluble polymer only in acetic medium; high purification cost | [89–95] |
| Dextran | Natural | Water solubility; biodegradability, biocompatibility; good proliferation and differentiation behavior; costly polymer; mechanically soft and unstable; necessitates alteration to enhance cell adhesion, used as immobilization support | [89, 96, 97] |
| Gelatin | Natural | Water-soluble biopolymer; obtained from various animal by-products; forms thermally revisable and high mechanical hydrogels; versatile biomedical applications; easily forms films and matrix hydrogels; biodegradable and microencapsulating agents; excellent biocompatibility and non-immunogenicity, lower thermal stability at high temperatures, can be modified at amino acids level, antimicrobial and antioxidant property | [24, 98–100] |

| | | | |
|---|---|---|---|
| Starch | Natural | Water-soluble and inexpensive biopolymer; biodegradable–biocompatible; easy to modify with other polymers; difficulties in crosslinking itself; mechanically weak; needs modification to enhance cell adhesion | [101, 102] |
| Poly(acrylic acid) (PAA) | Synthetic | Water-soluble polymer, biodegradable, also used as a thickening, suspending, and emulsion stabilizing agent for topical application, ideal drug delivery vehicle | [103, 104] |
| Poly(acrylamides) | Synthetic | A simple and economical polymer, non-toxic, chemically inert and stable over wide pH intervals, medium of choice for all electrophoretic techniques, acrylamide, a monomer of poly(acrylamide) is a known neurotoxin, support matrix for immobilization | [105] |
| Poly(vinyl alcohol) (PVA) | Synthetic | Non-toxic, water-soluble, biologically compatible, easily degradable, relatively low strength and thermal stability, excellent mechanical properties such as elastic modulus, strain, and toughness and flexibility in dry state, biomedical applications including eye wetting drops and cartilage replacements | [106–110] |
| Poly(vinyl pyrrol-idone) (PVP) | Synthetic | Water-soluble polymer; excellent wetting properties; swells rapidly; excellent film; non-toxic; biocompatible; wide application in blood plasma expander polymer; high storage stability; low mechanical and thermal stability | [111, 112] |
| Poly(ethylene glycol) (PEG) | Synthetic | Soluble in water, low intrinsic toxicity, forms temperature-responsive and chemically crosslinked hydrogels, used for storage of blood and organs | [112–115] |
| Poly(*N*-isopropyl acrylamide) (PNIPAM) | Synthetic | Water-soluble polymer; temperature-responsive polymer; good mechanical properties; biocompatible polymer for tissue engineering and controlled drug delivery; necessitates chemical modification to enhance culture surface for cell delivery; cytotoxic with lower thermal stability | [116, 117] |

**Table 7.3** Potential/proposed applications of some bacterial cellulose-based 'Green' composite materials. (Reproduced with permission from Ref. [16]).

| BC-based materials | Methodology | New/improved functionalities | Potential/proposed applications | References |
|---|---|---|---|---|
| BC/Chi/Alg | Molding | Physical, mechanical, biocompatibility | Wound dressing | [119] |
| BC-Vaccarino | Immersion | Physical, mechanical, biocompatibility | Wound dressing | [120] |
| BC–xGnP | Impregnation | Thermal properties and electrical conductivity | Biosensors, tissue engineering | [121] |
| BC-$Fe_2O_3$ | Immersion | Magnetic behavior | Magnetic paper, loud-speaker membranes | [122] |
| BC–HA | Immersion | Biocompatibility | Bone tissue regeneration | [123] |
| P(3HB)-*g*-BC | Laccase-assisted grafting | Thermo-mechanical strength | Bio-plastics, biomedical | [5] |
| AMPS-*g*-BC | Ultraviolet-induced polymerization | Conductivity, effective methanol barrier | Fuel cells | [124] |

| | | | | |
|---|---|---|---|---|
| BC-MMTs | Immersion | Antibacterial properties | Wound dressing, regeneration materials | [125] |
| BC/GO | Vacuum-assisted self-assembly | Thermal, mechanical, conducting properties | Biochemical and electrochemical devices | [116] |
| BC-PAni | Immersion | Electrical conductivity | Flexible electrodes, flexible display devices, biosensors etc. | [126] |
| BC-MMT | Impregnation | Physical and mechanical properties | Biomedical | [127] |
| PANI/BC | Oxidative polymerization | Thermal, mechanical, conductivity | Flexible electrodes, display, sensors | [128] |
| BC/Chi | Immersion | Physical, mechanical, biocompatibility | Wound dressing | [129] |
| ε-PL/BC | Immersion | Physical, antibacterial | Packaging | [130] |

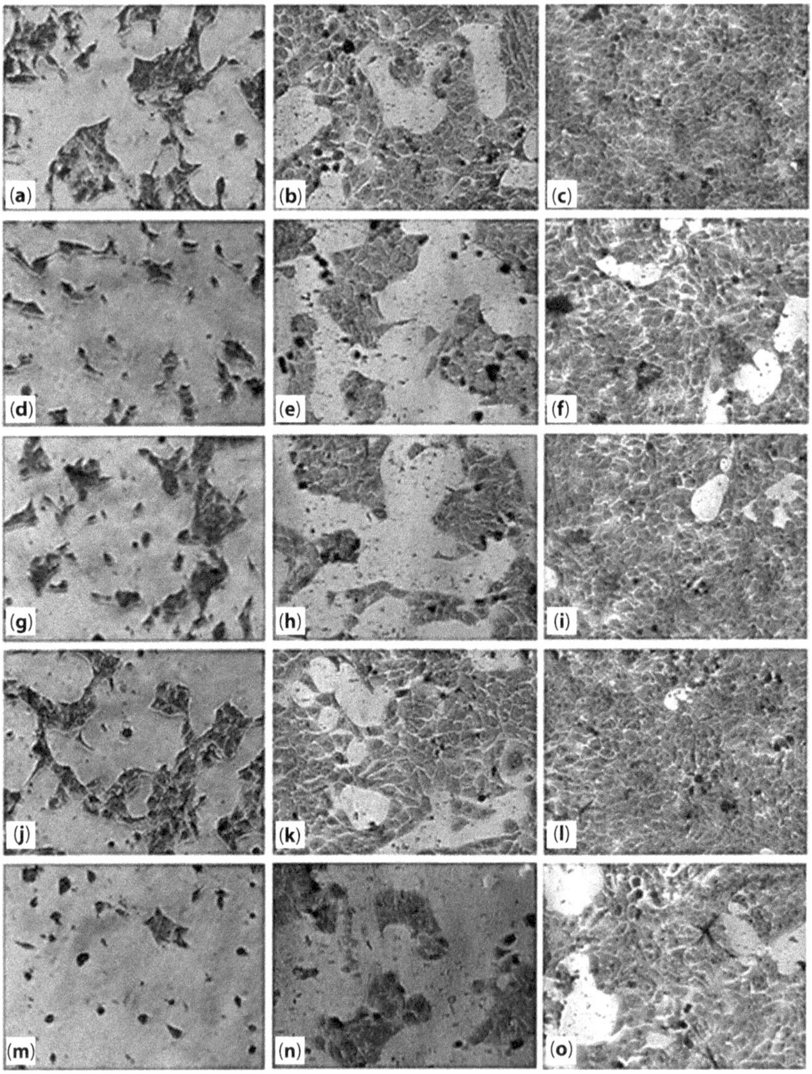


**Figure 7.10** Adherent morphology of stained images of human keratinocytes-like HaCaT cells seeded onto the composite surfaces. Images A, B, and C represent the HaCaT cells on native P(3HB)-EC composite (i.e. 0CA-*g*-P(3HB)-EC) after 1, 3, and 5 days of incubation, respectively; images D, E, and F represent the adhered HaCaT cells on 5CA-*g*-P(3HB)-EC composite after 1, 3, and 5 days of incubation, respectively; images G, H, and I represent the adhered HaCaT cells on 10CA-*g*-P(3HB)-EC composite after 1, 3, and 5 days of incubation, respectively; images J, K, and L represent the adhered HaCaT cells on 15CA-*g*-P(3HB)-EC composite after 1, 3, and 5 days of incubation, respectively, and images M, N, and O represent the adhered HaCaT cells on 20CA-*g*-P(3HB)-EC composite after 1, 3, and 5 days of incubation, respectively. All of the test samples were stained using neutral red dye (5 mg/mL) for 1 h followed by three consecutive washings with PBS at an ambient temperature. All images were taken at 100× magnification. (Reproduced with permission from Ref. [2, 10]).

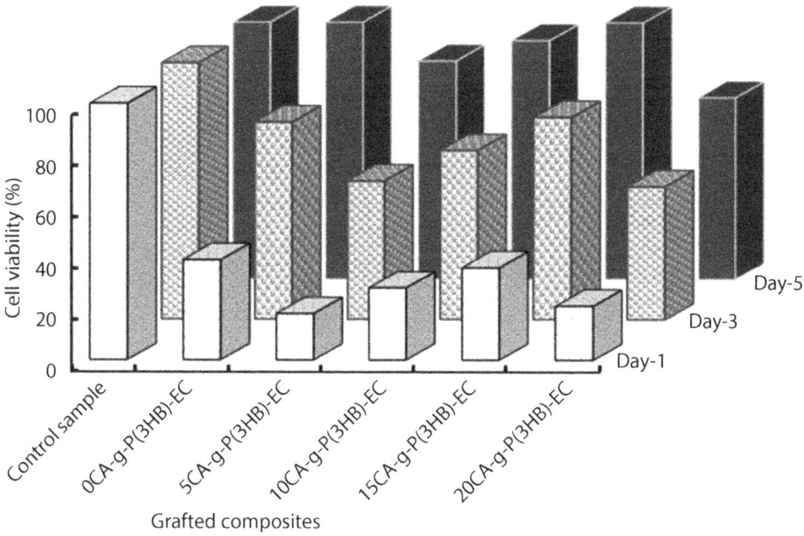


**Figure 7.11** Neutral red dye concentration-dependent percentage cell viability of human keratinocytes-like HaCaT cells after 1, 3, and 5 days of incubation onto the CA-*g*-P(3HB)-EC composite surfaces (mean ± SD, n = 3). (Reproduced with permission from Ref. [2, 10]).

## 7.6 Concluding Remarks and Future Considerations

In conclusion, through sophisticated design and novel characteristics, the polymeric materials can be modified to achieve an optimal infective capability. Such materials include but are not limited to the biodegradable and biocompatible composite, scaffolds, films, blends, and highly porous three-dimensional (3D) constructs. Various types of natural and synthetic materials are versatile for synthesizing novel constructs with multifunctional characteristics for potential applications in the biomedical sector. Moreover, a novel type of potent material could be designed for the management and skin regeneration/repair from injury, particularly burns and ulcers, where the risk of bacterial infection is high. Material structure and performance integrity need to be accessed using a range of analytical and imaging techniques. Considering this scenario, research, production, and commercialization of such novel materials have drawn global efforts from numerous transnational companies as well as highly skilled research groups from around the world and diverse research areas.

## Conflict of Interest

The authors declare no conflicting, competing, and financial interest in any capacity.

## Acknowledgements

The authors are grateful to the Shanghai Jiao Tong University, Shanghai 200240, China, Huaiyin Institute of Technology, Huaian 223003, China, and Tecnologico de Monterrey, Mexico for providing literature services.

## References

1. Iqbal, H.M.N., Kyazze, G., Keshavarz, T., Advances in the valorization of lignocellulosic materials by biotechnology: an overview, *BioResources*, 8(2), 3157–3176, 2013.
2. Iqbal, H.M.N., *Development of bio-composites with novel characteristics through enzymatic grafting*, Doctoral dissertation, University of Westminster, 2015.
3. Iqbal, H.M.N., Kyazze, G., Tron, T., Keshavarz, T., 'One-pot' synthesis and characterisation of novel P (3HB)–ethyl cellulose based graft composites through lipase catalysed esterification. *Polym. Chem.*, 5(24), 7004–7012, 2014.
4. Iqbal, H.M.N., Kyazze, G., Tron, T., Keshavarz, T., A preliminary study on the development and characterisation of enzymatically grafted P (3HB)-ethyl cellulose based novel composites. *Cellulose*, 21(5), 3613–3621, 2014.
5. Iqbal, H.M.N., Kyazze, G., Tron, T., Keshavarz, T., Laccase-assisted grafting of poly (3-hydroxybutyrate) onto the bacterial cellulose as backbone polymer: development and characterisation. *Carbohydr. Polym.*, 113, 131–137, 2014.
6. Iqbal, H.M.N., Kyazze, G., Locke, I.C., Tron, T., Keshavarz, T., *In situ* development of self-defensive antibacterial biomaterials: phenol-g-keratin-EC based bio-composites with characteristics for biomedical applications. *Green Chem.*, 17(7), 3858–3869, 2015.
7. Iqbal, H.M.N., Kyazze, G., Locke, I.C., Tron, T., Keshavarz, T., Poly (3-hydroxybutyrate)-ethyl cellulose based bio-composites with novel characteristics for infection free wound healing application. *Int. J. Biol. Macromol.*, 81, 552–559, 2015.
8. Iqbal, H.M.N., Kyazze, G., Locke, I.C., Tron, T., Keshavarz, T., Development of bio-composites with novel characteristics: evaluation of phenol-induced antibacterial, biocompatible and biodegradable behaviours. *Carbohydr. Polym.*, 131, 197–207, 2015.

9. Iqbal, H.M.N., Kyazze, G., Tron, T., Keshavarz, T., Laccase-assisted approach to graft multifunctional materials of interest: keratin-EC based novel composites and their characterisation. *Macromol. Mater. Eng.*, 300(7), 712–720, 2015.
10. Iqbal, H.M.N., Kyazze, G., Locke, I.C., Tron, T., Keshavarz, T., Development of novel antibacterial active, HaCaT biocompatible and biodegradable CA-gP (3HB)-EC biocomposites with caffeic acid as a functional entity. *Express Polym. Lett.*, 9, 764–772, 2015.
11. Iqbal, H.M.N., Kyazze, G., Tron, T., Keshavarz, T., Laccase from Aspergillus niger: a novel tool to graft multifunctional materials of interests and their characterization. *Saudi J. Biol. Sci.*, 25(3), 545–550, 2018.
12. Rizwan, M., Yahya, R., Hassan, A., Yar, M., Azzahari, A.D., Selvanathan, V., Abouloula, C.N., pH sensitive hydrogels in drug delivery: brief history, properties, swelling, and release mechanism, material selection and applications. *Polymers*, 9(4), 137, 2017.
13. Asgher, M., Ahmad, Z., Iqbal, H.M.N., Bacterial cellulose-assisted de-lignified wheat straw-PVA based bio-composites with novel characteristics. *Carbohydr. Polym.*, 161, 244–252, 2017.
14. Bedian, L., Rodríguez, A.M.V., Vargas, G.H., Parra-Saldivar, R., Iqbal, H.M.N., Bio-based materials with novel characteristics for tissue engineering applications – a review. *Int. J. Biol. Macromol.*, 98, 837–846, 2017.
15. Plackett, D., Andersen, T.L., Pedersen, W.B., Nielsen, L. Biodegradable composites based on L-polylactide and jute fibres. *Comp. Sci. Technol.*, 63(9), 1287–1296, 2003.
16. Gallegos, A.M.A., Carrera, S.H., Parra, R., Keshavarz, T., Iqbal, H.M.N., Bacterial cellulose: a sustainable source to develop value-added products – a review. *BioResources*, 11(2), 5641–5655, 2016.
17. Bilal, M., Rasheed, T., Iqbal, H.M., Li, C., Hu, H., Zhang, X., Development of silver nanoparticles loaded chitosan-alginate constructs with biomedical potentialities. *Int. J. Biol. Macromol.*, 105, 393–400, 2017.
18. Bilal, M., Rasheed, T., Iqbal, H.M.N., Hu, H., Zhang, X., Silver nanoparticles: biosynthesis and antimicrobial potentialities. *Int. J. Pharmacol.*, 13(7), 832–845.
19. Rasheed, T., Bilal, M., Iqbal, H.M., Li, C., Green biosynthesis of silver nanoparticles using leaves extract of Artemisia vulgaris and their potential biomedical applications. *Colloids Surf. B: Biointerfaces*, 158, 408–415, 2017.
20. Villalba-Rodriguez, A.M., Dhama, K., Iqbal, H., Biomaterials-based hydrogels and their drug delivery potentialities. *Int. J. Pharmacol.*, 13(7), 864–873, 2017.
21. Baniani, D.D., Bagheri, R., Solouk, A., Preparation and characterization of a composite biomaterial including starch micro/nano particles loaded chitosan gel. *Carbohydr. Polym.*, 174, 633–645, 2017.
22. Dou, Y., Zhang, B., He, M., Yin, G., Cui, Y., Savina, I.N., Keratin/polyvinyl alcohol blend films cross-linked by dialdehyde starch and their potential application for drug release. *Polymers*, 7(3), 580–591, 2015.

23. Aroguz, A.Z., Baysal, K., Adiguzel, Z., Baysal, B.M., Alginate/polyoxyethylene and alginate/gelatin hydrogels: preparation, characterization, and application in tissue engineering. *Appl. Biochem. Biotechnol.*, 173(2), 433–448, 2014.
24. Gómez-Guillén, M.C., Giménez, B., López-Caballero, M.A., Montero, M.P., Functional and bioactive properties of collagen and gelatin from alternative sources: a review. *Food Hydrocoll.*, 25(8), 1813–1827, 2011.
25. Wang, L., Auty, M.A., Kerry, J.P., Physical assessment of composite biodegradable films manufactured using whey protein isolate, gelatin and sodium alginate. *J. Food Eng.*, 96(2), 199–207, 2010.
26. Gómez-Estaca, J., De Lacey, A.L., López-Caballero, M.E., Gómez-Guillén, M.C., Montero, P., Biodegradable gelatin–chitosan films incorporated with essential oils as antimicrobial agents for fish preservation. *Food Microbiol.*, 27(7), 889–896, 2010.
27. Müller, C.M., Laurindo, J.B., Yamashita, F., Effect of cellulose fibers addition on the mechanical properties and water vapor barrier of starch-based films. *Food Hydrocoll.*, 23(5), 1328–1333, 2009.
28. Jabeen, S., Islam, A., Ghaffar, A., Gull, N., Hameed, A., Bashir, A.,... Hussain, T. Development of a novel pH sensitive silane crosslinked injectable hydrogel for controlled release of neomycin sulfate. *Int. J. Biol. Macromol.*, 97, 218–227, 2017.
29. Pawar, S.N. and Edgar, K.J., Alginate derivatization: a review of chemistry, properties and applications. *Biomaterials*, 33(11), 3279–3305, 2012.
30. Lee, K.Y. and Mooney, D.J., Alginate: properties and biomedical applications. *Prog. Polym. Sci.*, 37(1), 106–126, 2012.
31. Chen, S.C., Wu, Y.C., Mi, F.L., Lin, Y.H., Yu, L.C., Sung, H.W., A novel pH-sensitive hydrogel composed of N, O-carboxymethyl chitosan and alginate cross-linked by genipin for protein drug delivery. *J. Control. Release*, 96(2), 285–300, 2004.
32. Campo, V.L., Kawano, D.F., da Silva Jr, D.B., Carvalho, I., Carrageenans: biological properties, chemical modifications and structural analysis – a review. *Carbohydr. Polym.*, 77(2), 167–180, 2009.
33. Li, L., Ni, R., Shao, Y., Mao, S., Carrageenan and its applications in drug delivery. *Carbohydr. Polym.*, 103, 1–11, 2014.
34. Piyakulawat, P., Praphairaksit, N., Chantarasiri, N., Muangsin, N., Preparation and evaluation of chitosan/carrageenan beads for controlled release of sodium diclofenac. *Aaps PharmSciTech*, 8(4), 120, 2007.
35. Sannino, A., Demitri, C., Madaghiele, M., Biodegradable cellulose-based hydrogels: design and applications. *Materials*, 2(2), 353–373, 2009.
36. Lim, L.S., Ahmad, I., Lazim, M.A.S.M., pH sensitive hydrogel based on poly (acrylic acid) and cellulose nanocrystals. *Sains Malaysiana*, 44(6), 779–785, 2015.
37. Zhou, H., Zhu, H., Yang, X., Zhang, Y., Zhang, X., Cui, K., Yao, J., Temperature/pH sensitive cellulose-based hydrogel: synthesis, characterization, loading, and release of model drugs for potential oral drug delivery. *BioResources*, 10(1), 760–771, 2014.

38. Bostan, M.S., Senol, M., Cig, T., Peker, I., Goren, A.C., Ozturk, T., Eroglu, M.S., Controlled release of 5-aminosalicylicacid from chitosan based pH and temperature sensitive hydrogels. *Int. J. Biol. Macromol.*, 52, 177–183, 2013.
39. Atta, S., Khaliq, S., Islam, A., Javeria, I., Jamil, T., Athar, M.M., Ghaffar, A., Injectable biopolymer based hydrogels for drug delivery applications. *Int. J. Biol. Macromol.*, 80, 240–245, 2015.
40. Muxika, A., Etxabide, A., Uranga, J., Guerrero, P., de la Caba, K., Chitosan as a bioactive polymer: processing, properties and applications. *Int. J. Biol. Macromol.*, 105, 1358–1368, 2017.
41. Kammoun, M., Haddar, M., Kallel, T.K., Dammak, M., Sayari, A., Biological properties and biodegradation studies of chitosan biofilms plasticized with PEG and glycerol. *Int. J. Biol. Macromol.*, 62, 433–438, 2013.
42. Elgadir, M.A., Uddin, M.S., Ferdosh, S., Adam, A., Chowdhury, A.J.K., Sarker, M.Z.I., Impact of chitosan composites and chitosan nanoparticle composites on various drug delivery systems: a review. *J. Food Drug Anal.*, 23(4), 619–629, 2015.
43. Wu, T., Zivanovic, S., Draughon, F.A., Conway, W.S., Sams, C.E., Physicochemical properties and bioactivity of fungal chitin and chitosan. *J. Agric. Food Chem.*, 53(10), 3888–3894, 2005.
44. Siedenbiedel, F. and Tiller, J.C., Antimicrobial polymers in solution and on surfaces: overview and functional principles. *Polymers*, 4(1), 46–71, 2012.
45. Kulbokaite, R., Ciuta, G., Netopilik, M., Makuska, R., N-PEG'ylation of chitosan via 'click chemistry' reactions. *React. Funct. Polym.*, 69(10), 771–778, 2009.
46. Chen, S.H., Tsao, C.T., Chang, C.H., Lai, Y.T., Wu, M.F., Chuang, C.N., Hsieh, K.H., Assessment of reinforced poly (ethylene glycol) chitosan hydrogels as dressings in a mouse skin wound defect model. *Mater. Sci. Eng.: C*, 33(5), 2584–2594, 2013.
47. Rinaudo, M. (2006). Chitin and chitosan: properties and applications. *Prog. Polym. Sci.*, 31(7), 603–632, 2013.
48. Du, H., Liu, M., Yang, X., Zhai, G., The design of pH-sensitive chitosan-based formulations for gastrointestinal delivery. *Drug Discov. Today*, 20(8), 1004–1011, 2015.
49. Zhang, M., Li, X.H., Gong, Y.D., Zhao, N.M., Zhang, X.F., Properties and biocompatibility of chitosan films modified by blending with PEG. *Biomaterials*, 23(13), 2641–2648, 2002.
50. Su, F.Y., Lin, K.J., Sonaje, K., Wey, S.P., Yen, T.C., Ho, Y.C., Sung, H.W., Protease inhibition and absorption enhancement by functional nanoparticles for effective oral insulin delivery. *Biomaterials*, 33(9), 2801–2811, 2012.
51. Feng, C., Wang, Z., Jiang, C., Kong, M., Zhou, X., Li, Y., Chen, X., Chitosan/o-carboxymethyl chitosan nanoparticles for efficient and safe oral anticancer drug delivery: *in vitro* and *in vivo* evaluation. *Int. J. Pharm.*, 457(1), 158–167, 2013.

52. Gujarathi, N.A., Rane, B.R., Patel, J.K., pH sensitive polyelectrolyte complex of O-carboxymethyl chitosan and poly (acrylic acid) cross-linked with calcium for sustained delivery of acid susceptible drugs. *Int. J. Pharm.*, 436(1–2), 418–425, 2012.
53. Pacelli, S., Paolicelli, P., Casadei, M.A., New biodegradable dextran-based hydrogels for protein delivery: synthesis and characterization. *Carbohydr. Polym.*, 126, 208–214, 2015.
54. Simonsen, L., Hovgaard, L., Mortensen, P.B., Brøndsted, H., Dextran hydrogels for colon-specific drug delivery. V. Degradation in human intestinal incubation models. *Eur. J. Pharm. Sci.*, 3(6), 329–337, 1995.
55. Hovgaard, L. and Brøndsted, H., Dextran hydrogels for colon-specific drug delivery. *J. Control. Release*, 36(1–2), 159–166, 1995.
56. Chiu, H.C., Wu, A.T., Lin, Y.F., Synthesis and characterization of acrylic acid-containing dextran hydrogels. *Polymer*, 42(4), 1471–1479, 2001.
57. Kim, I.S. and Oh, I.J., Drug release from the enzyme-degradable and pH-sensitive hydrogel composed of glycidyl methacrylate dextran and poly (acrylic acid). *Arch. Pharm. Res.*, 28(8), 983–987, 2005.
58. Prabaharan, M., Prospective of guar gum and its derivatives as controlled drug delivery systems. *Int. J. Biol. Macromol.*, 49(2), 117–124, 2011.
59. Sinha, V.R. and Kumria, R., Polysaccharides in colon-specific drug delivery. *Int. J. Pharm.*, 224(1–2), 19–38, 2001.
60. Wong, D., Larrabee, S., Clifford, K., Tremblay, J., Friend, D.R., USP dissolution apparatus III (reciprocating cylinder) for screening of guar-based colonic delivery formulations. *J. Control. Release*, 47(2), 173–179, 1997.
61. Seeli, D.S. and Prabaharan, M., Guar gum succinate as a carrier for colon-specific drug delivery. *Int. J. Biol. Macromol.*, 84, 10–15, 2016.
62. Dumitriu, S. (Ed.), *Polysaccharides: Structural Diversity and Functional Versatility*, CRC Press, New York, NY, 2004.
63. Bueno, V.B., Bentini, R., Catalani, L.H., Petri, D.F.S., Synthesis and swelling behavior of xanthan-based hydrogels. *Carbohydr. Polym.*, 92(2), 1091–1099, 2013.
64. Das, S. and Subuddhi, U., pH-responsive guar gum hydrogels for controlled delivery of dexamethasone to the intestine. *Int. J. Biol. Macromol.*, 79, 856–863, 2015.
65. Arunbabu, D., Shahsavan, H., Zhang, W., Zhao, B., Poly (AAc-co-MBA) hydrogel films: adhesive and mechanical properties in aqueous medium. *J. Phys. Chem. B*, 117(1), 441–449, 2012.
66. Jabbari, E. and Nozari, S., Swelling behavior of acrylic acid hydrogels prepared by γ-radiation crosslinking of polyacrylic acid in aqueous solution. *Eur. Polym. J.*, 36(12), 2685–2692, 2000.
67. Chang, A., pH-sensitive starch-g-poly (acrylic acid)/sodium alginate hydrogels for controlled release of diclofenac sodium. *Iran. Polym. J.*, 24(2), 161–169, 2015.

68. Sheikh, N., Jalili, L., Anvari, F., A study on the swelling behavior of poly (acrylic acid) hydrogels obtained by electron beam crosslinking. *Radiat. Phys. Chem.*, 79(6), 735–739, 2010.
69. Yang, T.H., Recent applications of polyacrylamide as biomaterials. *Recent Patents Mater. Sci.*, 1(1), 29–40, 2008.
70. Bashir, S., Teo, Y.Y., Ramesh, S., Ramesh, K., Synthesis, characterization, properties of N-succinyl chitosan-g-poly (methacrylic acid) hydrogels and *in vitro* release of theophylline. *Polymer*, 92, 36–49, 2016.
71. Mukhopadhyay, P., Sarkar, K., Bhattacharya, S., Bhattacharyya, A., Mishra, R., Kundu, P.P., pH sensitive N-succinyl chitosan grafted polyacrylamide hydrogel for oral insulin delivery. *Carbohydr. Polym.*, 112, 627–637, 2014.
72. Jia, Y.T., Gong, J., Gu, X.H., Kim, H.Y., Dong, J., Shen, X.Y., Fabrication and characterization of poly (vinyl alcohol)/chitosan blend nanofibers produced by electrospinning method. *Carbohydr. Polym.*, 67(3), 403–409, 2007.
73. Bano, I., Ghauri, M.A., Yasin, T., Huang, Q., Palaparthi, A.D.S., Characterization and potential applications of gamma irradiated chitosan and its blends with poly (vinyl alcohol). *Int. J. Biol. Macromol.*, 65, 81–88, 2014.
74. Rafique, A., Zia, K.M., Zuber, M., Tabasum, S., Rehman, S., Chitosan functionalized poly (vinyl alcohol) for prospects biomedical and industrial applications: a review. *Int. J. Biol. Macromol.*, 87, 141–154, 2016.
75. Park, S.Y., Jun, S.T., Marsh, K.S. Physical properties of pvoh/chitosan-blended films cast from different solvents. *Food Hydrocoll.*, 15, 499–502, 2001.
76. Shahzad, S., Yar, M., Siddiqi, S.A., Mahmood, N., Rauf, A., Anwar, M.S., Afzaal, S., Chitosan-based electrospun nanofibrous mats, hydrogels and cast films: novel anti-bacterial wound dressing matrices. *J. Mater. Sci. Mater. Med.*, 26, 1–12, 2015.
77. Islam, A., Yasin, T., Bano, I., Riaz, M., Controlled release of aspirin from pH-sensitive chitosan/poly (vinyl alcohol) hydrogel. *J. Appl. Polym. Sci.*, 124(5), 4184–4192, 2012.
78. Kadajji, V.G. and Betageri, G.V., Water soluble polymers for pharmaceutical applications. *Polymers*, 3(4), 1972–2009, 2011.
79. Sohail, K., Khan, I.U., Shahzad, Y., Hussain, T., Ranjha, N.M., pH-sensitive polyvinylpyrrolidone-acrylic acid hydrogels: impact of material parameters on swelling and drug release. *Brazil. J. Pharm. Sci.*, 50(1), 173–184, 2014.
80. Pawar, R., Tekale, S.U., Shisodia, S.U., Totre, J.T., Domb, A.J., Biomedical applications of poly (lactic acid). *Recent Patents Regen. Med.*, 4(1), 40–51, 2014.
81. Gombotz, W.R. and Wee, S., Protein release from alginate matrices. *Adv. Drug Deliv. Rev.*, 31(3), 267–285, 1998.
82. Remminghorst, U. and Rehm, B.H., Bacterial alginates: from biosynthesis to applications. *Biotechnol. Lett.*, 28(21), 1701–1712, 2006.

83. Lee, K.Y. and Mooney, D.J., Alginate: properties and biomedical applications. *Prog. Polym. Sci.*, 37(1), 106–126, 2012.
84. Cavallaro, G., Gianguzza, A., Lazzara, G., Milioto, S., Piazzese, D., Alginate gel beads filled with halloysite nanotubes. *Appl. Clay Sci.*, 72, 132–137, 2013.
85. Zhang, Q., Yu, P., Li, Z., Zhang, H., Xu, Z., Li, P., Antioxidant activities of sulfated polysaccharide fractions from Porphyra haitanesis. *J. Appl. Phycol.*, 15(4), 305–310, 2003.
86. Yuan, H., Song, J., Zhang, W., Li, X., Li, N., Gao, X., Antioxidant activity and cytoprotective effect of κ-carrageenan oligosaccharides and their different derivatives. *Bioorg. Med. Chem. Lett.*, 16(5), 1329–1334, 2006.
87. Briones, A.V., Sato, T., Bigol, U.G., Antibacterial activity of polyethylenimine/carrageenan multilayer against pathogenic bacteria. *Adv. Chem. Eng. Sci.*, 4(02), 233, 2014.
88. Zia, K.M., Tabasum, S., Nasif, M., Sultan, N., Aslam, N., Noreen, A., Zuber, M., A review on synthesis, properties and applications of natural polymer based carrageenan blends and composites. *Int. J. Biol. Macromol.*, 96, 282–301, 2017.
89. Cascone, M.G., Maltinti, S., Barbani, N., Laus, M., Effect of chitosan and dextran on the properties of poly (vinyl alcohol) hydrogels. *J. Mater. Sci.: Mater. Med.*, 10(7), 431–435, 1999.
90. Chatelet, C., Damour, O., Domard, A., Influence of the degree of acetylation on some biological properties of chitosan films. *Biomaterials*, 22(3), 261–268, 2001.
91. Huang, M., Khor, E., Lim, L.Y., Uptake and cytotoxicity of chitosan molecules and nanoparticles: effects of molecular weight and degree of deacetylation. *Pharm. Res.*, 21(2), 344–353, 2004.
92. Kofuji, K., Qian, C.J., Nishimura, M., Sugiyama, I., Murata, Y., Kawashima, S., Relationship between physicochemical characteristics and functional properties of chitosan. *Eur. Polym. J.*, 41(11), 2784–2791, 2005.
93. Freitas, R.M., Spin-Neto, R., Spolidório, L.C., Campana-Filho, S.P., Marcantonio, R.A.C., Marcantonio Jr, E., Different molecular weight chitosan-based membranes for tissue regeneration. *Materials*, 4(2), 380–389, 2011.
94. Tsai, M.L., Chang, H.W., Yu, H.C., Lin, Y.S., Tsai, Y.D., Effect of chitosan characteristics and solution conditions on gelation temperatures of chitosan/2-glycerophosphate/nanosilver hydrogels. *Carbohydr. Polym.*, 84(4), 1337–1343, 2011.
95. Liu, Z., Ge, X., Lu, Y., Dong, S., Zhao, Y., Zeng, M., Effects of chitosan molecular weight and degree of deacetylation on the properties of gelatine-based films. *Food Hydrocoll.*, 26(1), 311–317, 2012.
96. Sarwat, F., Qader, S.A.U., Aman, A., Ahmed, N., Production & characterization of a unique dextran from an indigenous Leuconostoc mesenteroides CMG713. *Int. J. Biol. Sci.*, 4(6), 379, 2008.
97. Ullah, F., Othman, M.B.H., Javed, F., Ahmad, Z., Akil, H.M., Classification, processing and application of hydrogels: a review. *Mater. Sci. Eng.: C*, 57, 414–433, 2015.

98. Einerson, N.J., Stevens, K.R., Kao, W.J., Synthesis and physicochemical analysis of gelatin-based hydrogels for drug carrier matrices. *Biomaterials*, 24(3), 509–523, 2003.
99. Aewsiri, T., Benjakul, S., Visessanguan, W., Functional properties of gelatin from cuttlefish (Sepia pharaonis) skin as affected by bleaching using hydrogen peroxide. *Food Chem.*, 115(1), 243–249, 2009.
100. Xing, Q., Yates, K., Vogt, C., Qian, Z., Frost, M.C., Zhao, F., Increasing mechanical strength of gelatin hydrogels by divalent metal ion removal. *Sci. Rep.*, 4, 4706, 2014.
101. Kenawy, E.R., Kamoun, E.A., Eldin, M.S.M., El-Meligy, M.A., Physically crosslinked poly (vinyl alcohol)-hydroxyethyl starch blend hydrogel membranes: synthesis and characterization for biomedical applications. *Arab. J. Chem.*, 7(3), 372–380, 2014.
102. Alcázar-Alay, S.C. and Meireles, M.A.A., Physicochemical properties, modifications and applications of starches from different botanical sources. *Food Sci. Technol. (Campinas)*, 35(2), 215–236, 2015.
103. Ayers, D., Cuthbertson, J.M., Schroyer, K., Sullivan, S.M., Polyacrylic acid mediated ocular delivery of ribozymes. *J. Control. Release*, 38(2–3), 167–175, 1996.
104. Bromberg, L., Polyether-modified poly (acrylic acid): synthesis and applications. *Ind. Eng. Chem. Res.*, 37(11), 4267–4274, 1998.
105. Risbud, M.V. and Bhonde, R.R., Polyacrylamide-chitosan hydrogels: *in vitro* biocompatibility and sustained antibiotic release studies. *Drug Deliv.*, 7(2), 69–75, 2000.
106. Da Silva, E., Lebrun, L., Metayer, M., Elaboration of a membrane with bipolar behaviour using the semi-interpenetrating polymer networks technique. *Polymer*, 43(19), 5311–5320, 2002.
107. Wang, J., Wang, X., Xu, C., Zhang, M., Shang, X., Preparation of graphene/poly (vinyl alcohol) nanocomposites with enhanced mechanical properties and water resistance. *Polym. Int.*, 60(5), 816–822, 2011.
108. Gopishetty, V., Tokarev, I., Minko, S., Biocompatible stimuli-responsive hydrogel porous membranes via phase separation of a polyvinyl alcohol and Na-alginate intermolecular complex. *J. Mater. Chem.*, 22(37), 19482–19487, 2012.
109. Swapna, V.P., Suresh, K.I., Saranya, V., Rahana, M.P., Stephen, R., Thermal properties of poly (vinyl alcohol)(PVA)/halloysite nanotubes reinforced nanocomposites. *Int. J. Plast. Technol.*, 19(1), 124–136, 2015.
110. Gaaz, T.S., Sulong, A.B., Akhtar, M.N., Kadhum, A.A.H., Mohamad, A.B., Al-Amiery, A.A., Properties and applications of polyvinyl alcohol, halloysite nanotubes and their nanocomposites. *Molecules*, 20(12), 22833–22847, 2015.
111. Razzak, M.T. and Darwis, D., Irradiation of polyvinyl alcohol and polyvinyl pyrrolidone blended hydrogel for wound dressing. *Radiat. Phys. Chem.*, 62(1), 107–113, 2001.
112. Kadajji, V.G. and Betageri, G.V., Water soluble polymers for pharmaceutical applications. *Polymers*, 3(4), 1972–2009, 2011.

113. Bhadra, D., Bhadra, S., Jain, P., Jain, N.K., Pegnology: a review of PEG-ylated systems. *Die Pharmazie*, 57(1), 5–29, 2002.
114. Ruel-Gariepy, E. and Leroux, J.C., *In situ*-forming hydrogels – review of temperature-sensitive systems. *Eur. J. Pharm. Biopharm.*, 58(2), 409–426, 2004.
115. Knop, K., Hoogenboom, R., Fischer, D., Schubert, U.S., Poly (ethylene glycol) in drug delivery: pros and cons as well as potential alternatives. *Angewandte chemie international edition*, 49(36), 6288–6308, 2010.
116. Feng, Y., Zhang, X., Shen, Y., Yoshino, K., Feng, W., A mechanically strong, flexible and conductive film based on bacterial cellulose/graphene nanocomposite. *Carbohydr. Polym.*, 87(1), 644–649, 2012.
117. Qing, G., Li, M., Deng, L., Lv, Z., Ding, P., Sun, T., Smart drug release systems based on stimuli-responsive polymers. *Mini Rev. Med. Chem.*, 13(9), 1369–1380, 2013.
118. Fowler, P.A., Hughes, J.M., Elias, R.M., Biocomposites: technology, environmental credentials and market forces. *J. Sci. Food Agric.*, 86(12), 1781–1789, 2006.
119. Chang, W.S. and Chen, H.H., Physical properties of bacterial cellulose composites for wound dressings. *Food Hydrocoll.*, 53, 75–83, 2016.
120. Qiu, Y., Qiu, L., Cui, J., Wei, Q., Bacterial cellulose and bacterial cellulose-vaccarin membranes for wound healing. *Mater. Sci. Eng.: C*, 59, 303–309, 2016.
121. Kiziltas, E.E., Kiziltas, A., Rhodes, K., Emanetoglu, N.W., Blumentritt, M., Gardner, D.J., Electrically conductive nano graphite-filled bacterial cellulose composites. *Carbohydr. Polym.*, 136, 1144–1151, 2016.
122. Barud, H.S., Tercjak, A., Gutierrez, J., Viali, W.R., Nunes, E.S., Ribeiro, S.J.L., Jafellici, M., Nalin M., Marques, R.F.C., Biocellulose-based flexible magnetic paper. *J. Appl. Phys.*, 117(17), 17B734, 2015.
123. Duarte, E.B., Bruna, S., Andrade, F.K., Brígida, A.I., Borges, M.F., Muniz, C.R., Filho, M.M.S., Morais, J.P.S., Feitosa, J.P.A., and Rosa, M.F., Production of hydroxyapatite–bacterial cellulose nanocomposites from agroindustrial wastes, *Cellulose*, 22, 3177–3187, 2015.
124. Lin, C.W., Liang, S.S., Chen, S.W., Lai, J.T., Sorption and transport properties of 2-acrylamido-2-methyl-1-propanesulfonic acid-grafted bacterial cellulose membranes for fuel cell application. *J. Power Sources*, 232, 297–305, 2013.
125. Ul-Islam, M., Khan, T., Khattak, W.A., Park, J.K., Bacterial cellulose-MMTs nanoreinforced composite films: novel wound dressing material with antibacterial properties. *Cellulose*, 20(2), 589–596, 2013.
126. Shi, Z., Zang, S., Jiang, F., Huang, L., Lu, D., Ma, Y., Yang, G., *In situ* nano-assembly of bacterial cellulose–polyaniline composites. *RSC Adv.*, 2(3), 1040–1046, 2012.
127. Ul-Islam, M., Khan, T., Park, J.K., Nanoreinforced bacterial cellulose–montmorillonite composites for biomedical applications. *Carbohydr. Polym.*, 89(4), 1189–1197, 2012.

128. Hu, W., Chen, S., Yang, Z., Liu, L., Wang, H., Flexible electrically conductive nanocomposite membrane based on bacterial cellulose and polyaniline. *J. Phys. Chem. B*, 115(26), 8453–8457, 2011.
129. Kim, J., Cai, Z., Lee, H.S., Choi, G.S., Lee, D.H., Jo, C., Preparation and characterization of a bacterial cellulose/chitosan composite for potential biomedical application, *J. Polym. Res.*, 18(4), 739–744, 2011.
130. Zhu, H., Jia, S., Yang, H., Tang, W., Jia, Y., Tan, Z., Characterization of bacteriostatic sausage casing: a composite of bacterial cellulose embedded with ε-polylysine. *Food Sci. Biotechnol.*, 19(6), 1479–1484, 2010.

# 8

# Seaweed-Based Binder in Wood Composites

**Kang Chiang Liew*** **and Nur Syafiqah Nadiah Abdul Ghani**

*Faculty of Science and Natural Resources, Forestry Complex, Universiti Malaysia Sabah, Jalan UMS, Kota Kinabalu, Sabah, Malaysia*

***Abstract***
Synthetic binder from petroleum-based sources is being widely used nowadays. The depleting source of petroleum would see the need for the current adhesive industries to scout for alternative binder. Such a source could be from seaweed. In this study, seaweed was cleaned, cooked with chemical and prepared for binding with wood particles. The mixtures of different percentages of binder were hot-pressed to form particleboards. Results from physical and mechanical testing indicated that seaweed binder used was at par or better compared to the synthetic urea formaldehyde according to standards used.

***Keywords:*** Seaweed, binder, wood composite, physical, mechanical

## 8.1 Introduction

Marine macroalgae or seaweeds are plant-like organisms that are generally live attached to rock or other hard substrata in coastal areas [1]. A total of 42 countries in the world are involved in commercial seaweed activity. In Malaysia, Sabah is the only state that is commercially producing seaweed. Seaweed is cultivated in four districts on large scale which are in Semporna, Lahad Datu, Kudat and Kunak. The main cultivated species are *Kappaphycus alvarezii* and *Eucheuma spinosum* [2].

*Corresponding author: liewkc@ums.edu.my

Shakeel Ahmed and Chaudhery Mustansar Hussain (eds.) Green and Sustainable Advanced Materials, Vol. 1 (191–212) © 2018 Scrivener Publishing LLC

According to researchers in Korea's Silla University, some of the seaweed extracted can inhibit bacteria that are known to cause gingivitis and adhesive disease. Seaweed can also be manufactured artificially or extracted and used to achieve various levels of viscosity called as industrial adhesives. Seaweed industrial adhesives also known as hydrocolloids fall into three categories which are alginates, agars and carrageenans. Seaweed hydrocolloids market grew at annual rates of 3%–5% in the 1980s and 1990s [3]. In this study, the type of seaweed that was used was red seaweed species, *K. alvarezii*. This type of species is widely cultured and highly demanded in industry. Hayashi *et al.* [4] stated that *K. alvarezii* has changed the social and economic aspects of several countries in which commercial farms were well established in a relatively short period such as Philippine, Indonesia and Tanzania, and it is also one of the main seaweed cultivated in the world. According to the Food and Agriculture Organization (FAO) database [5], the production of *K. alvarezii* in 2009 was 1755 tons, worth almost US$ 203 million. Bixler and Porse [3] mentioned that carrageenan production reached 50,000 tons, worth US$ 527 million. If *K. alvarezii* could be used in particleboard making, this will also help to increase global industrial demand for carrageenan.

Studies said that nature offers an extensive range of biological species with adhesive capabilities that can be sourced to seek ideas and inspirations. Seaweed was the organism which has very elegantly achieved underwater adhesion. According to Mather [6], seaweed has an ability to adhere to marine surfaces and withstand large tidal force. This kelp produces an extremely strong water resistant glue anchoring it to the rocks. The study on the seaweed glue including their strength testing in sticking to a variety of materials including glass, stainless steel, plastics and wood. To get good synthetic glues that can resist water and withstand dirty environments, high energy processes or strong chemicals were needed. Seaweed adhesive is a type of bioadhesive that specifically secretes natural organisms in marine species. Adhesive secretion is used by organisms for attachment, defence and predation. Bioadhesives differ from synthetic adhesives. It is able to build durable interfaces between hard and soft materials, often of disparate scales and exhibit a certain number of characteristics. Bioadhesives are of commercial interest because they tend to be biocompatible. Bioadhesives are one of the good alternatives compared to synthetic adhesives if concerned about the environmental, health and safety issues relating to hazardous ingredients, volatile organic compound emissions and difficulties in recycling or remediating synthetic adhesives.

## 8.2 Methods and Techniques

### 8.2.1 Preparation of Raw Material

*Acacia mangium* particles were screened for size by using a vibrator screener machine. The oversized particles (>2 mm) were recycled for further reduction and screened particles of 0.5–2.0 mm were used for particleboard making.

### 8.2.2 Seaweed Adhesive Preparation

There are various methods of seaweed (S) adhesive solution preparation, all of which involved dissolution of the adhesive in water. The adhesive in this study was prepared according to a recipe by Kayano [7] for the preparation of seaweed adhesive. A quantity of 300 g seaweed was swelled in 200 ml cold water or in alcohol/water overnight and then was cooked at 80 °C with 6 g of sodium hydroxide for 30–60 min or until mucilage dissolved, stirring frequently until most of the seaweed dissolved. Many seaweeds are insoluble in cold water but readily dissolve in boiling water. Alkali was used because it caused a chemical change that leads to increased adhesive strength in the final product. In chemical terms, it removed some of the sulphate groups from the molecules in seaweed. To adjust the pH to 5–6, the cooked seaweed was treated with sulphuric acid (90 ml). Seaweed adhesive is a thermoplastic type which means it can be reheated after it turns to solid form when cold.

### 8.2.3 Blending and Mat Forming

Adhesives in the form of urea formaldehyde (UF) are generally used to bind together the particle mix. After 24 h drying process, the particles and resin were added into the mixture along with 0.3% to 1% ammonium chloride according to the ovendry weight of the particles. The resin content ranged between 6% and 9% of UF as control. A mattress of wood particles coat with adhesive was formed by dropping them on to metal plates. Glue blending was used with five different mixing ratios which were 100%UF, 100%S, 50%UF:50%S, 75%UF:25%S and 25%UF:75%S. The mixture was blended properly to get even blending and to avoid from glue spot when forming board. All the adhesive was blended evenly with the particles according to predetermined amounts.

The mixture was then poured and distributed as even as possible into a metal mould of the size 300 mm × 300 mm with the use of steel stopper at 10 mm and another metal plate on top. After the distribution process, the mat then was cold-pressed for 3 min until the particles and adhesive were evenly distributed, before the mat was loaded into the hot press machine at 165 °C for 15 min to cure the adhesive. Heat and pressure were applied to the press to activate the adhesive and bond the particles into a solid panel. The hot press process was repeated for other mixing ratios.

### 8.2.4 Conditioning

Figure 8.1 showed the particleboard produced after the hot press. The board then was left to condition in a room with temperature 20 ± 2 °C and relative humidity (RH) 65%. The conditioning process is to avoid degradation in adhesive bonding of particleboard.

Then the board was cut to size by using trimming saws for the testing. The test pieces were prepared according to Japanese Industrial Standard (JIS A 5908) Type 8 [8]. Test pieces were tested for three tests which were for physical properties, mechanical properties and dimensional stability properties. The tests performed in this study were accordance to [9] which were moisture content (MC), density (D), Modulus of Elasticity (MOE) and Modulus of Rupture (MOR), thickness swelling (TS) and water absorption (WA) (24 h submersion) and internal bonding (IB).

**Figure 8.1** Particleboard.

### 8.2.5 Data Analysis

All the data were analysed by using Analysis of Variance (ANOVA). The significant difference between the groups was examined using $p \leq 0.05$ as the p-value point. The data analysis was obtained to see the differences of physical properties, mechanical properties and dimensional stability properties between the particles with UF only, particles with UF and S, and particles with S only according to the specified ASTM standard [9].

## 8.3 Results and Discussion

### 8.3.1 Overview

Five types of particleboards were produced using five different percentages of adhesive mix which were 100%UF, 100%S, 50%UF:50%S, 25%UF:75%S and 75%UF:25%S. From five types of boards, 15 replicates of test pieces were produced for each testing. The results of physical properties measured in this study for MC were highest for 100%S and lowest for 75%UF:25%S, and D result showed high value for both 100%UF and 75%UF:25%S, and the lowest value was 100%S. For dimensional stability, both results for WA and TS showed the highest value for 100%S and lowest value for 75%UF:25%S. The mechanical properties test was static bending strength as in MOE and MOR where both showed the highest value for 75%UF:25%S and lowest value for 100%S, and for IB, 100%S showed the lowest value and 75%UF:25%S showed the highest value. The dimensional stability and mechanical properties of boards improved with increase in resin content but will generally be independent of the formulation of the adhesive. Detailed results are shown in Table 8.1.

### 8.3.2 The Physical Properties of *Acacia Mangium* Particleboard

#### *8.3.2.1 Moisture Content*

Based on the result, the mean values of MC of particleboard test pieces as shown in Figure 8.2 were 100%UF (9.47%), 100%S (11.17%), 50%UF:50%S (9.46%), 25%UF:75%S (8.74%), and 75%UF:25%S (4.37%). Based on the histogram, the mean of MC of 75%UF:25%S was the lowest compared to other variables and the higher mean of MC was 100%S. The sequence for mean of MC in accordance with the type of adhesive in ascending order was 75%UF:25%S (4.37%) < 25%UF:75%S (8.74%) < 50%UF:50%S (9.46%)

**Table 8.1** Table shows the comparison and analysis with mean, standard deviation of physical properties, dimensional stability and mechanical properties of different types of adhesives with different types of adhesive percentages in particleboard.

| Type of adhesive | Properties | | | | | | |
|---|---|---|---|---|---|---|---|
| | Moisture content (%) | Density ($g/cm^3$) | Water absorption (%) | Thickness swelling (%) | Modulus of elasticity ($N/mm^2$) | Modulus of rupture ($N/mm^2$) | Internal bonding ($N/mm^2$) |
| 100%UF | $9.47_a$ | $0.73_a$ | $65.01_a$ | $8.04_a$ | $1656.10_a$ | $9.79_a$ | $0.45_a$ |
| | ± (0.73) | ± (0.026) | ± (6.01) | ± (0.89) | ± (259.77) | ± (1.71) | ± (0.02) |
| 100%S | $11.17_b$ | $0.51_b$ | $158.55_b$ | $48.34_b$ | $1304.66_b$ | $0.33_b$ | $0.00_b$ |
| | ± (0.54) | ± (0.032) | ± (4.24) | ± (2.96) | ± (79.88) | ± (0.05) | ± (0.0002) |
| 50%UF:50%S | $9.46_a$ | $0.63_c$ | $55.58_c$ | $8.74_{a,c}$ | $1912.32_{c,d}$ | $11.14_c$ | $0.12_c$ |
| | ± (0.91) | ± (0.026) | ± (6.75) | ± (1.56) | ± (286.28) | ± (2.26) | ± (0.01) |
| 25%UF:75%S | $8.74_c$ | $0.68_d$ | $65.97_a$ | $9.66_c$ | $1901.32_d$ | $9.66_a$ | $0.15_d$ |
| | ± (0.42) | ± (0.022) | ± (0.65) | ± (0.78) | ± (119.70) | ± (1.43) | ± (0.009) |
| 75%UF:25%S | $4.37_d$ | $0.70_{e,d}$ | $28.79_d$ | $3.68_d$ | $2682.93_e$ | $14.69_d$ | $0.27_e$ |
| | ± (0.64) | ± (0.034) | ± (6.70) | ± (0.63) | ± (380.03) | ± (2.79) | ± (0.03) |

Note: The number that is not in parentheses is the mean value; the number in parentheses with '±' is the standard deviation value.
Value in the same column with different alphabets a, b, c, d, e for different types of adhesives for each property in the column indicates significant difference at $p \leq 0.05$ in Least Significant Difference (LSD).

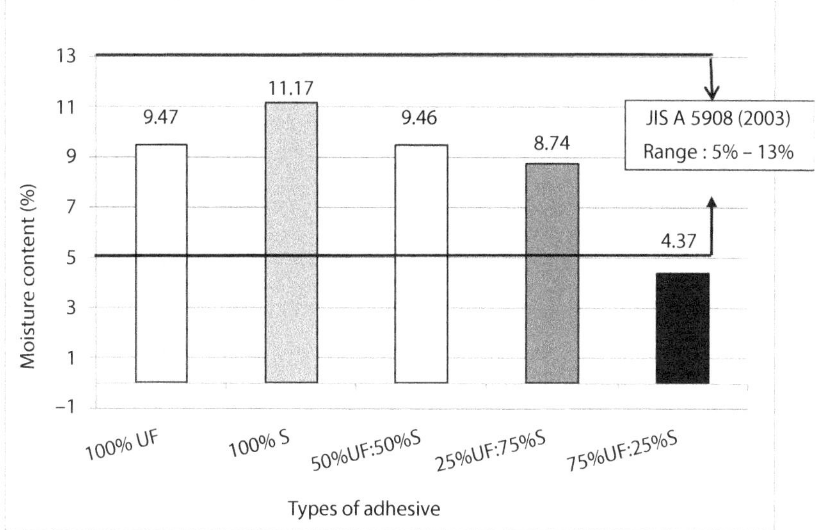


**Figure 8.2** Mean value of moisture content for particleboard bonded with different mixing ratios of adhesive.

< 100%UF (9.47%) < 100%S (11.17%). According to JIS A 5908 (2003) as shown in Figure 8.2, the mean of MC in particleboard is standardized between 5% and 13%. The significant difference of MC between groups was 0.000 which means that there is a significant difference in the MC between the groups. Based on Table 8.1, there was no difference between 100%UF with 50%UF:50%S in MC. That means in the particleboard manufactured, there is no difference in using either 100%UF or 50%UF:50%S. As stated by Smulski [10], the change in RH may affect the changes in MC of particleboard. Lower MC means that particleboard may extend the life span of board.

Based on Bowyer *et al.* [11], the MCs of wood products are in an equilibrium with the surrounding environment. MC is one the physical properties for particleboard. According to Widarmana [12], the rate of MC in particleboard will decrease with the increase of temperature and high amount of adhesive because the bonding between the particles will get stronger until water cannot enter into the particles. Particles as stated contain MC between 2% and 5% [12]. Moisture will increase by 4%–6% when mixed with adhesive until reaching between 5% and 6%.

### *8.3.2.2 Density*

Figure 8.3 shows the mean for density of 100%UF, 100%S, 50%UF:50%S, 25%UF:75%S and 75%UF:25%S. In this study, the mean of density from

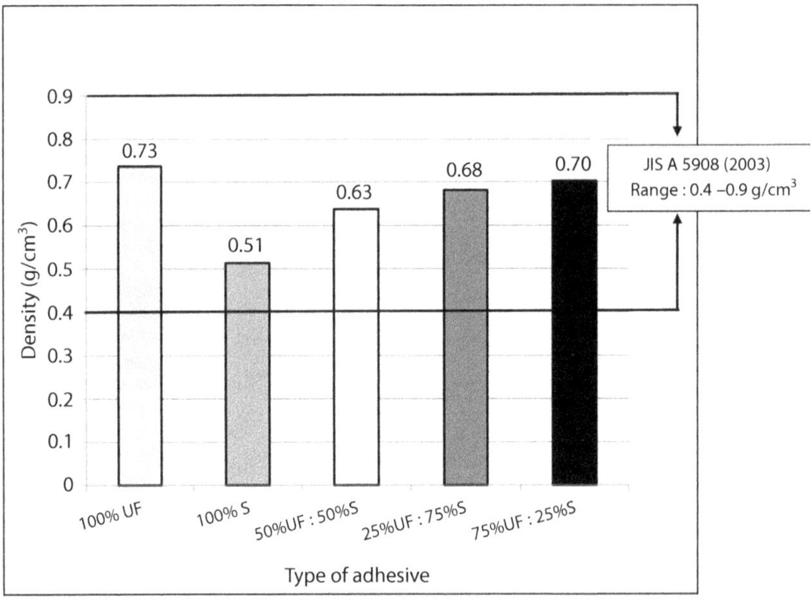


**Figure 8.3** Mean value of density for particleboard bonded with different mixing ratios of adhesive.

highest to lowest mean WA were 100%S (0.51 g/cm$^3$) < 50%UF:50%S (0.63 g/cm$^3$) < 25%UF:75%S (0.68 g/cm$^3$) < 75%UF:25%S (0.70 g/cm$^3$) < 100%UF (0.73 g/cm$^3$). Between the group, test pieces with 100% seaweed adhesive show the lowest mean of density and test pieces with 100%UF, the highest mean of density. The significant difference of density between groups was 0.000. The significant probability was $0.000 \leq 0.05$ which means that there was significant difference in the density between the groups. As stated by Maloney [13], higher density of particleboard will produce a higher value in physical properties, mechanical properties with better dimensional stability.

Density is a measure of the compactness of particles in a particleboard that is highly dependent on the density of the particles used and the pressure exerted during the compression process. Density is one of the important characteristics of particleboard. The density of a material varies with temperature and pressure. Increasing the pressure on a board decreases the volume of the board and thus increases its density. The mass and volume change proportionately. According to Widarmana [12], the higher the density of particleboard, the more the particles required to create a board of the same size. The higher the density, the better the strength. According to a Japanese Industrial Standard [8], the range of density of particleboard was 0.4–0.9 g/cm$^3$.

Based on Table 8.1, we can see that there was no difference between 25%UF:75%S adhesive with 75%UF:25%S adhesives in density. The difference in density of particleboard was affected by the amount of particles in mat, pressing pressure, the adhesive content and additives mix [14]. Bowyer *et al.* [11] stated that density of particleboard is higher than the original material components due to adhesive weight, additive and pressure during manufacturing.

According to [15], the strength of particleboard is largely determined by adhesive bond quality. The higher the overall density of the particleboard from a given raw material, the greater the strength. In this study, 100%S adhesive has the lowest mean value of density. This may be due to the low adhesive bond quality. The properties of the seaweed adhesive itself affect the bonding quality of particles that cause difficulties in producing boards with truly uniform density profiles, i.e. equal face and core density. Seaweed adhesive has lower tendency to spread or distribute evenly because it has high viscosity which was 11,000 mPa with contact angle at 63°(almost 90°) compared to UF 22°. The seaweed adhesive cannot penetrate easily into the particles and cause low internal bond strength. According to Fowkes [16], in applying liquid adhesive the viscosity should be as low as possible in order to increase the extent of adhesive flow through particles surface. This is also because as the press is closed, the surfaces of the mat heat first and to a higher temperature than the core. This softens the surface particles and allows them to become more densified than the core. When the temperature become high, there may result chemical changes in the seaweed adhesives or develop surface-chemical changes in the particles. This may cause the seaweed adhesives to be highly adsorbable and may cause the formation of new low-energy surface films on which the seaweed adhesive will not spread.

### 8.3.3 Dimensional Stability of *Acacia Mangium* Particleboard

#### *8.3.3.1 Water Absorption*

Figure 8.4 shows the mean for WA of 100%UF, 100%S, 50%UF:50%S, 25%UF:75%S and 75%UF:25%S. In this study, the mean of WA from highest to lowest mean WA were 100%S (158.55%) > 25%UF:75%S (65.97%) > 100%UF (65.01%) > 50%UF:50%S (55.58%) > 75%UF:25%S (28.79%). Between the group, test pieces with 100%S adhesive show the highest mean of WA and test pieces with 75%UF:25%S show the lowest mean of WA. Test pieces with 100% seaweed adhesive were absorbing water more than test pieces with 100%UF. This is because seaweed adhesive has weaker bonding

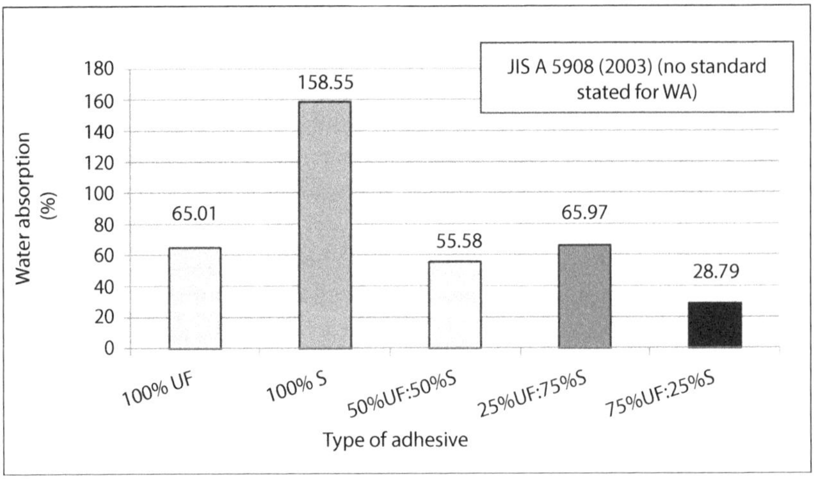


**Figure 8.4** Mean value of water absorption for particleboard bonded with different mixing ratios of adhesive.

than UF. The characteristics of the adhesive itself affected the bonding of the particles. But when both seaweed and UF mix together, this mixture produced better bonding and better surface compared to seaweed and UF only test pieces.

As we can see in the mean WA histogram, the mixture of seaweed and UF has greater value than 100%UF test pieces. It shows that WA is lower when adhesive content increases. This is due to more adhesive present in the particleboard that reduces the penetration of water molecules into the board.

According to Widarmana [12], the factors that can affect the WA were the volume of vacant space between particles that can hold water, the surface area of the particles which cannot be covered by the adhesive and the depth of penetration of the adhesive to the particles. The type of adhesive and the amount of the adhesive used also affect the amount of WA. Haygreen and Bowyer [17] stated that the higher the amount of adhesive used for producing wood panel, the lower the value of WA. The different type of adhesive used also shows the difference of WA as shown in Table 8.1. From the figure we can see the different amounts of water absorbed in the particles by looking at the change of thickness and measuring the weight of test pieces. More water absorbed shows that the type of adhesive is not good for board forming and bonding.

WA is a condition when the particle expands due to the absorption of moisture and water. WA can cause changes in shape, debonding or loss of strength in boards regularly exposed to moisture as stated by Tserki *et al.* [18]. The characteristic of the MC of the particleboard depends on the adhesive type, temperature, area exposed, content below the particles and permeability of the particles [19].

After the test pieces from each variable was submerged for 24 h, all the test pieces showed the significant difference in WA. The significant difference of WA between groups was $0.000 < 0.05$ meaning got difference in WA. Based on Table 8.1, it shows that there is no difference in WA between 100%UF and 25%UF:75%S but there is significance within the group type of adhesive in WA.

### 8.3.3.2 *Thickness Swelling*

Figure 8.5 shows the mean for TS of 100%UF (8.04%), 100%S (48.34%), 50%UF:50%S (8.74%), 25%UF:75%S (9.66%) and 75%UF:25%S (3.68%). After the 24 h submersion, there was significant difference of TS between groups $0.000 < 0.05$ which means that there is significant difference in the TS between the groups. Based on Table 8.1, it shows that there was no difference between 100%UF and 50%UF:50%S in TS. This means that both types of adhesives have the same water absorbing ability that can cause TS. However, there is no difference between 25%UF:75%S and 50%UF:50%S.

Based on the TS histogram, it shows that 100%S test pieces give the biggest value of mean compared to other type of adhesives. This is because seaweed adhesive does not cover all the particles and the amount of adhesive was less. This is because according to Handaya and Prayitno [20], the higher the amount of adhesive used, the lower the amount or value of TS. This is because the greater the amount of adhesive used, the bigger the particle surface covered with adhesive and the bonding between particles becomes stronger when pressure is applied. During exposure to water, cell walls were absorbing water which led to regaining the original shape or shape before pressing. This leads to release of compression stresses from pressing, hence swelling occurs. Halligan [21] said that regaining the original shape and straightness of particles in the presence of water caused the breakage of the glue bond, hence swelling occurs. This was what had happened to the 100%S particleboard test pieces when in contact with water. The weakness of the adhesive bonding led to the swelling occurring after 24 h submersion. As stated by Medved *et al.* [22], the interaction between moisture and the particleboard leads to the development of swelling stress that causes the failure of resin bond and the dislocation of the particles within the panel. All the changes before submersion and after submersion are shown in Table 8.2. Particleboard with 100%S has lower density

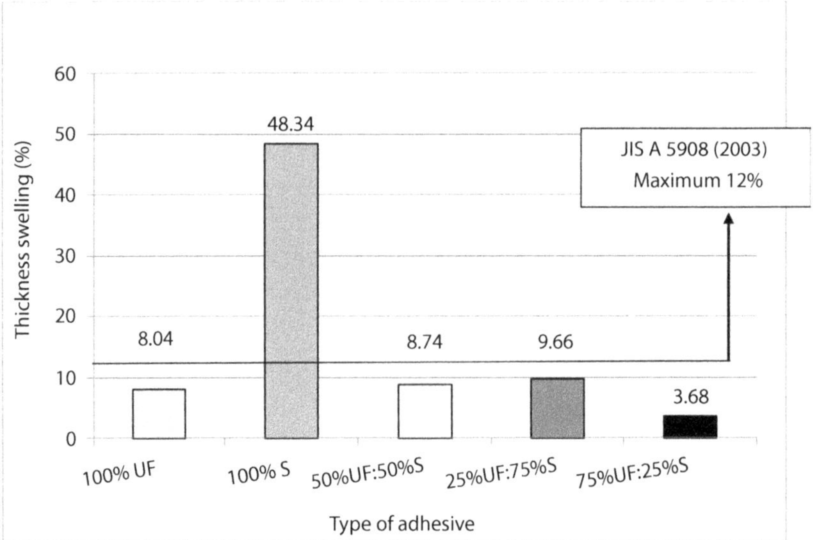


**Figure 8.5** Mean value of thickness swelling for particleboard bonded with different mixing ratios of adhesive.

compared to the other particleboards with different types of adhesive and thus affecting the TS.

Based on Table 8.2, we can see the difference in thickness of the test pieces before and after 24 h submersion. All the average thicknesses for 100%UF, 50%UF mix with 50%S, 25%UF mix with 75%S, 75%UF mix with 25%S are in the range between 5% and 12% except for 100%S which is 51.65% of average thickness. According to the standard of JIS A 5908 (2003) for particleboard, the maximum TS was 12%. So we can conclude that all the four types of adhesives except for 100%S are still in the standard range of TS.

The factor that may lead to these results was the characteristics of the seaweed adhesive itself. The viscosity of the seaweed adhesive is lower than the viscosity of UF which is 11,000 mPas and 17,000 mPas for UF.

According to Doyle [23], in order for an adhesive to bond or hold together the particles, the adhesive must wet the substrate which means that the adhesive must spread itself out to cover the surface of the particles. So in order for this to happen, the adhesive must have a low viscosity. Viscosity is temperature dependent, the application of a cold adhesive to particles or the application of adhesive to cold particles may result in poor wetting. Seaweed adhesive is a thermoplastic type of adhesive where it can be reheated after cold or in solid form but after the heating process it must be used immediately before it turned into solid form again. This is because when mixing the particles, the probability of the seaweed

**Table 8.2** The figure of test pieces before and after the 24 h submersion for water absorption and thickness swelling test.

| Type of adhesive | Before submersion (Top)/ after submersion (Bottom) | Weight (g) | Thickness (mm) |
|---|---|---|---|
| 100%UF | 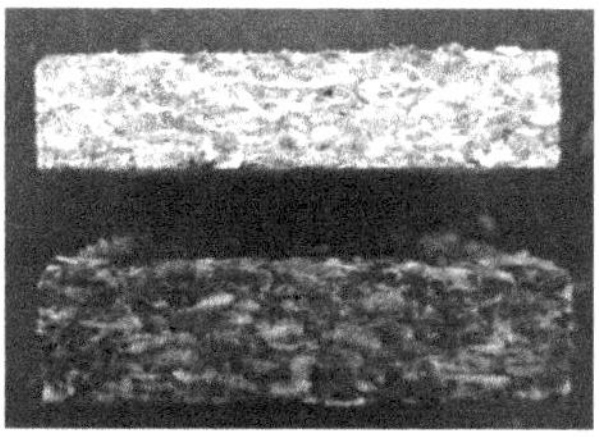 | Weight<br>B: 16.8<br>A: 26.07<br>AV: 55.17% | Thickness<br>B: 10.26<br>A: 11.66<br>AV: 7.79% |
| 100%S |  | Weight<br>B: 12.52<br>A: 33.49<br>AV: 167.49% | Thickness<br>B: 11.48<br>A: 17.41<br>AV: 51.65% |
| 50%UF:50%S |  | Weight<br>B: 16.00<br>A: 25.01<br>AV: 56.31% | Thickness<br>B: 9.73<br>A: 10.66<br>AV: 9.55% |
| 25%UF:75%S |  | Weight<br>B: 14.88<br>A: 24.93<br>AV: 67.54% | Thickness<br>B: 9.89<br>A: 10.69<br>AV: 8.08% |
| 75%UF:25%S |  | Weight<br>B: 15.34<br>A: 20.15<br>AV: 29.43% | Thickness<br>B: 9.75<br>A: 10.24<br>AV: 5.02% |

Note: B = Before, A = After, AV = Average.

adhesive to become cold is higher caused by the characteristics of the types of adhesive itself. These may lead to unfully covered surface of particles. The relative strengths of cohesive force between adhesive molecules and adhesive forces between the adhesive and particles are one of the factors in order for an adhesive to hold particles together. If the cohesive force among adhesive molecules is weaker than the adhesive force between the adhesive molecules and the surface of the particles, then the adhesive molecules will spread out over the particles and wet their surface. Thus, seaweed adhesive cannot be used alone without mixing with UF because the characteristics of the adhesive itself show that it cannot stand wet conditions. Moisture contamination is known to reduce the bond strength and is a common reason for bond failure. The characteristic of the seaweed red algae itself also contain polysaccharides which are their ability to form strong gels in aqueous solution and they may be amorphous or even insoluble in water.

### 8.3.4 The Mechanical Properties of *Acacia Mangium* Particleboard

#### *8.3.4.1 Modulus of Elasticity*

Figure 8.6 showed the mean of the MOE of 100%UF, 100%S, 50%UF:50%S, 25%UF:75%S and 75%UF:25%S were 1656.11, 1304.66, 1912.32, 1901.32 and 2682.93 N/mm$^2$ accordingly. From the mean MOE histogram, we can see that the highest mean MOE was 75%UF:25%S and the lowest mean MOE was 100%S. The sequence in ascending order was 100%S < 100%UF < 25%UF:75%S < 50%UF:50%S < 75%UF:25%S. The sequence shows the ability of the test piece to resist deformation or bending caused by the load from a different percentage and type of adhesive. From this figure, we can also see which particleboard has the higher density. The higher the density of the particleboard, the higher the MOE of the particleboard [11].

The results show that either the adhesive type or their contents present significant effect on MOE. The significant difference of MOE between groups is 0.000 and therefore there is significant difference in the MOE between the groups. Based on Table 8.1, it shows that there is no difference between 100%UF and 25%UF:75%S adhesives. The difference in MOE among the test pieces according to their types of adhesive and percentage may be caused by the difference in density, amount of adhesive in the particles, the amount of the MC with dimension and the orientation of the particles as stated by Tsoumis [24]. It was also stated by Maloney [13], that factors that affecting the particleboard properties were type of binder or adhesive, additive that being used to produce board that water resistant or

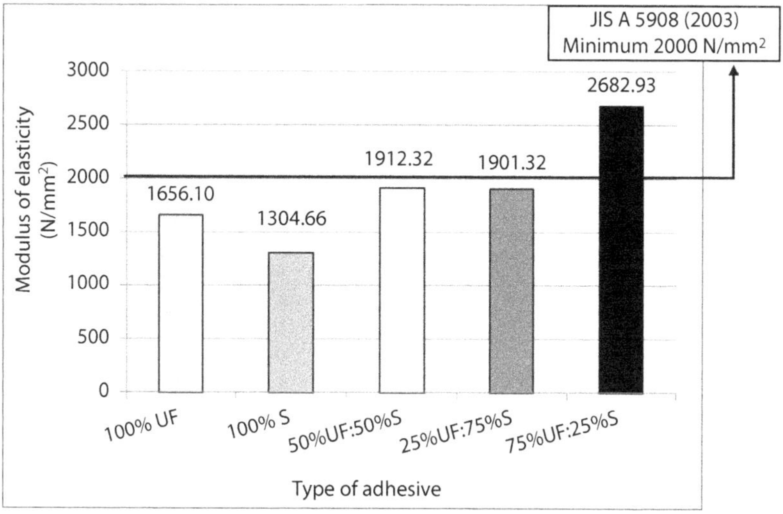


**Figure 8.6** Mean value Modulus of Elasticity for particleboard bonded with different mixing ratios of adhesive.

fire retardant, the particles alignment, the homogeneous of particles. The uneven distribution of adhesive can also affect the MOE of the test pieces. This is because, as stated by Yoesoef [25], the factors that affect the properties of the particleboard were during the process, the quality of production, mixing technique and even density of particleboard. So the uneven distribution of binder or adhesive caused by the mixing technique may affect the density and so the MOE. The MOE for all types of adhesives based on Figure 8.6 were following the standard except 100%S type because the MOE is lower and these types of adhesives are not suitable for construction panels but still can be used if there is no load applied directly on it. According to JIS A 5908 (2003) Type 8, the minimum value of MOE is 2000 N/mm$^2$.

### *8.3.4.2 Modulus of Rupture*

In Figure 8.7, the means of the MOR of 100%UF, 100%S adhesive, 50%UF:50%S, 25%UF:75%S and 75%UF:25%S were 9.79, 0.33, 11.14, 9.66 and 14.69 N/mm$^2$ accordingly. The highest mean MOR was for 75%UF:25%S and lowest for 100%S.

According to the Japanese Industrial Standard [8], the minimum MOR value for particleboard is 8.0 N/mm$^2$. The amount and type of adhesive play a significant role in improving the MOR values across the particleboard.

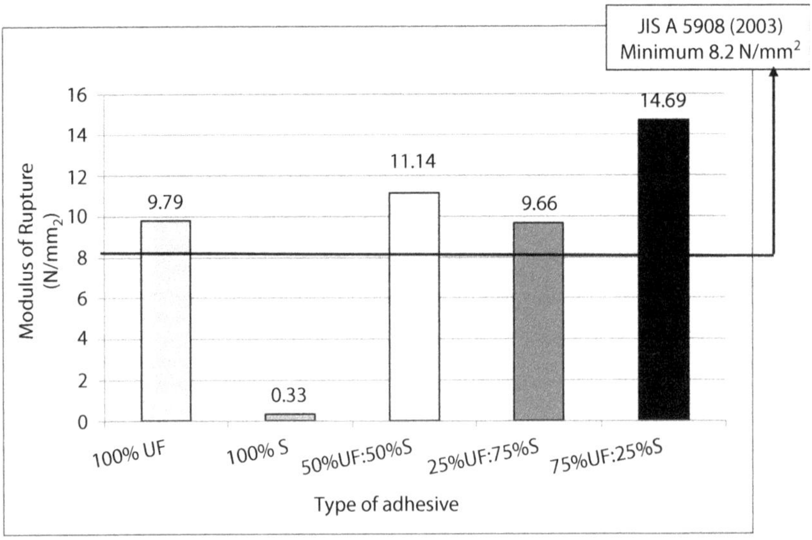


**Figure 8.7** Mean value of Modulus of Rupture for particleboard bonded with different mixing ratios of adhesive.

Adhesives can efficiently transfer and distribute stresses, thereby increasing the strength and stiffness of the particleboard. By the fact that UF has higher solid content which is 64.1%, thus as stated by Abdullah [26], the penetration of high viscosity UF adhesive probably would cause the board to withstand high load.

All four types of adhesive which were 100%UF, 50%UF:50%S, 25%UF:75%S and 75%UF:25%S adhesive had fulfilled the standard except for 100%S that had not fulfilled the standard. It was observed that the mean MOR values had significant difference with different percentage and type of adhesive. The highest mean MOR values were recorded from sample 75%UF:25%S with mean MOR 14.69 N/mm$^2$ compared to other type and percentage of adhesive. These results showed that the presence of seaweed adhesive has significant positive effect on bending strength especially on mixing UF with seaweed. The lowest mean MOR values were recorded from samples 100% seaweed adhesive with 0.33 N/mm$^2$. The significant difference of MOR between groups is 0.000. The significant probability was $0.000 \leq 0.05$ which means that there is significant difference in the MOR between the groups.

In a previous study by Kliwon and Iskandar [27], the authors used UF as the binder or adhesive and the MOR was between 117.78 and 137.00 N/mm$^2$; Korai and Lim [28] also did the same thing and the MOR was 105.50 N/mm$^2$. The higher the density, the larger the amount of particles used. Larger amount of particles were expected to reduce voids between particles in the particleboard and enhance the resistance to rupture. Since the MOR indicates the ability of a test piece to withstand a bending force perpendicular to its longitudinal axis, MOR results of the boards at higher density and resin contents were able to withstand such a force [29]. Test pieces with 100%S had lower mean MOR which may be caused by the weakness in the adhesive bonding and lower board density. The adhesive content in the board also influences the strength of the 100%S board. Seaweed adhesive has a lower solid content that caused it to not withstand the load that was applied on it. But when both seaweed and UF adhesive were blended together, it gave a better result than 100%UF. In general, it was observed that high MOR mean values were found for greater adhesive dosage applied regardless of adhesive type and density. The MOR results from the samples with high adhesive content and type had proven the fact that MOR indicates the ability of a test piece to withstand a transverse force perpendicular to its longitudinal axis. The adhesive content for use together with particles on strength performances had became an influencing factor where better adhesive content of board resulted better MOR mean values.

### *8.3.4.3 Internal Bonding*

Figure 8.8 shows the mean for IB of 100%UF, 100%S adhesive, 50%UF:50%S, 25%UF:75%S and 75%UF:25%S. The means were 0.4552, 0.00213, 0.12428, 0.15379 and 0.27401 N/mm$^2$ accordingly. The sequence of the testing in increasing order according to the mean of IB are 100%S < 50%UF:50%S < 25%UF:75%S < 75%UF:25%S < 100%UF. The significant difference of IB between groups was 0.000 where the significance probability was 0.000 < 0.05 which means that there was significant difference in the IB between the groups. The highest mean IB values were recorded from samples 100%UF and the lowest mean IB values were recorded from samples 100%S. Based on the Japanese Industrial Standard [8], the mean IB values according to that standard were minimum from 0.15 N/mm$^2$. 100%S and 50%UF:50%S did not fulfil the standard for mean IB values, thus suggesting lower bonding strength. The type and percentage of the adhesive affected the result of the IB. As we can see that seaweed adhesive was weak when seaweed adhesive was used alone but when seaweed adhesive was blended with UF, they become stronger. The viscosity of the adhesive also affected the bonding and strength. The amount and blending distribution of the adhesive can affect the IB values. Better blending and high amount of adhesive encouraged stronger interfacial bonding between particles in the boards, thus prolonging the ability of the boards to withstand the pulling force created through the test.

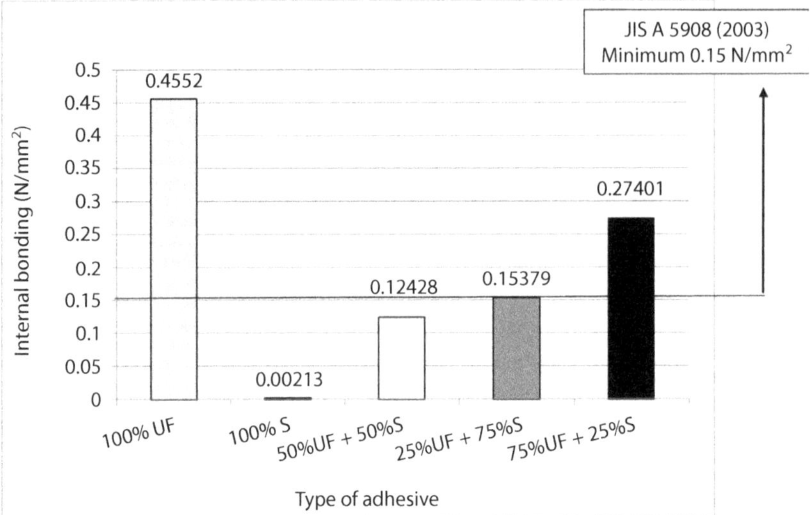


**Figure 8.8** Mean value of internal bonding for particleboard bonded with different mixing ratios of adhesive.

According to another standard [30], IB test was conducted to determine the IB strength between particles in the boards. Internal bond strength is an indication of the strength possible before breaking or split [17]. This parameter indicates how significant was the interaction performed between the particles and the adhesive used, and is directly related to all the other physical and mechanical properties evaluated in this study.

Maloney [13] said that on increasing the density of the particleboard, the particles will disintegrate during compression so that will increase the spread of adhesive per unit area, which ultimately resulted in a poor IB strength. The low value of IB may be related to the type and percentage of adhesive used which may have acted as a lack during adhesive blending and board forming. Low permeability of seaweed adhesive during adhesive application is also one of the probable influences why the adhesive had weak bonding and caused low IB values as mentioned by Youngquist *et al.* [31]. The low IB of seaweed specimens is due to adhesion and/or particle failure in the specimen.

## 8.4 Conclusion

Generally, it was found that the dimensional stability, physical and mechanical properties of the particleboard were significantly influenced by the different type of adhesive as in 100%UF, 100%S, 50%UF:50%S, 75%UF:25%S and 25%UF:75%S.

Based on this study, 100%UF and 75%UF:25%S showed high density which were 0.7373 and 0.7013 g/cm$^3$, respectively. The higher the density, the better the strength. 100%S adhesive produced a lower density board which was 0.1533 g/cm$^3$ but was still in the standard range for JIS A 5908 (2003) at 0.15 g/cm$^3$. The rate of MC in particleboard decreased with the increase of temperature and high amount of adhesive because the bonding between the particles will get stronger until water cannot enter into particles.

For WA, 100%S showed a higher mean value which was 158.55% and 75%UF:25%S showed the lowest mean value which was 28.79%. Meanwhile, for TS the highest mean value was for 100%S and the lowest mean value was for 75%UF:25%S. The higher the amount of adhesive used, the lower the amount or value of TS. This is because the greater the amount of adhesive used, the bigger the particle surface covered with adhesive and the bonding between particles becomes stronger when pressure is applied. During exposure to water, cell walls were absorbing water which led to regaining the original shape or shape before pressing. This leads to release of compression stresses from pressing, hence swelling occurs.

The mechanical tests done in this study, IB and the bending tests, showed that seaweed adhesive can be used as a binder but not for heavy duty purpose. It can be used as interior furniture. The results also show that seaweed adhesive has stronger bonding when mixed with UF compared to 100%S and 100%UF. Increasing of adhesive content, it will increases all the mechanical testing values. Seaweed adhesive also gives a nice appearance which gives a smooth surface and laminated look. The strongest board in this study is board with 75%UF:25%S since it has the highest mean value of MOR which was at 14.69 N/mm$^2$. The lowest strength was 100%S at 0.3333 N/mm$^2$.

## References

1. Michael, D.G., The seaweed site:. Information on marine algae, http://www.seaweed.ie/algae/seaweeds.php, 2000.
2. Cheryl, R.K. and Margaret, A., Seaweed culture and utilisation in Malaysia: Status, challenges and economic potential, Seminar on Developing the Seaweed Aquaculture Sector in Malaysia, Kuala Lumpur, 2009.
3. Bixler, H.J. and Porse, H., A decade of change in the seaweed hydrocolloids industry. *J. Appl. Phycol.*, 23. 321, 2011.

4. Hayashi, L., Hurtado, A.Q. *et al.*, A review of *Kappaphycus alvarezii* farming: prospects and constraints, in: *Seaweeds and Their Role in Globally Changing Environments*, A. Israel, R. Einav (Eds), pp. 255-283, Springer Dordrecht Heidelberg, London, 2010.
5. FAO, Global aquaculture production 19502009. http://www.fao.org/fishery/statistics/globalaquacultureproduction/ query/en, 2011.
6. Mather, A., *Bioadhesives: Seaweed could inspire a wet-resistant glue*, University of Canterbury, Christchurch, 2013.
7. Kayano, M., *Japanese scroll paintings: A handbook of mounting techniques*, Foundation of the American Institute for Conservation, Washington D.C., 1979.
8. Japanese Industrial Standard: Standard terminology of particleboards, JIS A 5908: 2003, 2003.
9. ASTM Standard test methods for evaluating properties of wood-base fiber and particle panel materials, ASTM D1037-99, 1994.
10. Smulski, S., *Humidity, temperature and wood moisture content*, Building Materials and Wood Technology Publication, Natural Resources Center University of Massachusetts, 1996.
11. Bowyer, J.L., Haygreen, J.G., Shmulsky, R., *Forest product and wood science*, The Iowa State University Press, Ames, 2003.
12. Widarmana, S., Panels derived from wood as a building material. Proceeding Persaki Seminar, 23–24 June, Bogor, 1977.
13. Maloney, T.M., *Modern particleboard and dry process fibreboard manufacturing*, Miller Freeman Inc., San Francisco, 1993.
14. Kelly, M.W., Critical literature review of relationship between processing parameter and physical properties of particleboard, U.S. Department of Agriculture Forest Service, Madison, 1977.
15. Wu, J.H., Particleboard. Department of Forestry, National Chung Hsing University, Taiwan, Taichung, 2014.
16. Fowkes, F.M., Attractive forces at interfaces. *Ind. Eng. Chem.*, 56, 40, 1964.
17. Haygreen, J.G. and Bowyer J.L., *Forest products and wood science: An introduction*, Iowa State University Press, Ames, 1989.
18. Tserki, V., Matzinos, P., Zafeiropoulos, N.E., Panayiotou, C., Development of biodegradable composites with treated and compatibilized lignocellulosic fibers. *J. Appl. Polym. Sci.*, 100, 4703, 2006.
19. Dhakal, H.N., Zhang, Z.Y., Richardson, M.O.W., Effect of water absorption on the mechanical properties of hemp fibre reinforced unsaturated polyester composites. *Comp. Sci. Technol.*, 67, 1674, 2007.
20. Handaya, R. and Prayitno, T.A., The factor type and amount of adhesive on particleboard properties Limbah Pasahan Sengon Wood, Proceeding Seminar Nasional VII MAPEKI, Makassar, 2004.
21. Halligan, A.F., A review of thickness swelling in particleboard. *Wood Sci. Technol.*, 4, 301, 1970.

22. Medved, S., Antonovic, A., Jambrekovic, V., Impact of resin content on swelling pressure of three layer particleboard bonded with urea-formaldehyde adhesive. *Drvna Ind.*, 62, 37, 2011.
23. Doyle, D.J., Criteria for proper adhesive selection: from application to viscosity. Society of Manufacturing Engineers Technical Paper AD90–450, 1990.
24. Tsoumis, G., *Science and technology of wood: Structure, properties, utilization*, Van Nostrand Reinhold, New York, 1992.
25. Yoesoef. Composite board. Universiti Gadjah Mada, Yogyakarta, 1977.
26. Abdullah, C.K., Impregnation of board using thermoset resins for structural applications, Universiti Sains Malaysia, 2010.
27. Kliwon, S. and Iskandar, M.I., Properties of particleboard from *Acacia mangium*. Research on Forest Product. 20. 195, 1999.
28. Korai, H. and Lim, N.P.T., Properties of *Acacia mangium* particleboard. Japan International Research Center for Agriculture, 1988.
29. Jacobs, J.A. and Kliduff, T.F., *Engineering material technology: Structure processing, properties and selection*, Prentice Hall Inc., 1994.
30. Anon., Particleboard and fibreboard – determination of tensile strength perpendicular to the plane of the board. *BS EN* 319: 1993.
31. Youngquist, J.A., Krzysik, A.M., English, E.W., Spelter, H.N., Chow, P., Agricultural fibers in composition panels, in: *Proceedings of the 27th International Particleboard/Composite Materials Symposium*, pp. 133-152, Washington State University, Washington, 1996.

9

# Green and Sustainable Textile Materials Using Natural Resources

**Pintu Pandit***, **Gayatri T.N.**, **and Saptarshi Maiti**

*Institute of Chemical Technology, Department of Fibers and Textile Processing Technology, Mumbai, India.*

***Abstract***

The challenge of developing green, sustainable materials to nullify or remediate the toxic effluents generated during the processing of textiles with conventional finishing agents has been a major concern for the textile manufacturing industry and academic research. Numerous innovative solutions have been developed to resolve these critical issues. This chapter critically reviews the different natural products derived from plant, animal and clay mineral sources such as protein, plant, starch, enzymes, different clays, chitosan, chicken feather, green coconut shell extract, banana pseudostem sap, *Sterculia foetida* fruit shell extract, *Delonix regia* stem shell waste, temple waste flower, natural essential oil, natural aroma compounds, etc. which have come into the domain of eco-friendly dyeing and functional finishing. These natural resources used to impart functional properties such as flame retardant, ultraviolet protection, antibacterial, colouration, aroma and mosquito repellency to textile materials are discussed.

***Keywords:*** Sustainable, eco-friendly, textile materials, natural resources, natural colouration, natural finishing

## 9.1 Introduction

The textile industry has to control the necessary inputs right from the raw material to the finished products and follows the 4R principle, i.e. reduce (chemical quantity), recycle (dyes/chemicals/fabrics), replace (harmful

**Corresponding author*: pintupanditict@gmail.com

Shakeel Ahmed and Chaudhery Mustansar Hussain (eds.) Green and Sustainable Advanced Materials, Vol. 1 (213–262) © 2018 Scrivener Publishing LLC

quantity) and reject (banned items), to make processing operations sustainable. Green chemistry using bio-sourced macromolecules has acquired incredible importance in the fibres and processing industry. Increasing global competition leading to rapid growth in value added textiles and green chemistry methods delivering functionality has created many opportunities for the application of newer kind of finishes. Among them, ultraviolet (UV) protection and antimicrobial textiles have a significant share in the market to fulfil the requirements of a healthy, hygienic and comfortable lifestyle. To date, numerous natural resources have been utilized for natural dyeing, antimicrobial finish, UV protection, flame retardancy, mosquito repellency and aroma finishing in textile materials and other natural polymers [1–11]. Tree species, tree part, age, season and location of the tree, are important factors affecting the chemical composition and amount of the extracts [12]. A lot of plant sources like tamarind seed coat, soya bean seed waste, flower waste from the temple and *Emblica Officinalis* G. fruit (amla) have been utilized for eco-friendly and natural dyeing [13–15]. Natural products derived from plants do not affect the natural ecological balance as their residues are easily degradable and thus it is a good alternative for control and maintenance of the ecosystem.

In the last few decades, a range of synthetic antimicrobial chemicals like triclosan, metallic salts, quaternary ammonium based products, etc. have been developed and widely used as finishing agents for making antimicrobial textile. However, these products have significant drawbacks such as their toxicity related to concentration and retention in living organisms, effluent problems, water pollution, etc. Present day consumers are more conscious about eco-friendliness in all the new generation functional textiles. Recently, there has been a lot of attention focused on producing a hygienic fabric with the products that are based on natural resources. On the basis of the sustainability parameters, the natural products have been used by many researchers for functional finishes giving value added textile materials. Herbal products such as tea tree oil, tulsi leaf, aloe vera, coconut shell extract (CSE), chitosan, *Sterculia foetida* fruit shell extract (FSE), *Delonix regia* stem shell extract (DSE) etc. [2, 4, 5, 8, 16, 17] have also been used to get the natural antibacterial effect. UV protective textile material has been reported using different plant sources such as manjistha, babool, annatto, CSE, *Delonix regia* (gulmohar) shell extract, *Sterculia foetida* FSE and ratan jot [1–4, 18–20].

Textile materials widely used in home furnishing, in hospital, in railway and in aircraft constitute a valuable sector of mass consumables. However, being cellulosic and lignocellulosic in nature, the textile material easily catches fire and thus is responsible for many serious accidents. Flame retardant textiles are increasingly in demand and numerous approaches have been taken to make textile material flame retardant. As far as the flame retardant effect of the natural textile fabric is concerned, borax and boric acid combination, phosphorous and nitrogen based condensate like di-ammonium hydrogen phosphate, urea, etc. are the most commonly used chemicals for imparting flame retardant property to textile material. Recently, researchers have also been advocating the use of nitrogen, sulphur based ammonium sulphamate for making flame retardant cellulosic and lignocellulosic textile materials [21–24]. Halogen based chemicals, also used for imparting flame retardancy by gas-phase mechanism, are however not widely used because of the release of toxic gases like polyhalogenated dioxins and furanes during processing. In the green chemistry approach, a few researchers have started exploratory work to use sustainable routes to make the materials flame retardant and revealed good research findings where they used different natural resources as flame retardant alternatives instead of synthetic chemicals [25–27]. In this direction researchers recently attempted to imparting flame retardance to the textile material using DNA, casein, whey proteins, hydrophobins, etc [28]. Starch as a source of the biomolecule was reported as a thermal stabilizing agent by the same group of researchers [29]. A research group in China has reported the use of chicken feather based protein biomolecule for thermal stability of cotton fabric [20]. Also, the use of natural green waste such as spinach leaves and banana pseudostem sap (BPS) was reported for making flame retardant finishing on textile [21–24]. Recently, Teli and Pandit [1, 2, 5], Basak and Pandit *et al.* [5, 6] have worked on the use of waste biomolecule such as green CSE, BPS, etc. and it was reported as an effective green flame retardant on cellulosic, lignocellulosic and proteinous textile material.

Clay is a fine-grained natural rock or soil material that combines one or more clay minerals with traces of metal oxides and organic matter. Geologic clay deposits are mostly composed of phyllosilicate minerals containing variable amounts of water entrapped into the mineral structure. Clays are generally plastic in nature due to the water content and become hard, brittle and non-plastic upon drying. Depending on the content of the soil, clay appears in various colours from white to dull grey or brown to deep orange-red. Clays form flat, hexagonal layered or sheet-like structure [35, 36]. Clay minerals are common weathering products and low-temperature hydrothermal alteration products. They are very common in soils, in fine-grained sedimentary rocks such as shale, mudstone and siltstone and in fine-grained metamorphic slate and phyllite. Clay finishing on textiles confers improved flame retardancy, thermal stability, peak heat release rate, fracture, and strength properties, while nano-clay finishes can be used to produce fibres that act as carriers for drugs, fragrances or other active agents and enable the controlled release of the incorporated species.

Our research group has also achieved functional properties from the waste natural plant resources such as colouration, antibacterial, UV protection and flame retardant properties using CSE, *Sterculia foetida* FSE, DSE, temple flower waste extract, etc. Thus, latest research findings of the natural resources on textile materials based on the usage of different plant, protein, starch, chitosan, BPS, CSE, FSE, stem shell extract, etc. derived natural biomolecules are critically summarized and reviewed in this chapter. Also, the review summarizes the comparison of the effects of different functional properties of natural resources using the plant, animal and clay on the performance properties of the textile materials.

## 9.2 Sustainable Colouration of Textile Materials Using Natural Plant Waste Resources

Natural colouration in textile materials mostly uses substances derived from flowers, roots, bark, etc. Waste was utilized as plant natural resources for colouration of textile materials using DSE, *Sterculia foetida* FSE, CSE and temple waste flower (TFW) on different textile materials.

### 9.2.1 Natural Dyeing with DSE on Silk Fabric

100% mulberry plain woven silk fabric having 110 ends/inch and 100 picks/inch with an areal density of 40 g/m$^2$ was used for this experiment. The plant specimen *Delonix regia* stem shell was collected from a local market of Mumbai, India (carpenter shop). Alum and myrobalan (harda) were used as mordants and the extraction procedure was as shown in Figure 9.1.

K/S values increased with an increase in DSE concentration indicating DSE has colouration properties. The efficacy of mordants in enhancing the K/S values of the dyeing was found to be in the following order: myrobalan

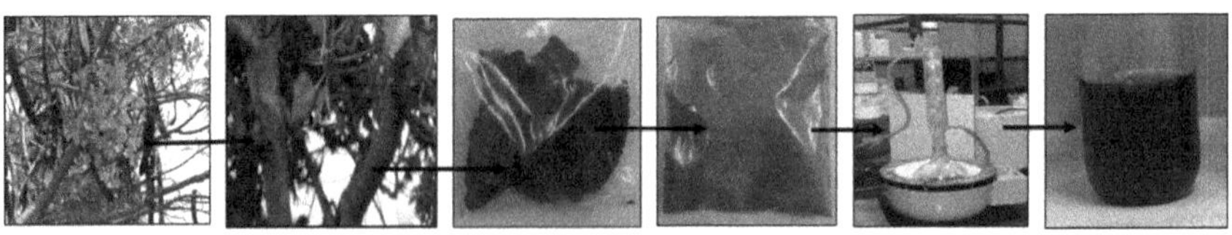

**Figure 9.1** Extraction procedure of *Delonix regia* stem shell extract.

> alum > without mordant. The highest K/S value was seen in the case of pre-mordanted silk fabric with myrobalan mordant as compared to another mordant. Maximum K/S value in this case was also seen with myrobalan as compared to other mordants. DSE in combination with alum and myrobalan (harda) mordant on silk fabric produced good improvement in colour depth (K/S) as they showed shifts in their tones resulting in a beautiful gamut of colours as compared to the dyeing obtained without using mordants. Myrobalan contains 30%–32% natural tannins and in combination with DSE, imparts enhanced colouration due to its synergistic effect. Alum mordant is well known for its ability to form coordination complexes, and they readily chelate with the dye. The colouring substance of DSE contains natural tannins, polyphenols and other phytochemical constituents. Dyeing of silk fabric with DSE is attributed to the presence of rich tannin content. Tannin contains phenolic compounds that can form hydrogen bonds with the carboxyl group of protein fibres. Additionally, there is a possibility that the anionically charged phenolic groups form an ionic bond with cationic (amino groups) in the silk protein substrate. The fastness ratings reported in terms of washing fastness of the silk fabrics dyed with DSE were good to very good (4 to 4–5) and the light fastness was of the grade good to very good (5 to 5–6) which indicates slight fading happened on silk fabric. The colour fastness to rubbing was found to be in the range of 4–5 to 5, i.e. very good to excellent, with the silk fabric dyed with or without a mordant. The good to very good fastness properties clearly indicate that dye fixed during exhaust dyeing is not just mechanically held by hydrogen bonding or van der Waals forces, but may be held by the formation of metal chelates in the presence of tannin, which was present to a significant extent in DSE. The tannins having phenolic structure contribute to the formation of metal chelate with different mordants. Hence, after mordanting, these tannins are insoluble in water, ultimately improving washing fastness [4].

### 9.2.2 Natural Dyeing of Textile Materials Using *Sterculia Foetida* Fruit Shell Waste Extract

The Latin name *Sterculia foetida* is a tropical Skunk tree, belonging to the Sterculiaceae family, with a straight trunk that can grow 50–70 feet tall, which has 2000 types of species also known as Javaolieve, Kelumpang, Bangar and Indian almond. The open segments of fruit look a lot like woody, heart-shaped bowls and are found in most of the countries. The plant specimen Sterculia fruit shell waste was collected from the Institute of Chemical Technology Campus, Mumbai, India (Figure 9.2).

**Figure 9.2** Extraction procedure of *Sterculia foetida* fruit shell [3].

Alum, myrobalan (harda) and copper sulphate were used as mordants. The colouring substances of *Sterculia foetida* fruit shell contain saponins, phenols, tannins, terpenoids, glycosides and flavonoids and the positive test results on the phytochemical constituents of the same have been reported earlier [37]. Tannins and flavonoids are considered to be very useful substances during the dyeing process because of their ability to fix dyes on fabrics. After dyeing K/S values on 100% mulberry silk fabric with an areal density of 40 g/m$^2$, cotton (80 g/m$^2$), organic cotton (60 g/m$^2$), worsted wool (188 g/m$^2$) and linen (124 g/m$^2$) fabrics are reported in Table 9.1 in terms of colour depth (K/S).

It was observed that K/S values of dyed silk fabric increased with an increase in dye concentration on mordant used and the efficacy in enhancing the K/S values of the dyeing was found to be in the order: myrobalan > $CuSO_4$ > Alum > without mordanting. The highest K/S value in the case of myrobalan and copper sulphate dyed sample with FSE was thus due to decreased values of L* (brightness) and relatively enhanced values of b*. With respect to a* (red), b* (yellow) values, FSE in combination with copper sulphate and myrobalan mordant onto silk fabrics produced good improvement and their values were positive and thus showed shifts in their tones resulting in beautiful colours as compared to the dyeing obtained without using mordant. The dyeing activity of worsted wool fabric both in the case of pre-mordanting and post-mordanting described by the enhancement in K/S values was in the order: myrobalan > ferrous sulphate > $CuSO_4$ > Alum > without mordanting for 10%, 30% and 50% dye shades. Tannin contains phenolic compounds that can form hydrogen bonds with the carboxyl group of protein fibres. Additionally, there can be possibilities involved that the anionically charged phenolic groups form an ionic bond with cationic (amino groups) in the protein substrate. It was observed from Table 9.1 that K/S values of the cotton fabric dyed with *Sterculiafoetida* FSE dye solution itself, i.e. without the use of mordant, increase with an increase in dye concentration. The mordant activity sequence in enhancing K/S values for the cotton was $CuSO_4$ > myrobalan > alum > without mordanting. FSE in combination with alum, copper sulphate and myrobalan mordant onto cotton fabric produced good improvement in colour depth (K/S). In the case of pre-mordanting, the enhancement in K/S values for the organic cotton

**Table 9.1** K/S values of dyed textile fabric using *Sterculia foetida* fruit shell waste extract.

| Type of mordants | Dye conc. (%) o.w.f. | Mulberry silk (K/S) | Cotton (K/S) | Organic cotton (K/S) | Worsted wool (K/S) | Linen (K/S) |
|---|---|---|---|---|---|---|
| Without mordant | 10 | 0.71 | 0.23 | 0.58 | 0.88 | 0.29 |
| | 30 | 1.38 | 0.32 | 0.63 | 1.32 | 0.30 |
| | 50 | 1.79 | 0.41 | 0.65 | 1.67 | 0.34 |

| | | Pre-mordanting (K/S) | | | | | Post-mordanting (K/S) | | | | |
|---|---|---|---|---|---|---|---|---|---|---|---|
| Type of mordants | Dye conc. (%) o.w.f. | Silk | Cotton | Organic cotton | Worsted wool | Linen | Silk | Cotton | Organic cotton | Worsted wool | Linen |
| Alum | 10 | 0.62 | 0.47 | 0.78 | 1.61 | 0.32 | 0.76 | 0.43 | 0.41 | 1.00 | 0.28 |
| | 30 | 1.41 | 0.71 | 0.93 | 2.30 | 0.36 | 1.73 | 0.65 | 0.67 | 1.45 | 0.39 |
| | 50 | 1.89 | 0.79 | 1.03 | 2.47 | 0.43 | 2.04 | 0.77 | 1.05 | 1.84 | 0.50 |
| Copper sulphate | 10 | 1.80 | 1.35 | 1.42 | 3.45 | 1.02 | 2.23 | 1.54 | 0.56 | 4.26 | 0.55 |
| | 30 | 4.67 | 1.57 | 2.59 | 5.58 | 1.27 | 4.87 | 3.58 | 0.83 | 5.63 | 0.90 |
| | 50 | 5.19 | 1.74 | 2.69 | 6.39 | 1.47 | 5.85 | 3.74 | 1.06 | 5.74 | 1.25 |
| Myrobalan | 10 | 4.31 | 1.11 | 1.78 | 7.35 | 0.93 | 5.37 | 1.24 | 1.59 | 6.87 | 1.30 |
| | 30 | 5.33 | 1.26 | 1.82 | 7.40 | 0.94 | 5.56 | 1.81 | 1.66 | 7.94 | 1.43 |
| | 50 | 5.48 | 1.31 | 1.87 | 9.05 | 1.23 | 5.94 | 1.88 | 1.68 | 8.00 | 2.25 |

fabric was in the following order: $CuSO_4$ > myrobalan > alum > without mordanting for 30% and 50% dye shades. The maximum K/S value was seen in the case of organic cotton fabric with copper sulphate mordant as compared to that with other mordants. In the case of post-mordanting, the sequence of enhancement in K/S values for the organic cotton fabric was myrobalan > $CuSO_4$ > alum > without mordanting for all dye shade percentages. FSE in combination with alum, copper sulphate and myrobalan mordants in organic cotton fabric produced good improvement in colour depth (K/S) and their values were in positive colour coordinates in terms of a* (red), b* (yellow) values. It was reported that the K/S values of the linen fabric dyed with *Sterculia foetida* FSE dye solution itself increased even without the use of mordant with an increase in dye concentration. However, the K/S values were, in general, quite low due to the absence of mordant. When mordant was used before or after dyeing, there was an increase in the K/S values, which is attributed to distinct chelation and complex formation of colouring compound with mordant, thus improving fixation on the fabric, giving enhanced K/S values. Different mordants, however, influenced this fixation of dye on fabric to a different extent. This is attributed to increasing the amount of colourant getting fixed on linen fabric. The mordant activity sequence in enhancing the K/S values for the linen is as follows: myrobalan > $CuSO_4$ > alum > without mordant. Copper sulphate and alum mordants are well known for their ability to form coordination complexes and to readily chelate with the dye. The colouring substance of *Sterculia foetida* fruit shell contains natural tannins and polyphenols of around 4%–5%. Thus, dyeing of the treated fabric with *Sterculia foetida* fruit extract may be attributed to the tannin richness. The washing fastness properties of treated textile fabric dyed without mordant and dyed with three different mordants was reported to be good to very good (4 to 4–5) and the light fastness was of the grade good to very good (4 to 5–6). The colour fastness to rubbing was found to be in the range of 4 to 5, very good to excellent, of the treated fabric. The tannins having phenolic structure contribute to the formation of metal chelate with different mordants. Hence, after mordanting, these tannins are insoluble in water, ultimately improving washing fastness. Thus, the overall fastness properties of treated textile materials were in the range of good to very good [3, 19, 20, 37–39].

### 9.2.3 Natural Dyeing of Textile Materials Using Green CSE

Plain woven bleached cotton (200 $g/m^2$), Hessian 100% scoured jute (281.6 $g/m^2$) and wool (120 $g/m^2$) were used as textile materials. Green coconut waste was collected from the local market at Mumbai, India. The mesocarp of green coconut was cut into pieces, and then sap was extracted out using a grinder. Original fresh CSE solution after extraction

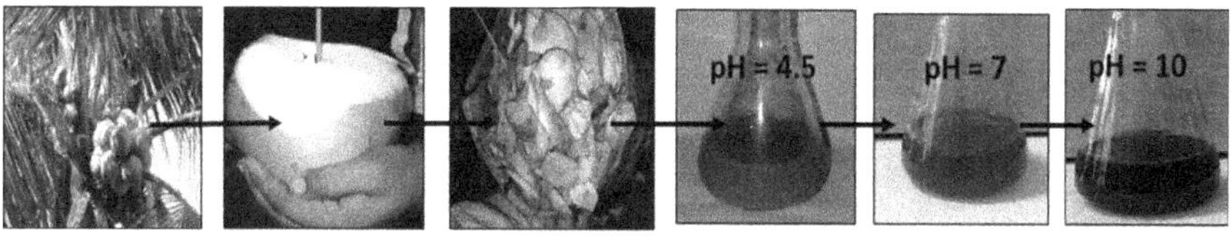


**Figure 9.3** Extraction process and colour of CSE at different pH.

was yellowish brown in colour and showed a pH value of 4.5. It was made neutral (pH 7) and alkaline (pH 10) by addition of anhydrous sodium carbonate ($Na_2CO_3$) as shown in Figure 9.3. Fresh scoured jute fabrics were then impregnated with different concentrations of CSE [original as it is extracted (CSE-A), double concentrated (CSE-B) by evaporating the extract to half of its volume] at 90 °C with material to liquor ratio of 1:15 for 60 min. Treated jute fabric was then dried in air at room temperature.

The K/S value and colour coordinates L*, a* and b* values obtained for CSE treated cotton and jute fabrics were determined at 420 nm (maximum wavelength). The results showed that there was an increase in brightness (L*) values for CSE-A as compared to CSE-B for treated samples at different pH. The highest K/S value was obtained in the case of pH 10 both for CSE-A and CSE-B. The results showed that there was an increase in brightness (L*) values for CSE-A dyed as compared to CSE-B dyed sample at different pH. When concentrated CSE-B was used, L* values decreased and hence a deeper tone was obtained (refer Table 9.2). The K/S values increased with an increase in pH and concentration as the colouring substances of CSE are the natural tannins and polyphenols present in it. Cotton and jute fabrics can be successfully coloured with CSE due to the tannin richness and the presence of polyphenols. Wash fastness ratings of the cotton and jute fabrics treated with CSE were found to be satisfactory. Light fastness was found to be good and colour rubbing fastness was found to be in the range of good to very good [2, 5].

CSE-A and CSE-B on treatment with wool fabric at pH 4.5 showed distinct colouration, and CSE-B showed 6.11% higher colour strength as compared to that of CSE-A. Colouration effect by CSE is responsible for the enhancement of K/S values (refer Table 9.2). Moreover, the bluish green colour observed when CSE was mixed with 2 mL of 2% solution of ferric chloride indicated the presence of tannin and phenols. Hence, it could be said that the colouration property of CSE may be due to its tannin richness. Tannin–protein complexation proceeds through the formation of highly crosslinked adducts at acidic pH. At low pH, reaction mechanism has been proposed in the literature, wherein quinone intermediates of the polyphenol formed adducts with the nucleophilic sites at $-COO^-$ and $-S^-$ on wool

**Table 9.2** Colour coordinates of dyed cotton, jute and wool fabrics with CSE.

| | | **CSE at its original concentration (CSE-A)** | | | **CSE concentration doubled by evaporation of water (CSE-B)** | | |
|---|---|---|---|---|---|---|---|
| **Types of fabrics used** | **Colour parameters** | **pH 4.5** | **pH 7** | **pH 10** | **pH 4.5** | **pH 7** | **pH 10** |
| Cotton | K/S | 1.05 | 6.48 | 8.76 | 1.14 | 7.95 | 11.92 |
| | L* | 71.63 | 69.74 | 73.13 | 69.48 | 69.48 | 68.05 |
| | a* | 7.64 | 15.58 | 11.74 | 9.39 | 16.60 | 13.48 |
| | b* | 17.43 | 17.21 | 20.17 | 18.21 | 18.18 | 15.10 |
| Jute | K/S | 2.57 | 3.78 | 5.47 | 17.60 | 4.98 | 9.53 |
| | L* | 60.33 | 59.41 | 60.40 | 53.72 | 60.72 | 56.94 |
| | a* | 6.12 | 9.64 | 13.33 | 14.06 | 9.06 | 15.71 |
| | b* | 19.96 | 19.84 | 22.71 | 13.21 | 22.03 | 19.84 |
| Wool (pH 4.5) | K/S | 1.31 | | | 1.39 | | |
| | L* | 80.5 | | | 81.00 | | |
| | a* | 1.74 | | | 2.52 | | |
| | b* | 11.10 | | | 11.76 | | |

protein. Non-covalent bonding between wool protein and polyphenols involves hydrogen bonds that are formed between electronegative atoms of nitrogen or oxygen, especially of hydroxyl (–OH) groups, amino (–$NH_2$) and a positively charged hydrogen atom from neighbouring hydroxyl or amino group of another polyphenol or protein molecules (wool in this specific case). In addition to hydrogen bonds that involve polar groups, protein and polyphenols may interact via hydrophobic, non-polar aromatic rings of polyphenols and aromatic amino acids (proline, phenylalanine, tyrosine, tryptophan, histidine). Wash fastness ratings of the wool fabrics dyed with only CSE-A and CSE-B were found to be good to very good. The rubbing fastness was also found to be very good to excellent. Light fastness was found to be good for treated wool fabric [1].

### 9.2.4 Colouration of Textile Materials Using Resources from Temple Flower Waste

Natural dyes are colourants derived from plants, insects or minerals. A survey report reveals that 40% of the total productions of the flower is unsold and wasted everyday, which is thrown in water or dumped and which also creates water pollution. Here in this survey, an attempt was made to isolate natural dyes from the flowers of hibiscus and marigold and apply them on cotton and cotton/silk blend fabrics with the help of different mordants like alum, harda and ferrous sulphate. Cotton and cotton/silk samples were dyed with yellow and orange marigold in fresh, dry and pulverized form. The depth of shade of marigold was studied using K/S values. Dry petals gave the highest colour yield as the K/S values obtained on the fabric were maximum. It was also confirmed that alum as metal mordant gave brighter shades with both yellow and orange marigold flowers, while harda and $FeSO_4$ gave darker and duller shades. Hibiscus flowers which were obtained as temple waste have also good colouration properties. The fastness properties of such dyed fabrics were found to be well within the accepted limits [40].

## 9.3 Sustainable Antibacterial Finishing of Textile Materials Using Natural Waste Resources

### 9.3.1 Antibacterial Activity of *Delonix Regia* Stem Shell Waste Extract on Silk Fabric

Silk fabric dyed with DSE inherently showed very good antibacterial properties both against *Staphylococcus aureus* and *Escherichia coli*. The percentage (%) reduction in bacteria increased on the application of different mordants. To test the durability to home laundering, the antibacterial activities of post-mordanted and without mordanted dyed silk fabrics with 30% shade were tested after 5 and 10 washes. It was seen that the highest colonies reduction percentage was observed in the case of myrobalan followed by alum mordants against *E. coli*. The reduction percentage in colonies against *S. aureus* and *E. coli* was more than 90%, even after 10 washes in the case of samples mordanted with alum and myrobalan of the dyed sample. Colouring component in DSE has strong potential for antibacterial activity not only in the presence of two selected mordants but also when no mordant was used. The observed antibacterial property of the treated silk fabric might be due to the presence of tannins and these polyphenolic compounds that bind to the proline-rich protein. Also, different metal ions (sulphur, phosphorus. magnesium, calcium, copper, chromium, iron) present in the DSE (as observed from energy dispersive x-ray spectroscopy (EDS) and Atomic Absorption Spectrophotometer (AAS) analysis), and the presence of tannin, terpenoid, saponin biomolecules jointly may rupture the negatively charged bacterial cell wall, finally killing the bacteria and providing antibacterial property. The antibacterial properties of *Delonix regia* bark extract may also be due to the presence of flavonoids, alkaloids, and phenolic compounds [4].

### 9.3.2 Antibacterial Textile Materials using Natural *Sterculia foetida* Fruit Shell Waste Extract

The quantitative analysis of the percentage reduction in gram-positive (G+, *S. aureus*) and gram-negative (G–, *E. coli*) bacteria was done with *Sterculia foetida* fruit shell waste extract (FSE) dyed without and with three different mordants of alum, copper sulphate and myrobalan. The results clearly indicate that the treated fabric dyed with FSE inherently showed excellent antibacterial properties both against *S. aureus* and *E. coli*. The percentage reduction in bacteria increased on the application of different mordants. It can be seen that the highest colonies reduction percentage was observed in the case of copper sulphate mordant followed by myrobalan and alum mordants against *E. coli* (refer Table 9.3). After 5 and 10 washes, this level of colonies reduction percentage gradually decreased. The reduction percentage in colonies against *S. aureus* was still very good for silk fabric even after 10 washes in the range of 93.87% to 97.33% as compared to *E. coli* reduction percentage in the range of 82.63% to 97.63%. The antibacterial property of the treated textile fabric might be due to the tannins, present in FSE which was found to be 4.71%, and these polyphenolic compounds

**Table 9.3** Reduction percentage of *E. coli* and *S. aureus* with FSE on dyed textile fabric.

| | *E. coli (G–)* | | | *S. aureus (G+)* | | |
|---|---|---|---|---|---|---|
| | **Colonies reduction (%)** | | | **Colonies reduction (%)** | | |
| **Types of mordants** | **Silk** | **Cotton** | **Organic cotton** | **Silk** | **Cotton** | **Organic cotton** |
| Without mordant | 97.45 | 97.37 | 87.82 | 99.91 | 96.50 | 82.63 |
| Alum | 98.64 | 97.66 | 86.52 | 100 | 96.84 | 84.95 |
| Myrobalan (Harda) | 99.82 | 98.65 | 94.78 | 99.9 | 97.91 | 93.05 |
| Copper sulphate | 99.90 | 99.26 | 97.82 | 100 | 98.81 | 95.37 |

FSE: fruit shell extract; *E. coli*: Escherichia coli; *S. aureus*: Staphylococcus aureus.

that bind to proline residues in silk have been shown to have antibacterial activity. The presence of flavonoids in the extract, as well as terpenoid and saponin, was also detected in FSE, out of which terpenoid acts as an antibacterial agent against both *S. aureus* and *E. coli* whereas saponins which are glycosides provide inhibitory effects on *S. aureus* bacteria. The EDS analysis also reported that FSE contains metal ions such as sodium, silicon, magnesium, calcium and potassium which might be responsible for getting good antibacterial property even after 10 washes for without and with three different mordant dyed textile fabric [19, 37–39].

### 9.3.3 Antibacterial Textile Materials Using Waste Green CSE

Antibacterial activity of untreated and CSE treated cotton fabric at different pH was quantitatively assessed to determine the percentage reduction in bacterial populations in liquid media for *E.coli* and *S. aureus*. Antibacterial activity of untreated and treated fabrics at different pH was qualitatively assessed (AATCC 147) to determine the zone of inhibition (mm) in the presence of *E. coli* and *S. aureus* media and results are shown in Table 9.4. Jute fabric treated with CSE showed good antibacterial properties against *S. aureus* and *E. coli*. Bacterial reduction percentage in the treated fabric for all pH is more than 96% (refer Table 9.4). Cotton and Jute fabrics treated with CSE showed very good antibacterial properties both against *S. aureus* and for *E. coli* having more than 90% bacterial reduction. CSE solution has antibacterial property, attributed to the presence of saponin, phenol, tannin, terpenoid, glycoside, flavonoids, etc. Also, various metallic salts and metal oxides also assist in enhancing the antibacterial properties of the

**Table 9.4** Antibacterial activity of *E. coli* and *S. aureus* on CSE treated fabric.

| Effect of CSE on different pH | Qualitative test (AATCC 147) | | Quantitative test (AATCC 100) | | | |
|---|---|---|---|---|---|---|
| | Diameter of zone of inhibition (mm) | | Bacterial reduction % in the treated cotton fabric | | Bacterial reduction % in the treated jute fabric | |
| | *S. aureus* | *E. coli* | *S. aureus* | *E. coli* | *S. aureus* | *E. coli* |
| pH 4.5 | 9 | 7 | 98.54 | 98.71 | 96.48 | 96.78 |
| pH 7 | 9 | 8 | 99.48 | 99.31 | 97.55 | 97.56 |
| pH 10 | 11 | 10 | 99.78 | 99.82 | 98.13 | 98.52 |

CSE. Tannin-like polyphenolic compounds are reported to have antibacterial activity. Flavonoids are hydroxylated polyphenolic compounds known to be produced by plants in response to microbial infections. Terpenoids although mainly used for their aromatic qualities have also been found to be a potential agent for inhibiting bacteria. Saponins, which are glycosides, have been found to have inhibitory effects on the gram positive organism. Green coconut shell (mesocarp) is rich in different types of plant secondary metabolites especially antioxidants, which are essential for microbial infection resistance to protect the coconut from biotic attack. Jute and cotton fibre have a strong ability to form chelating complexes with metal ion due to the presence of –OH and –COOH groups. Hydroxyl (–OH) group of jute and cotton fibre will form hydrogen bonding and minor van der Waals interaction with tannin. Moreover, metal ions, tannin and the hydroxyl group of the jute and cotton fibre form strong coordinate bonding between them on treatment with metallic salts. Besides, the large insoluble metal–tannin complex molecules are responsible for imparting antibacterial effect when they are inside the jute fibre [2, 5].

## 9.4 Sustainable UV Protective Textile Materials Using Waste Natural Resources

### 9.4.1 UV Protective Silk Fabric Using DSE

UV radiation is one of the primary causes of degradation of textile materials due to its very large surface volume ratio. Even though silk has poor swelling

**Table 9.5** Ultraviolet protection value of silk fabric dyed with DSE for post-mordanting.

| | **Ultraviolet protection factor (UPF) (290–400 nm)** | | |
|---|---|---|---|
| **Types of mordants used** | **Initial (no wash)** | **After 5 washes** | **After 10 washes** |
| Without mordant | 19.94 | 17.05 | 15.51 |
| Alum | 20.54 | 17.68 | 16.68 |
| Copper sulphate | 21.69 | 20.93 | 19.94 |
| Myrobalan | 29.38 | 25.36 | 23.41 |

properties, its very fine nature and a greater number of fibres in the cross-section of yarn result in higher swelling due to capillary absorption, and in turn less UV transmittance. The UV protection factor (UPF) of untreated silk fabric was 3.25, which has no protective abilities and causes transmission of solar radiation. There is an improvement in the UPF rating of silk fabric after dyeing with DSE in the case of without mordant and also with two different post-mordanted silk fabrics on 30% shade depth. Analysis of results reveals that silk fabric mordanted with myrobalan showed very good UPF rating. Maximum UV protection was found in the case of fabric mordanted with myrobalan. Even after 10 washes, DSE treated silk fabric showed good UPF property not only with two different mordants but also for without mordant (refer Table 9.5). Improvement in UV protection is mainly attributed to the synergistic effect of mordant and DSE used as natural dyes. Deep colour dyeing due to the presence of tannin also helps to increase the UPF property. Moreover, good to very good UPF property after dyeing with DSE might be due to the very thin coating layer on the dyed silk fabric as verified in the scanning electron microscopy (SEM) image [4].

### 9.4.2 UV Protective Textile Materials Using *Sterculia Foetida* FSE

Finishing treatments given to the fabrics to reduce swelling reduce the transmittance of UV rays. The UV protection abilities of the textile materials are considerably influenced depending upon the type of dye or pigment, the absorptive groups present in the dyestuff, depth after dyeing, the uniformity and additives. Dyes extracted from various natural resources also show the UPF within the range of 15–45 depending on the mordants used [7]. Untreated fabric, except worsted wool as reported in Table 9.6 has no protection from UV radiation. There is an improvement in the UPF rating

**Table 9.6** Ultraviolet protection value of textile fabric dyed and post-mordanted with FSE.

| | **Ultraviolet protection factor (UPF) (290–400 nm)** | | | |
|---|---|---|---|---|
| **Mordants used** | **Silk** | **Worsted wool** | **Linen** | **Organic cotton** |
| Untreated fabric | 3.14 | 81.71 | 3.21 | 2.97 |
| Without mordant | 22.89 | 218.60 | 8.37 | 8.64 |
| Copper sulphate | 22.64 | 651.37 | 21.67 | 15.51 |
| Alum | 16.74 | 716.42 | 15.51 | 11.92 |
| Myrobalan | 31.64 | 2355.82 | 25.42 | 17.68 |

FSE: Fruit shell extract.

of textile fabric after dyeing with FSE both in the case of without mordanting and mordanted with alum, myrobalan and copper sulphate after post-mordanting with 30% shade depth. It was also reported that mordanting of silk fabric with myrobalan and copper sulphate showed good UPF rating remaining mostly unaffected even after 10 washes. Maximum UV protection was found in the case of fabric mordanted with myrobalan. Improvement in UV protection is mainly attributed to the synergistic effect of mordant and *Sterculia foetida* FSE as a natural dye. Deep dyeing due to the presence of tannin also helps to increase the UPF property. EDS analysis showed the presence of elemental peaks of silicon, magnesium, calcium, potassium, sulphur, phosphorus and chlorine on treated fabric with FSE which also supports the UPF property of silk fabric [3, 19, 20, 38].

### 9.4.3 UV Protective Textile Materials Using Waste Green CSE

Ultraviolet protection abilities of textile materials are influenced by the functional groups (light absorbing) present in the dye or pigment, shade intensity, and additives. The UPF of untreated jute fabric was 4.64, not showing any protective abilities from solar radiations. There is an improvement in the UPF rating of jute fabric after application of CSE as the pH and concentration increased. Analysis of results showed a good UPF rating at pH 10 both in the case of CSE-A and CSE-B. The UPF of the untreated wool fabric of 20.38 showed inherently good ultraviolet protective abilities from transmission of solar radiation. There was an excellent improvement in the UPF rating of wool fabric after application of CSE-A and CSE-B at

**Table 9.7** Ultraviolet protection property of CSE treated jute and wool fabrics.

| | Jute CSE-B | | | | Wool CSE-A and CSE-B | | |
|---|---|---|---|---|---|---|---|
| **Parameters** | **Untreated** | **pH 4.5** | **pH 7** | **pH 10** | **Untreated** | **pH 4.5 CSE-A** | **pH 4.5 CSE-B** |
| UV-A (315–400 nm) | 23.35 | 9.09 | 8.40 | 6.47 | 14.11 | 1.55 | 0.94 |
| UV-B (290–315 nm) | 21.33 | 9.31 | 7.25 | 6.13 | 3.64 | 0.80 | 0.51 |
| UPF (290–400 nm) | 4.64 | 10.75 | 13.71 | 16.23 | 20.38 | 116.61 | 189.47 |

pH 4.5. It was observed that the UPF rating increased with the increase in the concentration of CSE-B. CSE behaves as a natural dye, and it imparts deeper colour when applied in CSE-B form as compared to CSE-A (refer Table 9.7). The UPF property increased with a deep colouration which might be due to the richness in tannin in the case of CSE-B. Moreover, excellent UPF property after treatment with CSE-A and CSE-B on jute and wool may be attributable to a coated layer formed on the dyed fabric as shown in the SEM image. EDS analysis also showed the presence of elemental peaks for sodium, potassium, magnesium, silicon, phosphorus, sulphur, calcium, etc. on the dyed jute and wool fabrics, which supports the improvement in the UPF property of the fabric. Polyphenols (phytochemical analysis) and anthocyanins (UV visible spectroscopy analysis) have been reported as potential effective agents to prevent UV radiation [1, 5].

## 9.5 Sustainable Green Flame Retardant Textile Materials Using Natural Resources

Horrocks and Bourbigot *et al.* have reported the thermal degradation and flammability behavior of the cellulosic cotton, wool and other synthetic textiles in detail [41–43]. Among the various functional finishing of textile material, eco-friendly flame retardant finishing is important as it protects from a major human health hazard.

In recent years, research in the field of green flame retardancy in textile materials has geared up to find a sustainable approach to reducing the carbon footprint. Efforts have been made, especially in the last 5–7 years, to use different eco-friendly biomolecules extracted from a plant (sap and grain based) and animal (protein and polysaccharide based) natural resources for making the textile materials thermally stable as shown in Figure 9.4. The major meaningful findings of recent research work in the area of natural flame retardant resources for application in textile materials are discussed in the following sections.

### 9.5.1 Flame Retardancy Imparted by Plant Based Waste Natural Resources

As some of the plant extract and wastage plant products contain phosphorous and other positive metal ions, inorganic salt biomolecules, metal oxides, etc., they can be utilized for flame retardant finishing of cellulosic, lignocellulosic and protein fibre based textile materials. In the recent years, flame retardancy effect with plant based biomolecules has been found with CSE (coconuts aqueous shell waste), BPS (an agricultural waste), spinach juice (SJ – a vegetable extract), starch (S – plant seed and modified stem extract) and pomegranate rind extract (PRE – fruit waste) applied mostly on cellulosic, lignocellulosic and proteinous textile materials.

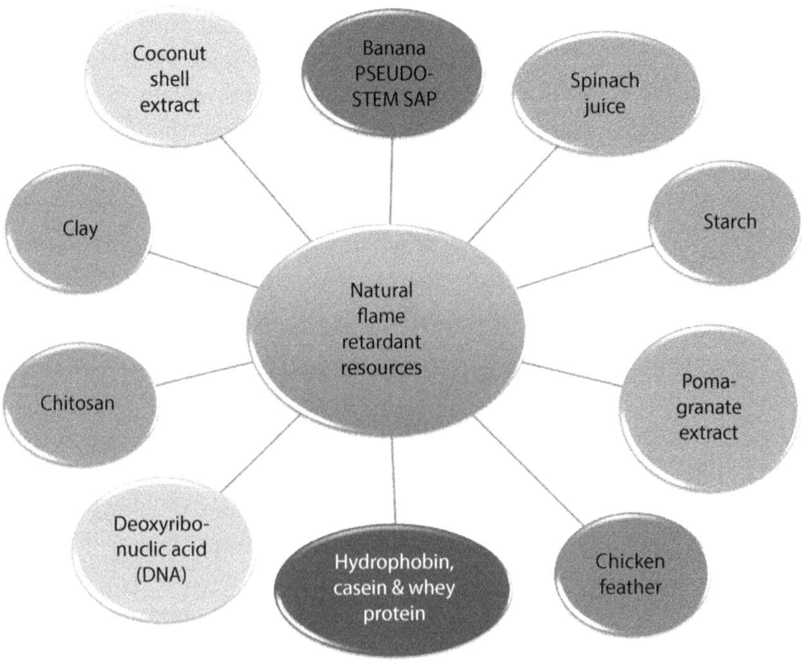


**Figure 9.4** Natural flame retardant resources for textile material.

### *9.5.1.1 Flame Retardant Textile Materials Using Green CSE*

The coconut tree is a very useful plant for its versatile uses in daily life, mostly found in the sea-shores and gardens as an ornamental plant. The green coconut water is consumed as a delicious, nutritional and refreshing natural drink. The green coconut shell waste was collected from a local market at Mumbai, India. The mesocarp of green coconut was cut into pieces, and then sap was extracted out using an extractor. Original fresh CSE solution after extraction appeared yellowish brown in colour and showed the pH value of 4.5. It was made neutral (pH 7) and alkaline (pH 10) by the addition of anhydrous sodium carbonate ($Na_2CO_3$) as shown in Figure 9.5. Cellulosic, lignocellulosic and proteinous fabrics were then impregnated with different concentrations of CSE [original as it is extracted (CSE-A), double concentrated (CSE-B) by evaporating the extract to half of its volume] at 90 °C with material to liquor ratio of 1:15 for 60 min. Treated cotton and jute fabrics were then dried in air at room temperature. Subsequently, samples were conditioned for 48 h at 65 ± 5% relative humidity at 27 ± 2 °C, before testing.

The limiting oxygen index (LOI) and flammability values of the control and the CSE treated cotton fabrics at different pH with different concentrations are given in Table 9.8. After application of the CSE, the LOI value was found to increase significantly. When the fabrics were treated with CSE-A, the LOI value was found to increase to 28.8%, 44.4% and 66.7% higher than that obtained for the control sample. When the CSE-B was

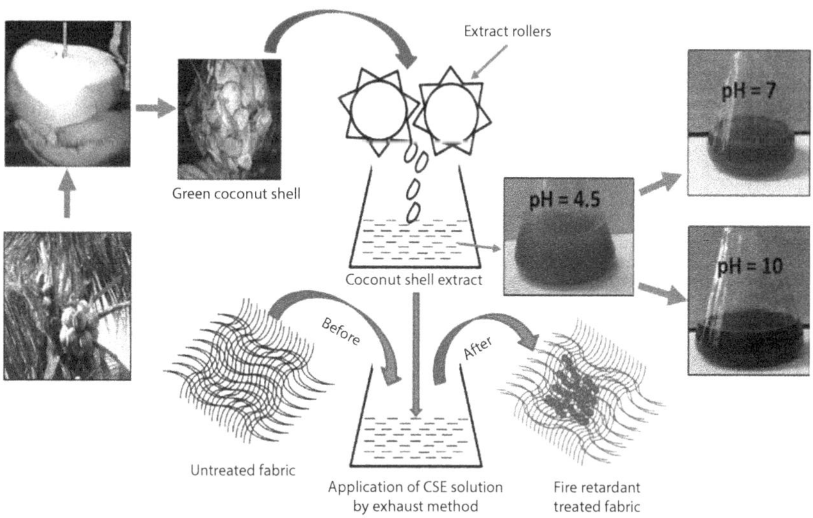


**Figure 9.5** Extraction process and colour of coconut shell extract (CSE) at different pH.

**Table 9.8** Vertical flammability test of the CSE treated cotton, jute and wool fabrics.

| **Textile fabric** | **Cotton** | | | | | | | **Wool** | | |
|---|---|---|---|---|---|---|---|---|---|---|
| Flammability parameters | C | pH 4.5 | | pH 7 | | pH 10 | | C | pH 4.5 | |
| Concentration of solution | – | CSE-A | CSE-B | CSE-A | CSE-B | CSE-A | CSE-B | – | CSE-A | CSE-B |
| Add-on (%) | – | 4.12 | 4.98 | 4.78 | 5.19 | 5.14 | 5.97 | – | 6.89 | 10.9 |
| LOI (%) | 18 | 23 | 25 | 26 | 28 | 30 | 31 | 25 | 30 | 36 |
| Burning with flame time (s) | 60 | 94 | 118 | 189 | 38 | 12 | 4 | 90 | 0 | 0 |
| Burning with afterglow time (s) after flame stops | 0 | 0 | 0 | 0 | 244 | 838 | 975 | 3–4 | 0 | 0 |
| Char length (mm) | Nil | Nil | Nil | Nil | Nil | Nil | Nil | Nil | 80 | 50 |
| **Textile fabric** | **Jute** | | | | | | | | | |
| Flammability parameters | C | pH 4.5 | | pH 7 | | pH 10 | | | | |
| Concentration of solution | – | CSE-A | CSE-B | CSE-A | CSE-B | CSE-A | CSE-B | | | |
| Add-on (%) | – | 4.92 | 6.15 | 5.18 | – | 6.25 | 9.18 | | | |
| LOI (%) | 21 | 25 | 27 | 27–28 | 21 | 31 | 38 | | | |
| Burning with flame time (s) | 30 | 40 | 26 | 25 | 30 | 8 | 0 | | | |
| Burning with afterglow time (s) after flame stops | 60 | 130 | 268 | 212 | 60 | 281 | 30 | | | |
| Char length (mm) | – | – | 122 | – | – | 50 | 12 | | | |

CSE: Coconut shell extract; C: Control.
Flame contact time = 12 s. Sample size = $(16 \times 4)$ $cm^2$ for cotton; $(25 \times 4)$ $cm^2$ for jute; $(30 \times 7.6)$ $cm^2$ for wool.

used, the LOI value increased by a magnitude of 38.9%, 55.6% and 72.2% as compared to that for the untreated jute. It may be noted that in all the CSE treated samples, as the LOI value increased significantly, the samples did not get ignited. Control jute fabric showed an LOI value of 21, reflecting easy flammability of the substrate. Any polymeric material showing an LOI value of equal or more than 26 can be considered to be flame retardant. When jute fabrics were treated with CSE at its initial concentration (CSE-A), the LOI value was found to increase to 19%, 33% and 48% higher than that obtained with the untreated sample. This is attributed to increase in swelling of jute as pH varied from acidic to alkaline pH, which consequently increased the add-on percentage of CSE and the LOI value. It was found that with the CSE-B the LOI value increased for all the cases at different pH. This is because as the add-on percentage increased, it caused a further increase in the LOI to 27, 34 and 38 for acidic, neutral and alkaline pH, respectively. Control wool fabric showed the oxygen index value of 25, which means it has some inherent flame retardant properties. However, the wool polymer cannot sustain prolonged high-temperature heating as it fails in the vertical flammability test. After application of the CSE-A and CSE-B, LOI values of the treated wool samples were found to increase to 30 and 36, respectively. This increase in LOI was about 20% in CSE-A and 44% in CSE-B when compared with that of untreated wool. These results indicate that the untreated wool fabric got burnt with flame within 90 s, whereas CSE-A and CSE-B treated wool fabrics showed no flame and no afterglow giving 8 and 5 cm char lengths, respectively.

The presence of alkaline metal ion containing inorganic salt molecules was observed in the ATR-FTIR (attenuated total reflection–Fourier transform infrared spectroscopy) analysis of CSE and in the peaks of the elements in the EDS analysis of CSE. It has been found that the amount of phosphorus and silicon is more in the alkaline CSE treated cotton and jute fabrics, and they showed greater thermal stability rather than the acidic CSE treated material. As these phosphates and chlorinated products can act as flame retardants for the lignocellulosic and cellulosic substrate, the flame retardancy effect of the CSE treated cotton, and jute fabric is attributed to the combined synergistic effect of the presence of different metallic salts, silicate, phosphate, nitrate and bound and unbound water molecules and various phytochemicals. Their combined presence has helped in the formation of more char and non-flammable gases like $CO_2$, $H_2O$, etc. as observed from the thermogravimetric (TG) analysis. Physical observation showed a thick intense black colour char mass formation after burning, earlier dehydration. SEM analysis indicated voluminous foamy nature of CSE. It can be thus said that CSE worked on the condensed phase intumescent mechanism. Flame retardancy property in the case of the CSE treated wool fabric was due to the presence of phosphate, sulphate, silicates, metal salts especially sodium and potassium chloride, etc. The increase in thermal stability for CSE-B treated wool was supported by the lower mass loss as compared to that of untreated wool. The flame retardant acts by altering the initial stage of wool pyrolysis favourably. The mechanism of flame retardancy proceeds by a decrease in the wool polymer volatilization causing char formation as well as the formation of a protective layer against the heat of the flame and retarding diffusion of combustible volatile compounds to the flame [1, 2, 5].

### *9.5.1.2 Flame Retardant Textile Materials Using BPS*

BPS is extracted from the pseudostem of the banana tree (*Musa cavendish*). It looks like colourless water immediately after extraction. However, with time it turns light khaki, natural colour due to the oxidation of the aromatic phenolic rings present in it. Under alkaline condition, the natural khaki colour of BPS becomes darker, and under acidic condition, it tends to become colourless [33, 44].

BPS was applied to pre-mordanted (5% tannic acid and 10% alum) cellulosic cotton textile at 90 °C by the exhaust method in different pH conditions [34]. On the other hand, the same BPS solution was applied in jute textile in different concentrations [33]. It is observed that BPS pick-up by the cellulosic cotton fabric results in a greater LOI value and a lower burning rate as shown in Table 9.9 in the vertical flammability test [34]. As far as flame retardancy is concerned, BPS treated jute fabric showed more thermal resistant behavior than the BPS treated cotton fabric as denoted by specific char length after the vertical flammability test.

It was confirmed from both the cone and bomb calorimeter results that the BPS treated cotton fabric produced less amount of heat energy compared to the control fabric during burning. This is attributed to the endothermic coating of the BPS which absorbs heat energy during burning [32, 34]. Concerning char morphology (char after the vertical flammability test) of the BPS treated textile, control cellulosic polymer showed bird nest–like open char structure through which flammable gas can easily flow out. However, BPS treated cellulosic polymer showed closed cell pockets where flammable gas cannot flow properly and cannot come in contact with the flame source. Since a flame represents a gas-phase combustion process, both the control and all the BPS treated wool fabrics (at different pH values) were tested in a vertical condition against the premixed flame of a Bunsen burner. The control wool fabric was found to burn with a flame

**Table 9.9** Different vertical flame retardant parameters for BPS treated cotton, jute and wool.

| Fabric | Cotton | | Jute | | Wool | | | |
|---|---|---|---|---|---|---|---|---|
| **Flammability parameters** | **C** | **BPS treated** | **C** | **Conc. BPS treated** | **C** | **BPS treated (pH = 4.5)** | **BPS treated (pH = 5.5)** | **BPS treated (pH = 7)** |
| Add-on (%) | – | 4.5 | – | 3 | – | 3 | 4 | 5 |
| LOI | 18 | 30 | 21 | 33 | 25 | 29 | 31 | 35 |
| Burning with flame time (s) | 60 | 4 | 60 | 5 | 90 | 110 | 125 | 0 |
| Burning with afterglow time (s) after flame stops | 15 | 900 | 10 | 600 | No | No | No | 2–3 |
| Char length (mm) | Nil | Nil | Nil | Nil | Nil | Nil | Nil | 6 |

BPS: Banana psedustem sap; C: Control.

within 80 s, whereas the neutral pH BPS treated wool fabric showed no flame and no afterglow, but a 6 cm char length. However, acidic pH (4.5 and 5.5) BPS treated wool fabric samples were burnt with a black colour, forming a heavy char mass [6]. High thermal stability of the BPS based biomolecule was attributed to the presence of various inorganic metals, metallic salts, metal oxides, phosphate, phosphite, aromatic phenolic OH groups etc. which is evident from the EDS, secondary ion mass spectroscopy (ToF-SIMS) and x-ray fluorescence (XRF) analysis [32, 33].

#### 9.5.1.3 Flame Retardant Textile Materials Using SJ

SJ is extracted from the fresh green leaves of the spinach plant (*Spinaciaoleracea*). SJ looks deep green in colour, and its pH is neutral immediately after extraction due to the presence of chlorophyll, flavonoids, etc. However, with time the deep green colour of SJ turns into blackish green in open condition at room temperature which is attributed to the photochemical degradation and air oxidation of the flavonoids and chlorophyll molecules [31].

SJ is composed of different nutrients, nitrogen, silicon, protein, potassium, calcium, sodium, vitamin K, vitamin A, vitamin B complex, antioxidant, polyphenolic aromatic rings, etc. [45]. Conventionally, it is used as a food material and depending on its chemical composition, it could find application in the medical sector. Thermal stability of dry SJ powder is attributed to the presence of silicate, iron, potassium and other inorganic salts as observed from the EDS and FTIR analysis of dry SJ powder [31]. Flame retardant functionality was imparted to the cellulosic cotton textile by waste spinach (*Spinaciaoleracea*) leaves extracts. Extracted green colour SJ was made alkaline by adding soda ash and applied to the bleached cotton fabric by exhaust method (at an elevated temperature for 30 min). Cotton fabric treated with SJ (as it is) showed an LOI value of 30 and did not catch flame (Table 9.10). It was reported that unlike cotton fabric, SJ treated fabric contains sodium, potassium, magnesium and silicon (confirmed by EDS analysis). These elements are present in the SJ in the form of inorganic metallic salts or oxides. FTIR analysis of the SJ also confirmed the presence of silicate [31].

#### 9.5.1.4 Flame Retardant Textile Materials Using Starch

Starch is a carbohydrate, obtained from most green plants, and consists of a large number of glucose units joined by glycosidic bonds. The common sources of starch are staple grain foods, potatoes, wheat, maize, rice,

**Table 9.10** Flammability details of the spinach juice (SJ) treated cotton fabric.

| Flammability parameters | Control | SJ treated (1:0) | SJ diluted (1:1) | Mordanted + SJ treated |
|---|---|---|---|---|
| Add-on (%) | – | 8 | 5.7 | 8 |
| LOI | 18 | 30 | 28 | 25 |
| Burning with flame time (s) | 60 | 0 | 4 | 100 |
| Burning with afterglow time (s) after flame stops | 15 | 345 | 400 | 100 |
| Total burning time (s) | 75 | 345 | 404 | 100 |
| Char length (mm) | Nil | Nil | Nil | Nil |
| Observed burning rate (mm/min) | 200 | 37.5 | 43.5 | 150 |

Vertical flammability: 12 s flame; sample size = (25×4) $cm^2$.

sorghum and cassava. Pure starch is colourless, a white powder composed of linear chain crystalline amylase (20%–25%) and branched chain amorphous amylopectin (75%–80%). The structure depends on the source from which it is extracted. The starch grains derived from rice (2 μm) is smaller than the starch extracted from potato (100 μm) [38]. Most of the starch varieties are slightly acidic in nature (cocoyam starch: pH 6.54, cassava starch: pH 5.5, corn starch: pH 6). Apart from carbohydrate contents, it has been revealed that starch also contains a minute amount of protein and other minerals. Protein present in the cocoyam, cassava and corn starch are around 7, 3.7 and 7.2 g/kg, respectively. A range of mineral matters like phosphorous, potassium, magnesium, zinc, iron, sodium and manganese are also present in starch [46, 47].

It was found that the TG profile of all starch biomolecules (cassava, corn, rice and potato) showed a major mass loss at a lower temperature of around 300 °C whereas the phenomenon happens at 350 °C in the case of cellulose polymer. Thus, a distinct difference is obtained in the case of degradation behavior of both the cellulose and starch biomolecule even though both possess similar glucosidic linkage in their structural backbone [48]. The amount of inorganic metallic ions varied depending on its source (cassava, corn, potato, rice, etc.) and thus the thermal degradation profile also changes [49]. Thermal stability increases with increasing the GSM of the fabric. A 400 $g/m^2$ (GSM) fabric showed a burning rate of 0.05 cm/s, against the 0.10 cm/s burning rate in the case of the control fabric.

Also, no afterglow was observed during the burning of the treated fabric. Char morphologies reported by the researchers showed that all the treated cotton fabric maintained structural integrity (the texture of the fabric) even after burning.

### 9.5.1.5 *Flame Retardant Textile Materials Using PRE*

PRE was used as a flame retardant agent on jute 200 GSM (areal density) material. The gray jute fabrics were impregnated separately in PRE solutions at three different pH (as it is, neutral and alkaline). During application of PRE in jute textile, the material to liquor ratio was maintained at 1:20. The fabric was dipped in these solutions for 30 min, followed by drying at 110 °C for 5 min and was used for analysis. PRE treatment increased the thermal stability of the treated fabric in terms of LOI and vertical flammability, irrespective of the pH used (refer Table 9.11). During flammability testing, it was observed that the control jute fabric was burnt completely within 180 s with the generation of high temperature, flame and afterglow whereas all the treated fabrics did not catch flame. It was observed that as the application pH of the PRE is increased (from 4.5 to 10), the length and the width, i.e. the intensity of the afterglow, are reduced with its rate of propagation throughout the fabric. The treated jute fibre showed the intrinsic intumescent behaviour of flame retardancy as evidenced by char morphology. TG curves of the treated fabric showed catalyzing dehydration and the propensity of the extensive char mass formation. Isothermal TG analysis at 500 °C of the control jute showed more than 30% weight loss against only 4%–5% weight loss of the alkaline PRE treated jute after 20 min heat exposure. Nitrogen and phosphorous present in the PRE assist

**Table 9.11** Vertical flammability of treated jute fabric using pomegranate rind extract.

| Flammability parameters | Control | PRE treated (pH = 4.5) | PRE treated (pH = 7) | PRE treated (pH = 10) |
|---|---|---|---|---|
| LOI | 22 | 33 | 35 | 38 |
| Burning with afterglow time (s) after flame stops | 80 | 1560 | 2400 | 700 |
| Char length (mm) | Nil | Nil | Nil | 60 |

Sample size: $(25 \times 4)\ cm^2$.

in catalyzing dehydration, char formation and restricting the flammable gas formation [50].

### 9.5.2 Flame Retardancy Imparted by Animal Based Natural Resources

#### *9.5.2.1 Flame Retardant Textile Materials Using Chicken Feather*

Chicken feather is composed of total 95 hydrophilic and hydrophobic amino acids, out of which the hydrophilic amino acid serine is the most predominant which helps to absorb more moisture than proteinous wool fibre from the air. Keratin protein is composed of covalently bonded carbon, nitrogen, oxygen, hydrogen, sulphur and ash material [51]. Wang *et al.* [30] synthesized a novel eco-friendly chicken feather protein (CFP) based phosphorus–nitrogen containing flame retardant for imparting flame retardancy to the cotton textiles. As per their report, raw chicken feather was collected from the slaughter house and dissolved in a mixture of sodium hydroxide (10 gpl) and urea (4 gpl) at elevated temperature. Regarding flammability, the 250 gpl CFP treated cotton fabric showed an LOI value of 30 and extinguished the flame in 7 s in the vertical flammability test. However, afterglow was present for 180 s, and the sample was burnt totally with a very lower burning rate. CFP treated cotton fibre showed thick, compact, voluminous structural coating on the fabric surface whereas CFP with boric acid and borax treated fabric showed the uniform dispersion of fine granules on the fibre surface with more thick and voluminous charred fibre structure. Indeed, phosphorous, nitrogen and sulphur present in the formulation synergistically come into play and form a thick fully intumescent less conductive layer on the fabric surface which absorbs heat and restricts the free flow of the flammable gases inside the polymeric structure and increases the formation of compact strong char mass [30].

#### *9.5.2.2 Flame Retardant Textile Materials Using Casein*

Casein as a source of protein is a major component of milk, generally obtained during the production of skim milk. Casein is composed of essential, non-essential and a few branched chain amino acids. Carosio *et al.* applied casein based protein biomolecule on the cotton fabric for improving its thermal stability. According to their report, the thermal stability of casein molecule derived from the presence of nitrogen and phosphate groups in its structure. The water suspension of the casein biomolecule was heated up to 80 °C in a thermostatic bath. Thereafter the solution was cooled to 30 °C and bath pH was maintained to 8.5 before applying to the cotton fabric. The prepared alkaline casein solution was coated on a cotton fabric surface by a spatula followed by drying. The casein treated cotton fabric showed a burning rate of 1 mm/s against the 1.5 mm/s burning rate of control cotton fabric in the horizontal flammability test (refer Table 9.12). Also, the total burning time is 30% more than that for the control fabric [27]. The flame retardant efficacy of casein powder on the synthetic polyester and the polyester–cotton blended fabric was also evaluated [28]. After complete burning, residual char of all the casein treated fabrics showed less dense, voluminous, globular structures with a sign of blown-up phosphorous-rich bubbles during combustion, representing an intumescent mechanism of flame retardancy.

#### 9.5.2.3 *Flame Retardant Textile Materials Using Whey Protein*

Whey protein is the major component of the milk protein, generally obtained during the production of skim milk [52]. Most of the amino acids present in the whey protein contain sulphur which is not generally found in casein [28]. Bosco *et al.* [26, 28] have reported the application of whey protein powder containing 93.5% protein on the cotton fabric for improving the thermal stability of the fabric. Whey protein powder biomolecules contain lipid, carbohydrate, ash, moisture and 93.5% protein. Whey protein powder was collected from a commercial chemical company and dissolved in distilled water for 10 min, at room temperature. As a result, folded whey protein was suspended. On the contrary, separately alkali was added into the folded whey protein powder solution to adjust its pH 7. After that, the solution was heated at 90 °C for 2 min and cooled to room temperature. At high temperature, the whey protein powder was unfolded, and denatured whey protein was suspended. Whey protein and denatured whey protein treatment reduced the burning rate of the fabric by 25% and 22.5%, respectively, compared to the control cotton fabric in the horizontal flammability test (methane flame: 3 s). The burning time for treated fabric was 126 (whey protein) and 133 s (denatured whey protein), respectively, whereas control cotton fabric was burnt in 70 s. It proves that whey protein treatment slows down the combustion process of the treated material. TG analysis in air atmosphere showed that whey protein treatment reduced the degradation temperature at every stage due to the formation of oligopeptides, amino acids bearing carboxylic groups, and also increased the amount of char mass production.

**Table 9.12** Horizontal flammability of the casein biomolecule treated textiles.

| Textile materials | Cotton | | | Polyester | | Polyester/cotton (PC) blend | |
|---|---|---|---|---|---|---|---|
| Flammability parameters | Control | Casein treated (pH 8.5) | Casein treated (pH 10) | Control | Casein treated | Control | Casein treated |
| Add-on (%) | – | 20 | 20 | – | – | – | 20 |
| LOI | 18 | – | 24 | 21 | 26 | 19 | 21 |
| Total burning time (s) | 78 | 100 | 75 | 57 | 54 | 104 | 171 |
| Char length (mm) | Nil | Nil | 30 | Nil | 30 | Nil | Nil |

### 9.5.2.4 *Flame Retardant Textile Materials Using Hydrophobin*

Hydrophobin protein contains both hydrophobic and hydrophilic surfaces. Class 1 hydrophobin contains water insoluble monolayer of amyloid fibrils because of their highly ordered structure whereas class 2 hydrophobin contains water soluble protein aggregate. Hydrophobins are non-toxic and contribute to the surface phenomena of the fungal cell in the environment. Also, it helps in the maturation of the cell wall and spore dispersion into the air [53, 28]. Protein based hydrophobin solution (without any dilution) was applied to the cotton fabric in a climatic chamber (30 °C) with maintaining the material to liquor ratio of 1:26 by dipping method for 1 min [54]. Here add-on percentage maintained on the treated fabric was 20%. Thermogravimetry analysis of the hydrophobin treated cotton fabric showed pyrolysis at reduced temperature and a lower rate of weight loss with rising temperature. This is due to the cleavage of the disulphide bonds (cystine linkage) present in the hydrophobin coating at 200 °C and release of hydrogen sulphide and sulphuric acid, which helps to enhance dehydration. TG data of the hydrophobin treated cotton fabric is almost similar to that of the casein treated cotton fabric in air atmosphere as degradation occurred in three main stages including the degradation of the crude protein.

### 9.5.2.5 *Flame Retardant Textile Materials Using Deoxyribonucleic Acid*

Deoxyribonucleic acid (DNA) a double helix single chain polymer consists of long chain nitrogen containing polymers composed of adenine, guanine, cytosine, and thymine. Deoxyribose units and phosphate groups are present as the outside layer of the backbone of the DNA biomolecule whereas nitrogen-containing bases predominate in the inside layer [28]. Thermogravimetry of DNA molecule composed of carbon, nitrogen, oxygen, phosphorus extracted from herring sperm showed three stages of weight loss which was attributed to the cleavage of the phosphodiester bonds between two adjacent neucleosides or due to the breakage of the phosphate group in the DNA ring itself. As a result, deoxyribose units of DNA dehydrate at a lower temperature which leads to the formation of voluminous foamy (intumescent) char [55]. DNA molecule contains nitrogen, oxygen and phosphorous. Phosphoric acid catalyzed dehydration and ammonia release, diluted oxygen at high temperature leading to thermal stability. The most efficient flame retardant treatment by using protein biomolecule on cotton fabric has been achieved by using DNA biomolecule. DNA powder gradually dissolved in distilled water at pH 5.5 at 30 °C over a period of 30 min, under stirring. Cotton fabric was impregnated with the solution and squeezed by the rotary drum. Also, DNA was applied in different concentrations (5%, 10% and 19%) to the cotton fabric. All the treated fabrics were characterized by horizontal and vertical flammability tests and also the LOI value was determined. The LOI value of the DNA treated cotton fabrics is 23 (5% DNA), 25 (10% DNA) and 28 (19% DNA), respectively [25, 55]. The reason behind the thermal stability is the presence of phosphate group in DNA which decomposes to phosphoric acid and helps initiate earlier dehydration or char formation process. It has been proposed that DNA works as a flame retardant by phosphorous–nitrogen synergistic action, which controls the degradation mechanism of the coated cotton fabric [56, 57]. Here the authors have applied DNA extracted from both the herring sperm and herring testis to the cotton fabric and have reported that DNA extracted from herring sperm showed more thermal stability (shown by TG analysis of both sperms and by the flammability test result) [56].

#### *9.5.2.6 Flame Retardant Textile Materials Using Chitosan*

Chitosan is a bio-based polymer derived from the *N*-deacetylation of chitin, a natural nitrogenous polysaccharide component obtained from crabs, shrimp shells, cuticles of insects and cell walls of fungi. Chemically, chitosan contains linear polyamine, reactive amino groups, the reactive available –OH group, etc. [41, 49, 58–60]. It has been reported that chitosan improved the wash durability of soluble phosphate treated cotton fabric. Indeed, after drying, chitosan formed a thin film of coating on the phosphate layer on the cotton fabric. Researchers have reported the use of chitosan phosphate for making flame retardant cotton textile. Chitosan has been added during the phosphorylation (di-ammonium hydrogen phosphate) process as a nitrogen source. By increasing the concentration of chitosan up to 2%, wash durability was improved along with thermal stability [59]. Very recently, researchers have used chitosan and ammonium polyphosphate via layer by layer (L-B-L) technique as an intumescent flame retardant on the cotton fabric [61].

## 9.6 Sustainable Textile Materials Using Clay as Natural Resources

### 9.6.1 Different Types of Clay and its Application in Textile Materials

Depending on the academic source, there are mainly three or four main groups of clays known as *Kaolinite*, *Montmorillonite-Smectite*, *Illite* and *Chlorite*. There are approximately 30 different types of 'pure' clays in these categories, but most of the 'natural' clay deposits are mixtures of these different types, along with other weathered minerals (Figure 9.6). Kaolinite is a clay mineral, part of the group of industrial minerals, with the chemical composition $Al_2Si_2O_5(OH)_4$ – a layered silicate mineral, with one tetrahedral sheet of silica ($SiO_4$) linked through oxygen atoms to one octahedral sheet of alumina ($AlO_6$). Rocks that are rich in kaolinite are known as kaolin or china clay. It has a low shrink swell capacity and a low cation-exchange capacity (CEC, 1–15 meq/100 g). Montmorillonite is a very soft phyllosilicate group of minerals that form when they precipitate from water solution as microscopic crystals, known as clay. Montmorillonite, a member of the smectite group, is a 2:1 clay, meaning that it has two tetrahedral sheets of silica sandwiching a central octahedral sheet of alumina. The individual crystals of montmorillonite clay are not tightly bound hence water can intervene, causing the clay to swell. The water content of montmorillonite is variable and it increases greatly in volume when it absorbs water. Chemically, it is hydrated sodium calcium aluminium magnesium silicate hydroxide $(Na,Ca)_{0.33}(Al,Mg)_2(Si_4O_{10})(OH)_2{\cdot}nH_2O$.

Illite is a non-expanding clay crystalline mineral. It is a secondary mineral precipitate phyllosilicate or layered aluminosilicate. Its structure is

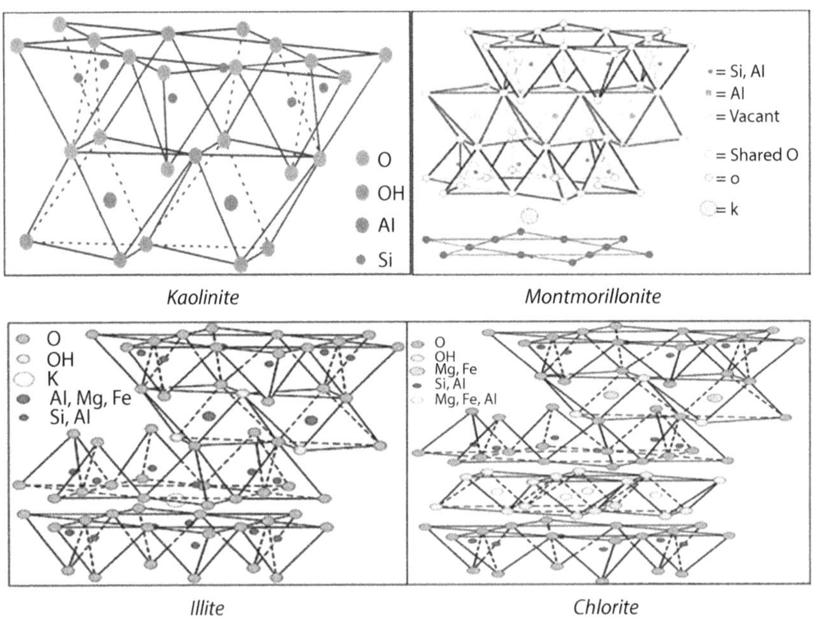


**Figure 9.6** Structure of different clays.

a 2:1 clay of silica tetrahedron–alumina octahedron–silica tetrahedron layers. The space between individual clay crystals is occupied by poorly hydrated potassium cations which are responsible for the absence of swelling. The chemical formula is given as $(K,H_3O)(Al,Mg,Fe)_2(Si,Al)_4O_{10}[(OH)_2,(H_2O)]$. The CEC of illite is smaller than that of smectite but higher than that of kaolinite, typically around 20–30 meq/100 g. Chlorites are not always considered to be a clay, but sometimes are classified as a separate group within the phyllosilicates. The typical general formula is $(Mg,Fe)_3(Si,Al)_4O_{10}(OH)_2{\cdot}(Mg,Fe)_3(OH)_6$. This formula emphasizes the structure of the group. Chlorites have a 2:1 sandwich structure (2:1 sandwich layer, i.e. tetrahedral–octahedral–tetrahedral). Unlike other 2:1 clay minerals, a chlorite's interlayer space (the space between each 2:1 sandwich filled by a cation) is composed of $(Mg^{2+}, Fe^{3+})(OH)_6$. Possible applications of different clays in textile materials are shown in Figure 9.7.

### 9.6.1.1 *Application of Clay in Nanocomposites*

Clay/polymer nanocomposites are typical examples of nanotechnology. This class of material uses smectite-type clays, such as hectorite, montmorillonite, and synthetic mica, as fillers to enhance the properties of polymers. Smectite-type clays have a layered structure. Each layer is constructed from tetrahedrally coordinated Si atoms fused into an edge shared

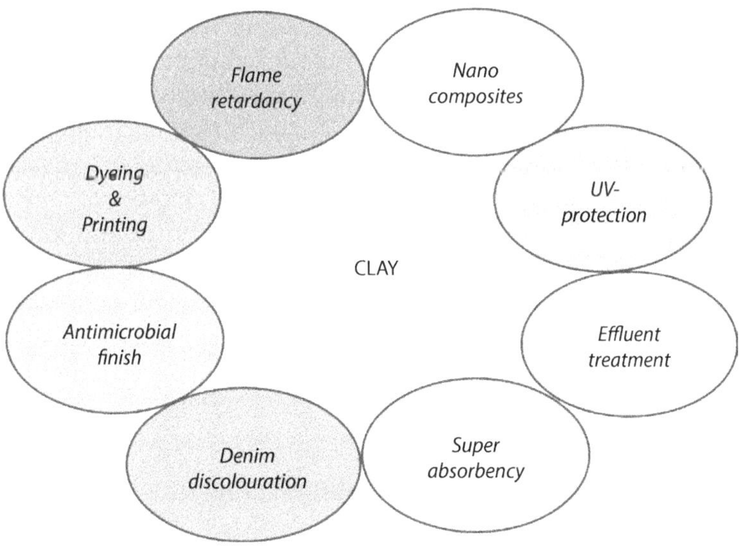


**Figure 9.7** Applications of clay in textile materials.

octahedral plane of either $Al(OH)_3$ or $Mg(OH)_2$. According to the nature of the bonding between these atoms, the layers should exhibit excellent material properties parallel to the layer direction. The principle used in clay/polymer nanocomposites is to separate not only aggregates but also individual silicate layers in a polymer by which the excellent mechanical properties of the individual clay layers can function effectively, while the number of reinforcing components also increases dramatically because each clay particle contains hundreds or thousands of layers. As a result, a wide range of engineering properties can be significantly improved with a low level of filler loading [62].

#### *9.6.1.2 Application of Clay in UV Protection*

The harmful effects of sunlight, especially solar UV radiation on human skin are well known. Chemically, sunscreens are compounds that absorb UV light and thereby decrease the amount of the solar radiation. Most sunscreen preparations in the market are appropriate only for UV-B rays, but protection from UV-A rays assumes increasing importance since it is more dangerous to human skin. The electronically excited sunscreen molecule can decay through different pathways: in the form of heat, by fluorescence, phosphorescence, interaction with neighboring molecules, or by undergoing photo-induced decompositions. The latter reactions decrease the sunscreen UV protective capacity during its use. Therefore, high photo-stability is a prerequisite for the effectiveness of sunscreen products. The photo-degradation of chemical sunscreens has been reduced by providing a protective film around the reacting molecule, by encapsulation in liposomes, microspheres and in glass beads, incorporation in nanoparticles or by complexation with cyclodextrins. Such intercalation in inorganic materials as hydrotalcite (HTlc) could represent a new strategy to improve sunscreen photo-stability. The advantages of sunscreen intercalation in hydrotalcite are sunscreen stabilization, absorption of UV lights in both the UV-A region and the UV-B region, and the absence of a close contact between skin and filter with consequent elimination of allergy problems. So, this concept of UV protection can be carried forward for the UV protective finishes on textiles by using a similar type of clay materials [63].

#### *9.6.1.3 Application of Clay in Effluent Treatment*

Health problems are caused by aluminium salts. Various reports have mentioned the direct and indirect toxic effects of metals in the form of tumors, cancers and allergies. In order to replace aluminium salts as adsorbents and coagulants, the bentonite and combination of bentonite have been used as coagulants in waste water treatment. The chemical oxygen demand (COD), turbidity, total suspended solids (TSS), and major cations are the main concerns of the characteristics of wastewaters. Generally, high levels of COD result in low dissolved oxygen (DO) in water and this can lead to mortality of aquatic life. In addition to that, suspended solids such as organic and inorganic materials can result in the water being contaminated with dirt and odour. Developing countries pay a high price to import chemicals including polyaluminium chloride and alum for water and waste water treatment. But polyaluminium chloride and alum add impurities, such as epichlodine, which are carcinogenic in nature. The use of clay mineral has undoubtedly become more popular and is widely used as adsorbent and ion exchange for water and waste water treatment applications. Clay minerals, such as bentonite and zeolite, are some of the potential alternatives, as they have large specific surface areas with a net negative charge. Their sorption capabilities come from their high surface areas and exchange capacities. It is a highly effective natural clay mineral, especially in granular form, used for the purification of waste water and sludge dewatering. The usage of natural clay minerals such as bentonite and zeolite for water and wastewater treatment is increasing because of their abundance, low price, and adsorption capabilities, as well as ion exchange. Bentonite is a natural material that contains essential compounds such as aluminium, iron and clay materials which are useful for the treatment of waste water [64]. Moreover, bentonite is cheaper than previously used chemicals and it fulfils the economic benefits of the operators as well as environmental concerns.

#### 9.6.1.4 Application of Clay in Superabsorbency

Superabsorbent polymers are loosely crosslinked networks that can absorb a large amount of water in a short time and retain water under pressure. Although superabsorbent polymers have been abundantly used in the disposable diaper industry for the past 30 years, their applications are still being extended to many fields including agriculture and horticulture, sealing composites, artificial snow, drilling fluid additives, medicine for drug delivery system and so on. The preparation of polymer/clay superabsorbent composites has received great attention because of their relatively low production cost and high water absorbency. The water absorbency of the poly(acrylic acid)/attapulgite superabsorbent composite in distilled water is greatly improved as compared with crosslinked poly(acrylic acid) superabsorbent polymer, but the water absorbing ability of superabsorbent composite in saline solutions only slightly increases [65]. It is well known that the water absorbing ability of superabsorbent composite in saline solutions is also very important for many applications. Therefore, the synthesis of the new superabsorbent composite which has high water absorbency both in distilled water and saline solution and developing their applications still represent interesting research areas. Attapulgite, as a good substrate for superabsorbent composite materials, is a layered aluminium silicate with respective –OH groups on the surface. In order to reduce costs and improve the comprehensive water absorbing properties of the superabsorbent materials, to graft acrylic acid and acrylamide copolymer onto attapulgite and fabricate a composite consisting of polymer and attapulgite micropowder can be a priority.

#### 9.6.1.5 *Application of Clay in Discolouration of Denim*

Nano-clays can be co-applied in the bio-stone washing of denim with cellulase to obtain denim garment with a new look along with the by-products clear effluent. Denim fabric usually has a weave of indigo dyed warp and white weft yarns. The abrasion process in stone washing is done with a pumice stone, with or without an oxidizing agent such as potassium permanganate. But this process results in dye removal as well as fabric strength loss to an undesirable extent. The stone washing or abrasion effect can be created on denim fabric either by using cellulase individually or along with pumice stone. The morphology of nano-clay is described as platelet with a thickness of less than 1 nm and width between 100 nm and 1 μm. Due to the huge aspect ratio, nano-clay has a very large surface area of about 700 $m^2/g$. This unique morphological structure provides possibilities for excellent sorption capability. Nano-clay can be modified with organic quaternary ammonium to provide the required interaction between clay and other organic molecules [66].

#### 9.6.1.6 *Application of Clay in Antimicrobial Finish*

Montmorillonite is the basic component of bentonite clays. Bentonite is a clay containing not less than 70% of a mineral of montmorillonite group. Montmorillonite is a fine-grained layered aluminium silicate carrying an extra negative charge because of non-stoichiometric replacements of cations of the crystal lattice, which compensates for exchange cations located in the interlayer space. This determines high hydrophilic properties of bentonite. Bentonites with montmorillonite containing mainly exchangeable sodium cations have the best technological properties. Elemental plates of clay are charged negatively; counterions $Na^+$ are necessary for the stabilization of the structure. Metal cations found in montmorillonite can be replaced with other ions of metals through ionic exchange that allows modifying clay by various ions and substances. This process is called intercalation. All clay minerals possess a certain capacity for cation exchange. This measure is an important characteristic of a mineral and indicates the quantity of exchange cations (in mg-equivalents) capable of being replaced by cations of another type in 100 g of clay. Montmorillonite possesses the highest capacity of cation exchange among clay minerals (upto 150 meq/100 g). The intercalated montmorillonite is a source of metal ions and metal oxides. Such intercalated montmorillonite particles coated on textile substrate do not allow for microbial growth and proliferation. This can be a suitable approach for antimicrobial finishes on textile materials [67].

### *9.6.1.7 Application of Clay in Flame Retardancy*

Nano-clay particles can play an important role in the flame retardant textile coating [1]. The exact mechanism of the combustion inhibition caused by the nano-clays is not yet known, but it is presumed that while the polymer matrix is burned and gasified during combustion, the incorporated nano-clays accumulate at the surface and form a barrier to oxygen diffusion, thereby slowing down the burning process. Besides, these nano-clays do slow down the burning process and enhance the char formation, but they have no effect on the ignition tendency and after flaming properties. The integration of nano-clay composites alone is not sufficient to provide a fabric with reliable combustion protection. Recent research indicates that optimized properties can be achieved by combining the nanocomposites with low concentrations of conventional flame retardants. For the integration of the described nanocomposites into textile fabrics mainly two process pathways are possible. For both application pathways, nano-clay or nanoparticle is melt blended with the polymer material. Before this process, the nano-clays have to be modified by replacing the (sodium-) cations found between the clay layers with other, sufficiently organophilic ions to allow the integration and dispersion of the nano-clays in the polymer matrix. One possibility to incorporate these nanocomposites into textiles is by melt spinning them into yarns which can subsequently be knitted or woven into textile fabrics. The second possibility is to apply the produced polymer nano-clay composite as a coating to finished textiles, as for example cotton or polyester fabrics.

#### *9.6.1.8 Application of Clay in Dyeing and Printing*

Clay can be used as an exhausting agent in the dyeing of textiles replacing the most commonly used exhausting agents like salts, thus helping to reduce the effluent load. Apart from that, it is also used as a thickening agent replacing sodium alginate in reactive printing and synthetic thickener in pigment printing. This replacement in printing yields a high colour value and comparable fastness property. Also, it eliminates high amount of water used in removing the unfixed thickener from the fabric, ultimately saving time, energy and cost. Clay being widely available in the earth can be a suitable material, whose use may be explored in various fields of science and technology.

## 9.7 Sustainable Application of Aroma Finishing in Textile Materials Using Natural Resources

### 9.7.1 Different Natural Sources of Aroma and Technology for Microencapsulation

The aroma quality or principle of plants or other substances constitutes their fragrance or agreeable odour. Clothing styles and popular apparel fashions are constantly in flux, and the fashion world is continuously inundated with runway innovations and fly-by-night fads. We have made an attempt to present comprehensive analysis done by fragrance finishing using natural resources. Adding fragrances to textiles has been carried out for many decades in the form of fabric conditioners in the wash and during tumble drying. However, no matter the quality of the technique used to incorporate the fragrance, its effect is relatively short-lived, only surviving one or two wash cycles. The encapsulated active agent is released from the textile by breaking the shell; when the size of microcapsules is higher, easier is the release of fragrance. The microcapsules can be applied by stamping works, exhaustion dyeing, impregnation, spraying, and coating or by direct incorporation in the fibre without modifying its touch and colour [68, 69].

Microencapsulation is a natural phenomenon and the examples of ideal microcapsules are found in nature, i.e. spores, seeds, eggs, pollen, etc. Due to numerous applications of microencapsulation, there are a number of processes developed to encapsulate a galaxy of materials to suit the individual applicability. In the broadest sense, microencapsulation provides a means of packaging, separating and storing solid and liquid materials in a microscopic scale for a delayed, controlled release. The fragrance compound and the essential oil are volatile substances. The most difficult task in preparing the aromatherapy textile is how to prolong its lifetime of odours. Microencapsulation is an effective technique to solve this task [70]. Microcapsules are minute containers that are normally spherical if they enclose a liquid or gas, and roughly of the shape of the enclosed particle if they contain a solid. It can be considered as a special form of packaging, in that particulate matter can be individually coated for protection against the environment and release the volatile substance from the enclosed capsule as required. The key to aromatic textile is how to make microcapsules of fragrance compounds and essential oils without omitting any ingredient in order to ensure its pharmaceutical effects. In addition, using a low-temperature polymer binder to attach a perfume microcapsule to the surface of the textile is also an important part of preparing an aroma textile. At the same time, durability in laundering and a soft handle should be carefully considered [71]. Although there are many effective approaches to microencapsulation for decreasing fragrance release, cyclodextrins are the best regarding safety to the human body, because β-cyclodextrin has no skin irritation, no skin sensitization and no mutagenic effect [72].

### 9.7.2 Preparation of Recipe and Method of Application for Aroma Finishing

The fragrance with β-cyclodextrin inclusions was formed by a mixture solution containing alcohol and distilled water (1:3). The solution was emulsified with a high-speed mixer at a speed of about 10,000 rpm for 5 min. The emulsified system was transferred into a flask. The fragrance alcohol solution was added into the emulsified solutions over 30 min and stirred at a temperature of 40 °C for 2 h. In the exhaust method, firstly β-cyclodextrin using acrylic acid, sodium hypophosphite and potassium persulphate was applied by exhausting and the fragrance was sprayed by means of a spray gun. The fabric is padded with the above recipe, fragrance having 10% of the weight of fabric under a padding pressure of 2 $kg/cm^2$, dried and cured at 70–120 °C (1–5 min). The recipe was prepared with water up to the desired concentration for application by exhaust at 40 °C for 20–30 min with fragrance 5%–10% of the weight of the fabric [72].

### 9.7.3 Fragrance Release Property of Aroma Finishing

There were several differences to the cyclodextrin's molecular encapsulation compared with other widely used encapsulation processes. The enclosed substance was totally or partially included in the cyclodextrin cavity. This system was reversible and was not sealed completely. The forms of the microcapsule were able to notably decrease the rate of fragrance release. The different rates of fragrance release of various perfume microcapsules were related to the nature of the fragrance oils themselves. The higher the detection threshold of the fragrance compounds, the lower the rate of fragrance release of the inclusion compound. Aroma therapy textiles based on lavender, lemon, chamomile, rose, cardamom, clove or jasmine have a sedative effect, which is fit for bed gown, underwear, sheets, curtains, carpets, etc. Basil, chamomile and lavender can serve to restrain bacterium growth on textiles. We may increase work efficiency after inhaling jasmine, lavender and mint scent deposited on aroma therapy garment. The antimicrobial/bacteriostatic product may be used on underwear, whereas outerwear such as a sweater is suitable to fulfil other pharmaceutical effects of essential oils.

### 9.7.4 Applications of Aroma Finishing in Textile Materials

Microencapsulated skin moisturizers, vitamins and provitamins are applied in garments and known as cosmetic-textiles, designed for wear in contact with skin. These are claimed to promote a younger look, counteracting the effects of skin ageing, e.g. as a result of exposure to UV radiation. There are also microencapsulated preparations for skin cooling. The uses of aromatherapy textile are diverse. Interior textiles such as sheets, quilt-covers, curtains, carpets and bed gowns are suitable for the attachment of lavender, chamomile, citrus or cinnamon microcapsules, which are good for hypnogenesis and eliminating fatigue. Patients suffering from high blood pressure feel sedation when they use a pillow made of fabric treated with lavender, basil and lemon or fennel microcapsules. The tired office clerk wearing clothing with a scent of lemon rose or jasmine oil may find his work efficiency improved. Household textiles such as curtains, sofas, cushions, sheets, as well as apparel items such as gloves, socks and ties may also be treated with microencapsulated fragrance and deodorizing finishes. The carpets can be finished with fragrances of different kinds and can be widely used in home textiles as well as automotive textiles. As the 'second skin' of the human body, all types of textiles are excellent media for transferring fragrance compounds and are essential to people in sports according to their preference for them. The type of fragrance necessary for sports wear may be orange, lemon which will keep them energized on the ground for a longer period. The scents of lavender rose, citrus or vanilla were encapsulated into fabrics, which proved to be a good way of meeting important psychological and emotional needs, as well as those of a purely physical and sensorial nature. Thus, the fabrics finished with the above fragrances can be used in apparels, leisure wear, daily wears, party wears, etc. Microencapsulation can effectively control the release rate of the fragrance compounds and essential oils as required, which ensures the storage life of volatile substances [73].

## 9.8 Sustainable Mosquito Repellent Textile Materials Using Natural Resources

### 9.8.1 Different Types of Repellent Insecticides

Mosquito repellents come under the category of Insecticides. Insecticides are of two types: Repellent Insecticides and Contact Insecticides. Repellent Insecticides are the substances that are mainly used to repel insects and pests rather than causing death. Several mosquito repellents are available in today's market, of which an estimated 110 million people safely use DEET (Di-Ethyl Meta Toluamide) worldwide each year. Contact Insecticides are the substances that contain neuro-toxins for pests and insects. They make the insects and pests unconscious when they come in contact with them [74].

### 9.8.2 Natural Resources of Mosquito Repellents

There are various natural mosquito repellents, which are as follows: citronella oil, castor oil, clove oil, cedar oil, rosemary oil, peppermint oil, lemongrass oil, geranium oil, chrysanthemum (refer Table 9.13). Among all the mentioned natural repellents, chrysanthemum was found to be the best. The most remarkable feature about chrysanthemum is that it does not lose its action towards mosquitoes for a long period of time even when it gets exposed to the environment. This is not so in the case of other natural repellents which are mentioned above because they deteriorate when exposed to sun, heat and rain. Some of the chemical repellents are DEET, allethrin, permethrin (synthetic analogue of pyrethrum) and malathion.

### 9.8.3 Mosquito Repellency Evaluation

Mosquito repellency activity was assessed using the test cage described in the American Society for Testing and Materials (ASTM) Standard E951-83 Laboratory testing of non-commercial mosquito repellent formulations on the skin. The test cages were positioned securely on the arms of each volunteer with Velcro tapes to ensure that only the test areas were exposed

**Table 9.13** Natural resources of aroma and its odour categories.

| S. no. | Natural resources | Representative odour categories |
|---|---|---|
| 1 | Citrus Lemon | Citral, Citronellal |
| 2 | Orange | Mandarin Oil, Decyl acetate |
| 3 | Floral Carnation | Phenethyl salicylate |
| 4 | Gardenia | Nonyl acetate |
| 5 | Geranium | Citronellol |
| 6 | Lilac | Anisylacetate |
| 7 | Lily | Hydroxycitronellal |
| 8 | Rose | Rose absolute |
| 9 | Violet | Costus Oil, Methyl-2-nonenoate |
| 10 | Fruity Apple | Benzyl acetate |
| 11 | Apricot | Allyl butyrate |
| 12 | Banana | Amyl acetate |
| 13 | Grape | Isobutyl isobutyrate |
| 14 | Peach | Allyl butyrate |
| 15 | Strawberry | Benzyl benzoate |
| 16 | Herbaceous Clove | Eugenylacetate |
| 17 | Minty | l-carveol, l-Carvone, l-Menthol |
| 18 | Sweet Anise | Ethyl acetate, Methyl sorbate |
| 19 | Cinnamon | Cinnamaldehyde |
| 20 | Honey | Allyl phenoxyacetate |
| 21 | Sweet | Acetanisole |
| 22 | Vanilla | Anisylacetate |

to mosquito bites. Fifteen female mosquitoes between 4 and 7 days-old were introduced into each cage and the number of biting was recorded at the end of 120 s [72]. There are different types of testing methods for evaluation of mosquito repellent textiles such as cone test, wire-ball test, cylinder test, field test, etc. In the case of cone test, exposed mosquitoes might spend more time resting on the cone not on the treated surface. In the case of wire-ball test, the treated fabric, which is to be tested, is wrapped around a wire frame and mosquitoes are introduced into this ball. In the case of cylinder test, the inside of a test cylinder is covered with the textile. The field test is the most meaningful evidence for the efficacy of treated textile. The field test is especially performed at such a location where the floodwater mosquitoes are rich.

### 9.8.4 Method of Application of Mosquito Repellency

In a process for applying microcapsules to textile materials, microcapsules can be initially introduced into the textile material without a crosslinking agent; a dispersant is introduced to disperse the microcapsules around and through the textile material, and thereafter the crosslinking agent is added to promote the adhesion of microcapsules to the textile material. Alternatively, microcapsules can be applied during the finishing process of textiles fabrication using a foulard in which the textile to be treated is impregnated by means of a finishing bath containing microcapsules, a softener and a self-crosslinking agent. Taking into account the type of final application, the microcapsules have to fulfil specific requirements, such as resistance to abrasion and to dry cleaning wash cycles.

For apparel designed to be worn in tropical climates where mosquitoes abound, there is clearly a market for insect repellent treatments that have a long-lasting effect. Microencapsulation of insect repellent agents can be employed to provide a longer lasting controlled release effect. Thor SARL (Salaise sur Sanne, France) has developed a method of microencapsulating a synthetic pyrethroid, permethrin, which is a known insecticide. A particular area of use is in clothing for the military serving in tropical climates. Textiles treated with tea tree oil Sensory Perception Technology (SPT) microcapsules have been tested by the Carrol Loye Biological Research Institute (CA, USA) and shown to demonstrate significant mosquito repellence properties. The natural organic ester plant oils act effectively as a nerve agent on insects, but more importantly, recent research has shown that these have an unusual effect on mosquitoes which is termed as 'jamming'. This effect results in the 'black box' which helps the mosquito search for victims being turned off. As a result, the microencapsulated release of these plant oils has been shown to significantly decrease the bites per minute in test boxes from 50 to virtually none. Another area is that of microencapsulated insect-resist treatments for textiles containing natural protein fibres like wool and silk. Such finishes could provide long-term protection against the ravages wrought by the larvae of moths and particularly many types of beetles.

### 9.8.5 Applications of Mosquito Repellency in Textile Materials

The mosquito repellent textiles can be used as antimosquito garment, bed sheets, wide-mesh netting jackets, head nets, bands and anklets, repellent ropes, bed nets, bed curtains, window curtains and bathroom curtains. During the past 10 years, a considerable amount of research has been conducted in the United States, Europe and Japan to measure not only the effects of fragrance on feeling, moods and emotions but also upon several areas of response like electrical activity in the brain, physiological parameters such as the heart rate and skin conductance, cognitive functions, and voluntary and involuntary behavior [72].

## 9.9 Conclusion

The colouration of textile materials is possible with the different natural resources of plants. DSE, *Sterculia foetida* FSE, and CSE can be successfully employed as a natural mordant as well as with mordants for dyeing of textile materials as a natural source of colourant. The dyed textile materials showed excellent antibacterial activity against both *S. aureus* and *E. coli* bacteria as per AATCC 100 test method. Even after 10 washes, the dyed sample showed excellent antibacterial and good UV protection properties. The concept of usage of naturally occurring resources from plants, animal and clay in the flame retardant finishing of textile materials is relatively new. The present chapter discusses the flame retardancy on cellulosic, lignocellulosic, polyester and proteinous textile materials. After application of plant based biomolecule at different pH and concentrations, the thermal stability of the fabric was found to increase. Flame retardancy in the plant based biomolecule treated textile material is attributed to the presence of inorganic metal salts, metal oxides, phenolic groups, etc. which helped in the production of more char and non-flammable gases. Reported animal based flame retardant finishes by DNA, casein, and hydrophobin-like biomolecules are also very expensive, and their availability in larger quantity is also a great challenge in the perspective of commercial implementation. Just to highlight, washing durability is one of the major challenges to address to retain thermal stability after repeated washing. Green coconut shell and banana pseudostem as raw material are widely available and mainly a waste product. This kind of sustainable research approach might lead to an excellent value addition in the recycling of the waste material for treatment of textile materials. Thus, the use of plant and animal based natural resources for imparting flame retardancy to textile substrates can find application in various areas like military tents, bags, public halls, theatre, sofa cover, table lamp, curtains, and other decorative products as home textile and protective textile segments. Clays can be a new perspective for a sustainable approach in textiles looking into its excellent natural properties. Such phenomenal characteristics as the flame barrier, good controlled release, biosorption capacity, cation exchange capacity, intercalation with metal ions, etc., offer a very good approach to enhance the engineering applications of textiles. Aroma finishing and mosquito repellency using natural resources are proving to be a new era of sustainable textile materials. Aroma is increasingly popular as one of the many approaches to healing with natural substances, which are favoured by the public. This chapter addresses the rationale behind aromatherapic and sustainable textile materials being the first choice of people who want to stay healthy in their daily life, and the reason such textiles will become a fashion in the near future.

## References


1. Teli, M.D. and Pandit, P., Novel method of ecofriendly single bath dyeing and functional finishing of wool protein with coconut shell extract biomolecules. *ACS Sustain. Chem. Eng.*, 5(9), 8323, 2017.
2. Teli, M.D. and Pandit, P., Development of thermally stable and hygienic coloured cotton fabric made by treatment with natural coconut shell extract. *J. Ind. Text.*, 48(1), 87, 2018.
3. Teli, M.D. and Pandit, P., A novel natural source Sterculia foetida fruit shell waste as colourant and ultraviolet protection for linen. *J. Nat. Fib.*, 15 (3), 337, 2018. http://dx.doi.org/10.1080/15440478.2017.1328327.
4. Teli, M.D., Pandit, P., Multifunctionalised silk using delonixregia stem shell waste. *Fib. Polym.*, 18(9), 1679, 2017.
5. Teli, M.D., Pandit, P., Basak, S., Coconut shell extract imparting multifunction properties to ligno-cellulosic material. *J. Ind. Text.*, 47 (6), 1261, 2018.
6. Basak, S., Samanta, K.K., Chattopadhyay, S.K., Pandit, P., Maiti, S., Green flame retardant finishing and combined dyeing of proteinous wool fabric. *Colour. Technol.*, 132, 135, 2016.
7. Sarvanan, D., Lakshmi, S.N.S., Raja, K.S., Vasahi, N.S., Biopolishing of cotton fabric with fungal cellulose and its effect on the morphology of cotton fibers. *Ind. J. Fib. Text. Res.*, 38, 156, 2013.
8. Joshi, M., Ali, S.W., Rajendran, S., Antibacterial finishing of polyester cotton blend fabric using neem: a natural bioactive agent. *J. Appl. Polym. Sci.*, 106, 793, 2007.
9. Ali, S.W., Rajendran, S., Joshi, M., Aloe vera – a potential eco-friendly biopolymer to impart antibacterial activity to cotton textiles. *Cellulose*, 2, 2063, 2014.

10. Teli, M.D., Viveki, S., Pandit, P., Standing bath technique for cost reduction in cotton dyeing with marigold. *J. Text. Assoc.*, 77, 378, 2017.
11. Teli, M.D., Sahoo, M.R., Pandit, P., Antibacterial and uv protective cotton fabric made by herbal synthesized silver nanoparticles, *Int. Res. J. Eng. Technol.*, 4, 1310, 2017.
12. Bikovens, O., Roze, L., Pranovich, A., Reunanen, M., Telysheva, G., Chemical composition of lipophilic extractives from grey alder (Alnusincana). *BioResources*, 8, 350, 2012.
13. Teli, M.D., Sheikh, J., Kamble, M., Ecofriendly dyeing and antibacterial finishing of soyabean protein fabric using waste flowers from temples. *Text. Light Ind. Sci. Technol.*, 1, 1, 2013.
14. Samanta, A.K., Kar, T.R., Mukhopadhyay, A., Shome, D., Konar, A., Eco-friendly salt-free reactive dyeing of cotton (muslin) fabric after cationization with amino acid from soya. *Text. Res. J.*, 86, 2179, 2016.
15. Prabhu, K.H., Teli, M.D., Waghmare, N.G., Eco-friendly dyeing using natural mordant extracted from Emblica officinalis G. Fruit on cotton and silk fabrics with antibacterial activity. *Fib. Polym.*, 12, 753, 2011.
16. Joshi, M., Ali, S.W., Purwar, R., Rajendran, S., Ecofriendly antimicrobial finishing of textiles using bioactive agents based on natural products. *Ind. J. Fib. Text. Res.*, 34, 295, 2009.
17. Hebeish, A., Sharaf, S., Farouk, A., Utilization of chitosan nanoparticles as a green finish in multifunctionalization of cotton textile. *Int. J. Biol. Macromol.*, 60, 10, 2013.
18. Chattopadhyay, S.N., Pan, N.C., Roy, A.K., Saxena, S., Khan, A., Development of natural dyed jute fabric with improved colour yield and UV protection characteristics. *J. Text. Inst.*, 104, 808, 2013.
19. Teli, M.D., Pandit, P., Ugale, P., Sterculia Foetida fruit shell for multiple wellness properties to organic cotton. *Asian Dyer*, 13, 31, 2016.
20. Teli, M.D., Pandit, P., Ugale, P., Sterculia Foetida fruit shell used for dyeing and ultra-violet protection properties to worsted wool. *J. Text. Assoc.*, 77, 27, 2016.
21. Horrocks, A.R, Flame retardant challenges for textiles and fibers: new chemistry versus innovatory solutions. *Polym. Degrad. Stab.*, 96, 377, 2011.
22. Horrocks, A.R., Flame retardant finishing of textiles. *Colour. Technol.*, 16, 62, 1986.
23. Schindler, W.S. and Hauser, A.H., Flame retardant finishes, in: *Finishing of Textiles*, A.H. Hauser (Ed.), p. 98, Woodhead Publishing, London, 2004.
24. Samanta, A.K. and Bhattacharya, K., Simultaneous dyeing and fire-retardant finishing of jute fabric using an acid dye and selective FR finishing chemicals. *Text. Light Ind. Sci. Technol.*, 4, 1, 2015.
25. Alongi, J., Carletto, R.A., Di Blasio, A., Cuttica, F., Carosio, F., Bosco, F., Malucelli, G., Intrinsic intumescent-like flame retardant properties of DNA-treated cotton fabrics. *Carbohydr. Polym.*, 96, 296, 2013.
26. Bosco, F., Carletto, R.A., Alongi, J., Marmo, L., Di Blasio, A., Malucelli, G., Thermal stability and flame resistance of cotton fabrics treated with whey proteins. *Carbohydr. Polym.*, 94, 372, 2013.

27. Carosio, F., Di Blasio, A., Cuttica, F., Alongi, J., Malucelli, G., Flame retardancy of polyester and polyester–cotton blends treated with caseins. *Ind. Eng. Chem. Res.*, 53, 3917, 2014.
28. Malucelli, G., Bosco, F., Alongi, J., Carosio, F., Di Blasio, A., Mollea, C., Cuttica, F., Casale, A., Biomacromolecules as novel green flame retardant systems for textiles: an overview. *RSC Adv.*, 4, 46024, 2014.
29. Carosio, F., Fontaine, G., Alongi, J., Bourbigot, S., Starch-based layer by layer assembly: efficient and sustainable approach to cotton fire protection. *ACS Appl. Mater. Interfaces*, 7, 12158, 2015.
30. Wang, X., Lu, C., Chen, C., Effect of chicken-feather protein-based flame retardant on flame retarding performance of cotton fabric. *J. Appl. Polym. Sci.*, 131, 2014.
31. Basak, S., Samanta, K.K., Chattopadhyay, S.K., Fire retardant property of cotton fabric treated with herbal extract. *J. Text. Inst.*, 106, 1338, 2015.
32. Basak, S., Samanta, K.K., Chattopadhyay, S.K., Narkar, R.S., Mahangade, R., Flame retardant cellulosic textile using banana pseudostem sap. *Int. J. Cloth. Sci. Technol.*, 27, 247, 2015.
33. Basak, S., Samanta, K.K., Chattopadhyay, S.K., Narkar, R., Self-extinguishable ligno-cellulosic fabric using banana pseudostem sap. *Curr. Sci.*, 108, 372, 2015.
34. Basak, S., Samanta, K.K., Chattopadhyay, S.K., Fire retardant property of cotton fabric treated with herbal extract. *J. Text. Inst.*, 106, 1338, 2015.
35. Ghosh, A., Nano-clay particle as Textile Coating. *J. Eng. Technol.*, 11, 34, 2011.
36. Uddin, F., Clays, nanoclays and montmorillonite minerals. *J. Metall. Mater. Trans.*, 39, 2804, 2008.
37. Teli, M.D., Pandit, P., Jangle, A., Antibacterial property of Sterculia foetida fruit shells. *Asian Dyer*, 12, 40, 2015.
38. Teli, M.D. and Pandit, P., Application of Sterculia foetida fruit shell waste biomolecules on silk for aesthetics and wellness properties. *Fib. Polym.*, 19(1), 41, 2018.
39. Teli, M.D. and Pandit, P., A novel natural source Sterculia foetida fruit shell waste as colourant and antibacterial agent for cotton. *J. Text. Assoc.*, 76, 293, 2017.
40. Teli, M.D., Valia, S.P., Kolambkar, D., Flower waste from temple for dyeing of cotton and cotton/silk. *J. Text. Assoc.*, 74, 210, 2013.
41. Bourbigot, S., Richard-Campisi, L., Le Bras, M., Delobel, R., The thermal behaviour of a blend of cotton and fire-retardant polyester fibers: a kinetic study. *J. Text. Inst.*, 88, 64, 1997.
42. Horrocks, A.R. and Price, D. (Eds), *Fire Retardant Materials.* Cambridge, UK: Woodhead Publishing, pp. 128–178, 2001.
43. Horrocks, A.R., Moss, W.A., Edwards, N.C., Price, D., The spontaneous igniting behaviour of oil-contaminated cotton. *Polym. Degrad. Stab.*, 33, 295, 1991.
44. Feriotti, D. and Iguti, A., Proposal for use of pseudostem from banana tree (Musa cavendish). In: *International Congress of Engineering and Food*, 2012.
45. Lomnitski, L., Bergman, M., Nyska, A., Ben-Shaul, V., Grossman, S., Composition, efficacy, and safety of spinach extracts. *Nutr. Cancer*, 46, 222, 2003.

46. Pathania, D. and Sharma R. Synthesis and characterization of graft copolymer of methacrylic acid onto gelatinized potato starch using chromic acid inhibitor in presence of air. *Adv. Mater. Lett.*, 3, 142, 2012.
47. Horrocks A.R. (Ed.), *Textiles in Fire Retardant Materials*, pp. 128–178, CRC Press, Boca Raton, 2000.
48. Tian, Y., Li, Y., Xu, X., Jin, Z., Starch retrogradation studied by thermogravimetric analysis (TGA). *Carbohydr. Polym.*, 84, 1165, 2011.
49. Dutta, P.K., Dutta, J.D., Tripathi, V.S., Chitin and chitosan: chemistry, properties and applications. *J. Sci. Ind. Res.*, 63, 20, 2004.
50. Basak, S. and Ali, S.W., Leveraging flame retardant efficacy of pomegranate rind extract, a novel biomolecule, on ligno-cellulosic materials. *Polym. Degrad. Stab.*, 144, 83, 2017.
51. Niekraszewicz, A., Potocka, D.A., Mik, T., Pałczyńska, M., Fibrous composites based on keratin from chicken feathers. *Fib. Text. East. Eur.*, 19, 89, 2011.
52. Keri Marshall, N.D., Therapeutic applications of whey protein. *Altern. Med. Rev.*, 9, 136, 2004.
53. Basak, S. and Ali, S.W., Sustainable fire retardancy of textiles using bio-macromolecules. *Polym. Degrad. Stab.*, 133, 47, 2016.
54. Alongi, J., Carletto, R.A., Bosco, F., Carosio, F., Di Blasio, A., Cuttica, F., Antonucci, V., Giordano, M., Malucelli, G., Caseins and hydrophobins as novel green flame retardants for cotton fabrics. *Polym. Degrad. Stab.*, 99, 111, 2014.
55. Alongi, J., Carletto, R.A., Di Blasio, A., Cuttica, F., Carosio, F., Bosco, F., Malucelli, G., Intrinsic intumescent-like flame retardant properties of DNA-treated cotton fabrics. *Carbohydr. Polym.*, 96, 296, 2013.
56. Bosco, F., Casale, A., Mollea, C., Terlizzi, M.E., Gribaudo, G., Alongi, J., Malucelli, G., DNA coatings on cotton fabrics: effect of molecular size and pH on flame retardancy. *Surf. Coat. Technol.*, 272, 86, 2015.
57. Alongi, J., Cuttica, F., Di Blasio, A., Carosio, F., Malucelli, G., Intumescent features of nucleic acids and proteins. *Thermochim. Acta*, 591, 31, 2014.
58. Corazzari, I., Nisticò, R., Turci, F., Faga, M.G., Franzoso, F., Tabasso, S., Magnacca, G., Advanced physico-chemical characterization of chitosan by means of TGA coupled on-line with FTIR and GCMS: thermal degradation and water adsorption capacity. *Polym. Degrad. Stab.*, 112, 1, 2015.
59. El-Tahlawy, K., Chitosan phosphate: a new way for production of eco-friendly flame-retardant cotton textiles. *J. Text. Inst.*, 99, 185, 2008.
60. Fang, F., Zhang, X., Meng, Y., Gu, Z., Bao, C., Ding, X., Li, S., Chen, X., Tian, X., Intumescent flame retardant coatings on cotton fabric of chitosan and ammonium polyphosphate via layer-by-layer assembly. *Surf. Coat. Technol.*, 262, 9, 2015.
61. Alongi, J., Cuttica, F., Carosio, F., Bourbigot, S., How much the fabric grammage may affect cotton combustion. *Cellulose*, 22, 3477, 2015.
62. Bitinis, N., Hernández, M., Verdejo, R., Kenny, J.M., Lopez-Manchado, M.A., Recent advances in clay/polymer nanocomposites. *Adv. Mater.*, 23, 5229, 2011.

63. Grancarić, A.M., Tarbuk, A., Kovaček, I., Nanoparticles of activated natural zeolite on textiles for protection and therapy. *Chem. Ind. Chem. Eng. Q.*, 15, 203, 2009.
64. Pala, A. and Tokat, E., Colour removal from cotton textile industry wastewater in an activated sludge system with various additives. *Water Res.*, 36, 2920, 2002.
65. Li, A. and Wang, A., Synthesis and properties of clay-based superabsorbent composite. *Eur. Polym. J.*, 41, 1630, 2005.
66. Maryan, A.S., Montazer, M., Rashidi, A.S., Nanoclay as new tool for discolouration of dyed denim garment with indigo. *Prog. Colour, Colour. Coat.*, 6, 25, 2013.
67. Grethe, T., Bidu, J., Mahltig, B., Haase, H., Antimicrobial finishing of textiles by modified clay minerals. *Melliand Int.*, 20(3), 2014.
68. Gordon, N., Microencapsulation in textile finishing, *Rev. Prog. Colour.*, 31, 57, 2001.
69. Rowe, D. (Ed.), *Chemistry and Technology of Flavours and Fragrances*, John Wiley & Sons, 2009.
70. Mei, W.P., Application of microencapsulation technology in textile colouration and finishing. *China Text. Inst. J.*, 5, 188, 1995.
71. Schaab, C.B., Impregnating nonwoven fabrics with microencapsulated components. *Nonwovens Ind.*, 16, 14, 1985.
72. Aggarwal, A.K., Dayal, A., Kumar, N., Microencapsulation processes and applications in textile processing. *Colourage*, 45, 15, 1998.
73. Wang, C.X. and Chen, S.L., Aromachology and its application in the textile field. *Fib. Text. East. Eur.*, 13, 41, 2005.
74. Holme, I., Innovative technologies for high performance textiles. *Colour. Technol.*, 123, 59, 2007.

# 10

# Green Engineered Functional Textile Materials

**Pravin Chavan[1], Shahid-ul-Islam[2], Akbar Ali[3], Shakeel Ahmed[4,5] and Javed Sheikh[2*]**

*[1]Department of Fibres and Textile Processing Technology, Institute of Chemical Technology, Mumbai, India*
*[2]Department of Textile Technology, Indian Institute of Technology Delhi, New Delhi, India*
*[3]Department of Chemistry, Jamia Millia Islamia, New Delhi, India*
*[4]Department of Chemistry, Government Degree College Mendhar, Jammu & Kashmir, India*
*[5]Higher Education Department, Government of Jammu & Kashmir, Jammu, India*

***Abstract***
Textile materials are one of the basic needs of human beings and mainly developed to provide protection against climatic conditions. The desirable properties required for textiles however changed dramatically and a lot of functions are nowadays desired for demanding consumer market place. The main driving factors for the development of functional textiles are the increasing awareness and requirement of protection. Numbers of functional textiles are developed and are also being continuously developed using chemicals some of which might be a cause of concern. Hence, there is an urgent need to develop green engineered functional materials. The chapter summarizes some of the recent advances in functional textiles mainly in the development of mosquito repellent and aroma finishes.

***Keywords:*** Green, materials, textile, functionality, green textile, advanced materials

## 10.1 Introduction

The primary function of textiles, being the basic human need, is protection of human beings from the adverse environmental conditions. With

**Corresponding author*: jnsheikh@iitd.ac.in

Shakeel Ahmed and Chaudhery Mustansar Hussain (eds.) Green and Sustainable Advanced Materials, Vol. 1 (263–288) © 2018 Scrivener Publishing LLC

the evolution of human beings, environmental conditions changed drastically and hence the expectations out of the textile materials. Textiles, barrier materials between human body and the environment, are expected to protect human beings from the adverse environment as well as from the elements which can be dangerous to the user. Hence, textiles are now becoming a functional need of human beings.

All clothing is known to perform various functions ranging from very basic needs for protection to aesthetic values. Therefore, functional textiles and clothing can be considered as a generic term to include all types of clothing or fibrous assemblies that are specifically designed to deliver a pre-defined performance to users, over and above its normal functions [1]. Sometimes the term 'Technical textiles' is also used to denote functional textiles. Functional textile can include many technical textile fields like protective textiles, medical textile, sports textile, agro textile, geotextile, etc. In this textile field, textile material can be seen as more functional if it has special properties. Functional finishing of textile material is a very vast theme with each part of the fabric having its own specifications, processes and technologies. Most of the technology used is among the best sophisticated available at any given point of time. End-use applications of the fabric are also diverse and quite difficult, such as from the lifesaving material to the royal life style.

The properties included are insect repellent, aroma, water repellent, UV (ultraviolet) protective, flame retardant, antibacterial and easy care, etc. These finishes make the fabric functional and provide strength to survive against the changing environment. The applications of textiles are diverse and the meaning of the word 'protection' varies with the agency against which it needs to act. Eventually, it varies with the environment to which the user is exposed. Biological agencies like microbes are a big threat, and the antimicrobial textiles either inhibit or kill the microbes and protect human beings from the perspective threats. Other biological agencies like mosquitoes are the biggest challenge. Mosquito bites can be dangerous as it may transmit the vector disease such as dengue or malaria to humankind. Protection of users from UV and other high energy radiations is essential in order to prevent their harmful effects on the human body. Fire accident causes severe loss of lives and flame retardant textiles can minimize/prevent these losses. Even though existing functional finishing concepts are capable of creating the textiles with advanced functionalities, they are incompetent to provide the specific demand of futuristic textile products with more advanced features coupled with the environmental responsiveness, i.e. sensitivity and the compatibility to surrounding circumstances.

Green chemistry is a buzzword in the entire material science and textile is no exception. Toxicity of some of the chemicals used for functionalization of textiles is always a question. As clothing material makes contact with skin, the user ecology of the chemical used for functionalization becomes important. Apart from this, huge quantity of effluent is discharged from textile industries, and the presence of chemicals which are potentially toxic for aquatic life or difficult to treat in effluent treatment plant becomes unacceptable. Water scarcity is going to be another big issue and many countries are already witnessing it. In order to overcome the problem of water scarcity, sustainable methods are required and upcoming technologies should have a green approach.

The chapter deals with the various green engineering aspects of textiles to meet the requirements of new generation functional clothing. It covers the present techniques used for functionalization of textiles via green approach. In addition, future trends and additional sources of information are provided.

### 10.1.1 Green Chemicals

Ecological issues like global warming, pollution, disposal of solid waste, carbon footprints, etc. need to be addressed by all industries including textiles. Nowadays, many chemicals used in textile industries have harmful side effects on various types of ecologies like production, user and disposal ecology. The need of the hour for textile industries is to improve their performance and become more eco-friendly. The term 'Green Chemicals' implies that these chemicals are easily and quickly biodegradable and do not have harmful side effects. They are derived from plants and vegetables and make use of the chemical and colloidal properties of certain plant extracts, and these chemicals ascertain the safety in all forms of ecologies.

### 10.1.2 Functional Finishing of Textiles: The Expectations

Functional finishing of textiles is aimed at meeting customer's requirement of functionality out of the textile product. Various synthetic chemicals are available for imparting a particular type of functional finish onto textile materials. Desirable requirements for functional finishing include effective protection and control, absence of toxic effects, durability of activity to laundering, compatibility with other finishing agents, no adverse effect on handle or other physical properties of the fabric, applicability using other textile machinery, no environmental problems, etc. [2, 3].

There are well-known synthetic chemicals available in each category of finishing with their varying durabilities towards washing, but most of them are not eco-friendly. It necessitates the need of green finishing chemicals for textiles.

## 10.2 Different Finishes Applied onto Textiles: Present Techniques vs. Green Methods

Extensive investigations have been carried out to functionalize different textile substrates using both natural and innovative chemical finishing agents [4–13]. Among all functional properties, it is noteworthy to mention that textiles with self-cleaning, magnetic, antibacterial, anti-wrinkle, anti-oxidant, waterproof, UV-protection, mosquito repellent, flame retardant, etc. are widely preferred for various application fields (Figure 10.1) [14].

A lot of synthetic chemicals such as metal salts, surfactants, organometallic compounds, phenols, quaternary ammonium salts, nitro compounds, heterocyclic compounds with anionic groups, antibiotics, urea and related compounds, formaldehyde derivatives, amines and synthetic dyes are extensively employed to produce multifunctional textile surfaces [15]. Apart from these agents, plants, insects and minerals have been recently discovered as sustainable sources of agents for functional modification of textiles. It is pertinent to mention that natural products are rich sources of phytochemicals including phenols, terpenoids, essential oils (EO),

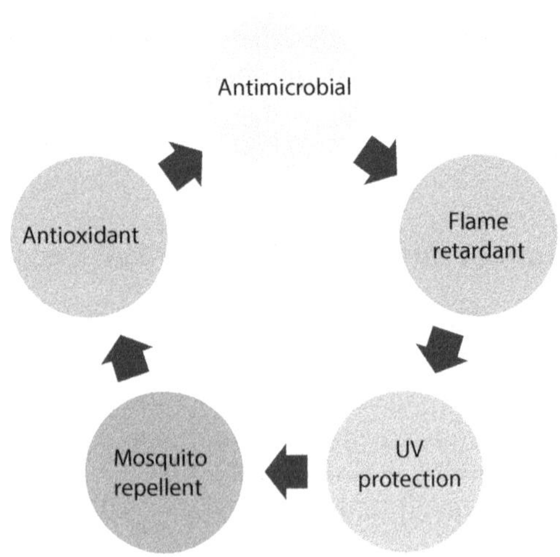


**Figure 10.1** Some important functional finishing properties imparted to textiles.

alkaloids, polypeptides and polyacetylenes [16]. This has led to their extensive use in the development functional clothing and textiles. Considering the importance of functional textiles, this chapter is focused on highlighting the mosquito repellency and aroma finishing using synthetic and eco-friendly agents.

### 10.2.1 Mosquito Repellent Finish

Mosquitoes are the most dangerous insect, which has been a cause of worry to mankind for many years. It deals with deadly diseases to humans and shows its presence in the nature as an important vector. They are known for spread of many diseases like malaria, dengue, chickengunya, etc. This use of mosquito repellent is one of the ways to deliver protection from these vectors. Various insect repellent products like liquid vaporizers, lotions, aerosol sprays, wristbands, etc. [17–19] are available in the market to fight against these vectors, but still there is 40% of the world's population, mostly those living in the world's poor nations, which is at risk of these deadly diseases. According to the World Health Organization (WHO), South East Asia has the second most affected region in the world. India followed by Indonesia and Myanmar has the highest malaria burden [20]. As human beings are fighting against these deadly diseases, researchers are playing a prodigious role in controlling such diseases, which is clearly reflected by a WHO report on malaria reporting around 37% decline in malaria incidence, which accounts for about 60% reduction in mortality rate globally between 2000 and 2015 [21].

Mosquito repellent textiles are the textile material which exhibits the property of repelling mosquitoes. Textiles play a big role in human life from covering of body to fashion, which is the reason to develop a mosquito repellent textile as a need in the sense of defence from mosquitoes in the areas which are highly inhabited by mosquitoes and are prone to disease, like malaria. To impart mosquito repellency, the textile material is given an anti-mosquito finish with a chemical agent. This agent exhibits the competency of being used on textile materials without pampering their natural characteristics and has good washing fastness. DEET (*N*,*N*-Diethyl-meta-toluamide) based chemicals are most commonly used for mosquito repellency. Initially, people used to apply mosquito repellents on their skin directly as a lotion. It remains effective for a few hours and that too most of them were toxic to humans because of direct contact with the skin. These lotions have side effects like skin itching and can get adsorbed inside body through skin pores. In the market, there are many more mosquito repellent pads, vaporizers available, but they are effective only for a short duration. Due to all these reasons mosquito repellent textiles are now preferred; mosquito repellent finishing of clothing showed effectiveness up to a week. The efficacy of the clothing can be extended by preventing the evaporation of the repellent agent, which can be obtained via sealing the impregnated fabric in a container or airtight bags when the material is not in usage. The repellent infused with clothing usually retains its effect for a longer period than on skin directly because there is:

- No abrasion loss;
- No skin absorption loss;
- No effect of sweating on the active compound;
- Slower evaporation rate because of lower temperature, except when clothing is exposed to sunlight;
- Better adherence to cotton and synthetic fibres.

Mosquito repellency property on textiles can be achieved by the application of many chemicals. However, most of these chemicals showed detrimental side effects on humans and the environment, and are thus banned by the WHO. Therefore, now technocrats are shifting towards some natural mosquito repellents, which also satisfy WHO requirements.

In general, Cotton and Nylon fabrics are mostly treated with mosquito repellents since both of them were commonly used in bed sheets and mosquito nets, respectively. Among the chemical repellents, DEET is the most familiar and well-known used insect repellent. The insects are found to locate humans through chemical, visual and thermal clues. Researchers proposed that DEET can function as an insect repellent via masking the host odour mainly for lactic acid and carbon dioxide, which act as the main attractants released by the human body. Although DEET prevents the tracking of people by insects; however, something more was expected in the function of DEET effectiveness [22]. Mosquitoes detect DEET through olfaction, and this is the main reason for their avoidance actions towards DEET treated materials [23]. However, only a few centimetres apart from DEET treated skin is susceptible to mosquito bites [24].

Various efforts have been reported to advance the mosquito repellent characteristic of synthetic and cellulose based fibres by applying various additives and crosslinkers. In these developments, polypropylene was prepared with good disperse dyeability and 100% mosquito repellency [25]. DEET is also used in combination with the other repellents like permethrin adsorbed into β-cyclodextrin empty cavities and grafted onto cotton [26]. Furthermore, the mixing up with other synthetic repellents too gives more enhanced insect repellency [27]. To obtain more durable mosquito protection, DEET was encapsulated *in situ* during graft copolymerization of butyl acrylate onto chitosan and the aqueous solution was sprayed on to the fabrics [28]. In one of the attempts, DEET was modified to impart amine group and the modified DEET (MD) was utilized for preparation of mosquito repellent cotton. Mosquito repellent textiles were also prepared through cotton fabric padding with modified DEET using a conventional pad–dry–cure method. The mosquito repellent achievement was evaluated by the cage test method, wherein a human arm covered with the treated textiles was exposed to *Aedes aegypti* mosquitoes. Fabrics finish with modified DEET displayed a higher and long lasting protection from mosquitoes assuring over 90% and 70%–80% mosquito repellency for unwashed and 10 times washed fabrics, respectively [29]. This modified DEET was also applied on to cotton fabric by diazotization process where modified DEET was used as the base material for synthesis of azo group. This dyed cotton fabric has imparted mosquito repellent finish in a single step process of dyeing cum finishing. The dyed cotton fabric using napthols and MD showed very good and durable mosquito repellent activity [30].

It is to be noted that DEET or any other repellents like IR3535 are acquiring properties of mosquito repellency; however, their use has some limitations as these chemicals cannot be used beyond certain concentrations. To overcome these problems, there is a need of eco-friendly mosquito repellents and nature is a great source of these kinds of repellents which are naturally protecting humans from these vectors even before the invention of chemical repellents.

## 10.2.2 Green Approach

EO are the main reason behind the distinctive odour of plants. Approximately, 3000 EO are known, with 10% having commercial importance in the field of cosmetics, food and pharmaceutical industries, which are generally documented as safe by the Food and Drug Administration (FDA). The composition of EO may differ significantly between different species and varieties of aromatic plants, and also with the geographic areas within the same variety [31–33].

There are numerous natural plant products which exhibit mosquito repellent properties. Extracts from different parts of plants such as roots, stem, leaves, flower, fruits and seed as well as from various species exhibit mosquito repellent properties. As EO are volatile in nature, they evaporate easily at room temperature or as per changing environment temperature. However, the efficacy of mosquito repellence is also dependent upon the nature of EO. Therefore, it should have a controlled release mechanism during its use so that it can last longer which is a major challenge for researchers to achieve. Naturally occurring EO such as chrysanthemum [34], eucalyptus, lavender, citronella, etc. are major natural sources which have properties of fighting against various insects. Among the category of these EO, pine, guggul and rosewood oils also have mosquito repellent property. These oils were applied on textiles and the outcome of the study showed that EO of pine and guggul have higher mosquito repellency than rosewood oil when applied on textile substrate [35].

These EO can be applied onto cotton fabric by using direct application, emulsion application or by microencapsulation application method. Among these methods, microencapsulation method is very popular among researchers because of its controlled release mechanism. The invention of microencapsulation has solved many problems of industries such as pharma, food, perfumery and textile. Although microencapsulation has a very small role to play in the textile industry, it has solved many problems of dyeing [36], printing [37] and finishing of textile material.

In particular, the EO extracted from *Cymbopogonnardus* (citronella) has been reported to show effective repellent activity against *Aedes aegypti* females for 2 h after the direct application of pure oil on the skin of human subjects. The release rate of these volatile EO is generally affected by the application conditions. At much lower concentration, it is found to be ineffective and if it is released in excess, it can cause uncomfortable feelings. Hence, it is necessary to have controlled release of such volatile EO for sustained period in various application environment conditions. Microencapsulation technology is evaluated as one of the best methods to achieve this goal of constant and controlled release [38].

Citronella oil also possesses properties like mood-lifting during depression, deodorizing, sterilizing, bug repelling and it is used as the core material for encapsulation. The citronella oil encapsulated by using a simple coacervation method and these microcapsules prepared where gelatine was used as shell material, were subjected to drying using different techniques such as oven drying, spray drying and freeze drying. The study showed that the drying condition of microcapsules also plays a great role. The encapsulation efficiency of citronella oil containing microcapsules dried by freeze drying technique was maximum followed by that of spray dried and oven dried microcapsules. Hence, the fabric treated with freeze dried microcapsules showed good aroma rating even after 10 washes. The maximum mosquito repellency was also observed in the case of the freeze dried microcapsule treated cotton fabric. The spray dried and oven dried microcapsule treated cotton fabric also displayed mosquito repellency. There was a slight increase in the bending length, which may be due to the polymeric binder present in the application system which forms a film on curing [39].

Using microencapsulation technique, extracts of Lemon grass was encapsulated using sodium alginate as shell material. These capsules were then applied onto 100% polyester fabric by using a pad–dry–cure technique. Polyester fabric treated with microcapsules showed the highest mosquito repellent activity (~92%) when compared to that of fabric treated with methanolic extract and aqueous extract of Lemon grass [40]. It can also be seen that some of the biopesticides, like Citriodiol, exhibit insect repellency property. In fact, it has been tested and registered by the United States Environmental Protection Agency (EPA) and included in the European Biocidal Products. It is a natural complex compound, containing a mixture of both cis- and trans isomers (about 64%) of p-menthane-3,8-diol (PMD), obtained from eucalyptus citriodiora oil. It is a monoterpene, and has lower volatility as compared to other plant derived compounds, thereby resulting in a longer protection time. The microcapsules containing two biopesticides, namely citronella EO and citriodiol were studied towards their efficacy in the preparation of mosquito repellent cotton and it was reported that cotton fabrics treated with microencapsulated citriodiol can provide 100% repellency for more than 30 days [41].

Mosquito repellent fabrics have been modified using monochlorotriazenyl-β-cyclodextrine (MCT-β-CD) then treated with different concentrations of pesticides like permethrin [42], bioallethrin [43], cypermethrin, prallethrin [44] and limonene [45] to impart insect repellency to the fabric. Modified fabrics showed excellent mosquito repellency with minimal changes in the physical properties of fabric along with good durability of mosquito repellency against washing.

The prallethrin application in to the fabric was carried out with a solution containing prallethrin along with a polymeric binder and a crosslinking agent, or by surface coating of the fabric with a polymeric binder containing prallethrin. The treated fabrics showed repellent action, knockdown (KD) action and killing action against mosquitoes, which were found to increase with the concentration of prallethrin in the treated fabric within the range studied (250–1500 mg/m$^2$) and the exposure time [46]. In another method, three different mosquito repellent agents such as permethrin, prallethrin, and Eucalyptus oil were fixed directly or via a binder or crosslinker on a fabric and repellency efficiency was evaluated before and after washing. All repellent-treated fabrics showed satisfactory levels of mosquito repellency in the unwashed state, but significant losses in efficiency were observed after washing [47].

Permethrin and bifenthrin were used as mosquito repellents and applied on to the Australian military shirt fabrics. Bifenthrin showed mosquito repellency higher at around 100% after initial treatment whereas permethrin showed 94%–100% repellency. However, the repellency was reduced to 58% and 66% after one wash indicating loss of repellent chemicals from the test fabric [48].

The microcapsules containing fragrance or insecticide can be applied which will get broken under pressure and release the active component which will show the desirable function. This phenomenon of application of microencapsulated products in military clothing is common for tropical climates with abundant mosquitoes. The textiles treated with microcapsules containing tea tree oil showed significant mosquito repellent properties. The encapsulated insect-resistant materials are also suitable for textiles containing protein fibres such as wool as they provide long-term protection against the larvae of moths [3, 49]. Moretti *et al.* prepared gelatine microcapsules containing EO such as Rosemary and Thymus via a phase separation method. The microencapsulation indicated a high yield over 98% with both EO. The prepared gelatine microcapsule was entrapped in the body hair of larvae which were destroyed after 2 h [50]. Microcapsules containing citronella oil were prepared by Specos *et al.* using the complex coacervation method, which were applied on the cotton fabric using pad–dry to obtain the mosquito repellent fabric. The fragrance smell was found without scratching after 28 d hanging in the open air, and also could only be smelt upon scratching the cotton fabric. The treated fabric presented the highest and longer lasting protection against insects than the fabric sprayed with citronella oil [51].

Some herbs like *Andrographis paniculata* plant also have high-repellent activity against mosquitoes, insects and snakes. Ramya and Maheshwari [52] prepared sodium alginate microcapsules containing Andrographolide extracted from the mentioned plant and applied on the bamboo/cotton fabric through exhaustion. The mosquito repellent efficiency was 94% for the fabric finished with microcapsules and 96% for the finished fabric with the direct application. The retention of the repellent activity on the fabric finished with microcapsules was higher than the direct application. A bamboo/Tencel (50:50) blended fabric was finished by Geethadevi and Maheshwari [53] using microencapsulated EO to obtain a mosquito repellent fabric. They prepared the fragrance microcapsules using sodium alginate, Acacia Arabica and Moringaoleifera gum as the wall materials and thyme, cypress oil and grapefruit oils as the core materials and applied on the bamboo/Tencel blended fabric through exhaustion method.

Moringaoleifera microcapsules indicated a good mosquito repellent activity, high UV protection and high durability without allergic reaction on the skin. The capsules enhanced the retention of mosquito-repellent action to 60% even after 30 washes [54].

## 10.3 Methods of Application of Microcapsules on Textiles

The application of microcapsules to textile materials is a three step process, starting with the introduction of microcapsules in the textile material, followed by dispersion of microcapsules around and through the textile material using dispersing agent, and finally the addition of crosslinking agent to support the adhesion of microcapsules to textile materials. Microcapsules can also be applied during the finishing process of textiles using padding mangle and a finishing bath containing microcapsules, along with self-crosslinking agent or binding agent as per requirement and feasibility of application.

The process presented above is a step forward on perfume microencapsulation technology and provides a viable alternative to systems that are traditionally used in textile applications (Figure 10.2). The majority of the microencapsulated application systems follow this finishing process.

## 10.4 Release Mechanism of Core Material from Microcapsules

The release mechanism of the materials from the capsule depends on the structure of polymeric wall including density, permeability and biological properties. It also depends on external environment. In general, there are four mechanisms for releasing the core materials from the capsules like mechanical stimuli, chemical stimuli, thermal stimuli and diffusion [55, 56].

## 10.5 Chemistry of EO

EO, also known as volatile oil, is a complex mixture of volatile components synthesized by a living organism. Due to the presence of volatile organic components, EO shows mosquito repellency property. They are composed of hydrocarbons and their oxygenated derivatives; besides these

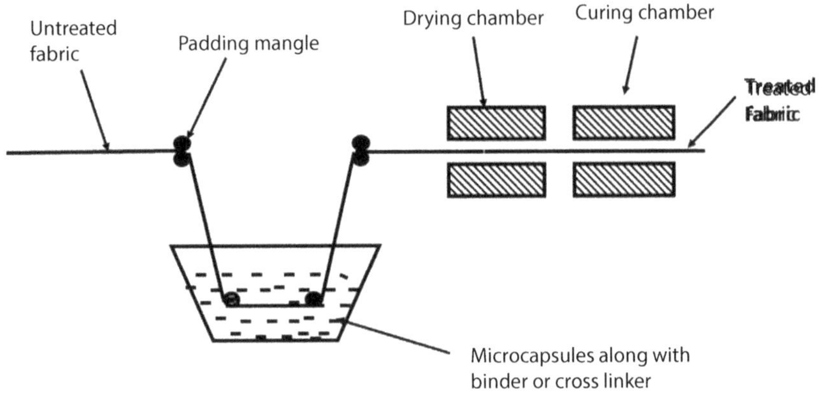


**Figure 10.2** Schematic representation of the process of microcapsules application in fabrics.

some may contain sulphur and nitrogen derivatives also. Hydrocarbons include terpenes and sesquiterpenes, and oxygenated derivatives such as ethers, alcohols, aldehydes, esters, lactones, ketones, phenols and phenol ethers, etc. Some monoterpenes such as cineole, eugenol, α-pinene, terpnolene, citronellat, thymol, limonene, camphor, citronemmam, etc. are found as common constituents of EO offering mosquito repellent activity. Among sesquiterpenes, β-caryophyllene is said to be a strong repellent against *Aedes aegypti* [57]. Chemical structures of some EO are given in Figure 10.3. The repellent properties of many EO regularly appear to be connected with the presence of sesquiterpenes and monoterpenoids. Some authors [58] also have reported that phytol has high repellent activity against *Anopheles gambiae*. Odalo *et al.* [58] assessed the repellent activities of different plants in Kenya against *Anopheles gambiae* (Diptera), and some pure metabolites extracted from them. They found the compounds like perillyl alcohol, *cis*-carveol, *cis*-verbenol, citronellal, geraniol, perillaldehyde, caryophyllene oxide, carvacrol, 4-isopropylbenzenemethanol, thymol, 3-carene and myrcene to be the most promising repelling chemicals. The oxygenated compounds such as β-citronellol, phenylethyl alcohol, cinnamyl alcohol, geraniol and α-pinene, isolated from the EO of *Dianthus caryophyllum*, showed strong repellency against ticks (nymphs of *Ixodes ricinus*) [59].

Most of the arthropod-repellents reported are oxygenated compounds (such as cinamyl alcohol, phenylethyl alcohol, α-pinene, geraniol, etc.), having the hydroxyl group linked to a primary, secondary or aromatic carbon. However, it is noteworthy that in some metabolites in which the hydroxyl group is linked to tertiary carbon (e.g. α-terpineol, linalool and

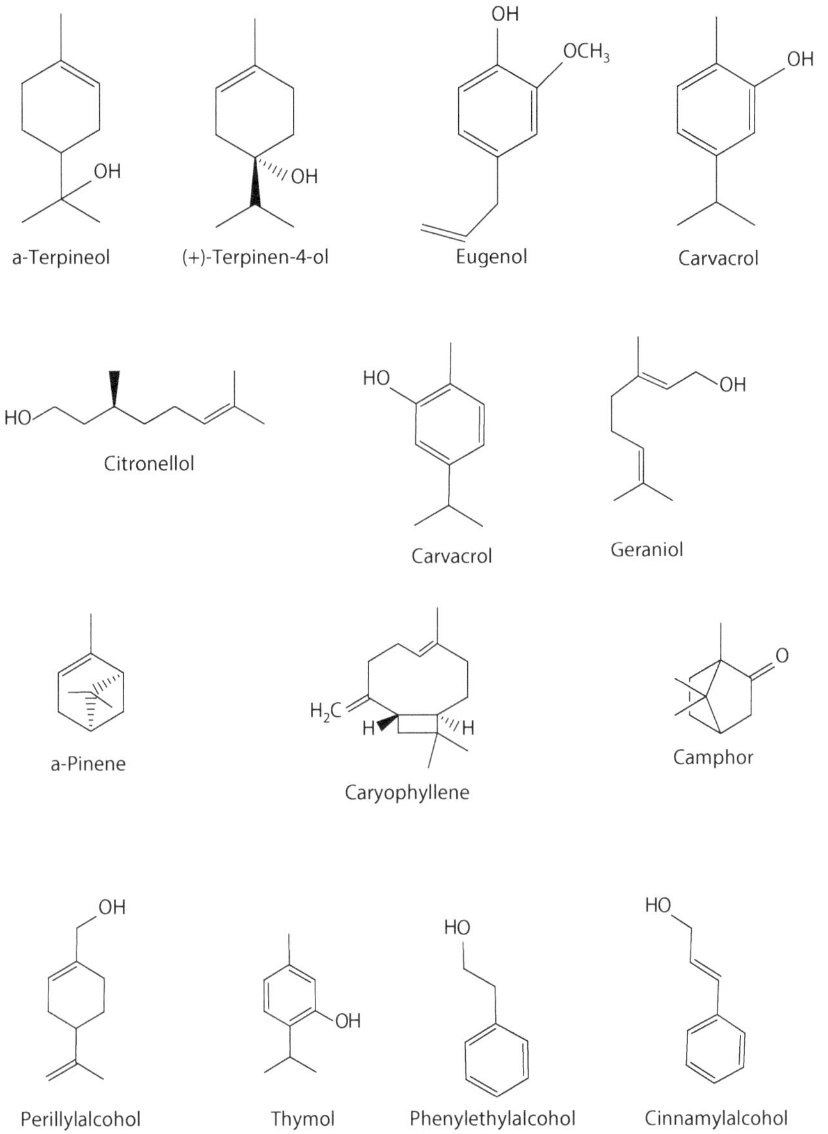


**Figure 10.3** Chemical structures of some essential oils (EO).

limonene) the activity is suppressed against *Anopheles gambiae*, suggesting the compounds in which hydroxyl group was substituted modulate repellency. Some studies further demonstrated that terpenoids having two functional groups are biologically active acting as mosquito repellents [60].

## 10.6 Evaluation of Mosquito Repellency

### 10.6.1 American Society for Testing and Materials (ASTM) Standard E951–83

This is a method usually used for non-commercial mosquito repellent preparation on the skin and the repellency activity is evaluated using the test cage as shown in Figure 10.4. The test cages are placed firmly on the arms with Velcro tapes ensuring the only exposure of test area to mosquito bites. Fifteen female mosquitoes, between 4 to 7 days-old were introduced into every cage and the number of biting was noted and recorded at the end of 120 s [61].

### 10.6.2 Screened Cage Method

In this test method, forearm treated with mosquito repellent is inserted into the cage (the hand portion is protected from mosquito bites by wearing a glove) (Figure 10.5). This technique is based on the idea of total number of mosquitoes that land and probe the skin in the given time (3 min), which is observed and recorded. The processes can repeat for every 30 or 60 min. Generally, two bites in one 3 min test or one bite in one 3 min test, followed by one or more bites after 30 min in a second test ends the test for the repellent activity. A second cage containing mosquitoes is used as positive or negative control. At the end of observation, the final data obtained through this bioassay technique can be used to estimate complete protection time (CPT).

### 10.6.3 WHO Cone and Field Test Method

WHO cone test was used to check the mosquito mortality rate. In this test, susceptible non-blood-fed mosquitoes aged between 2 and 5 days were introduced into WHO cones for a period of 3 min. After exposure, mosquitoes were placed in glass cylinders with sucrose solution in it for 24 h

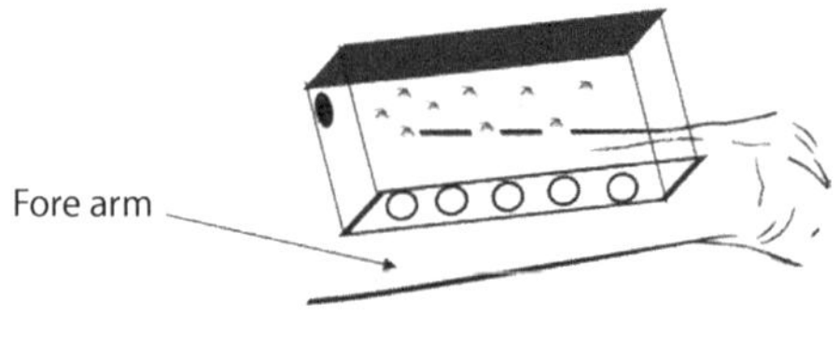


**Figure 10.4** Mosquito repellency evaluation by ASTM Standard E951–83.

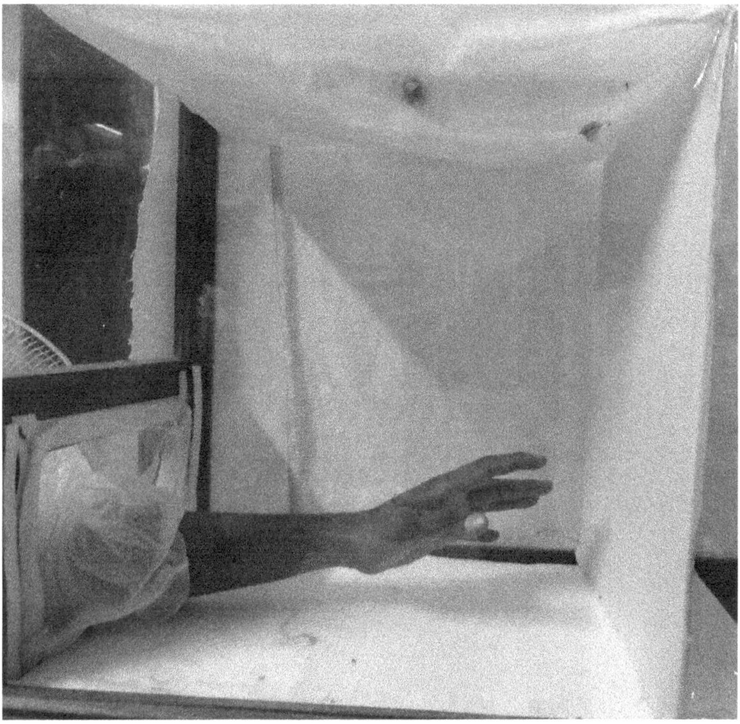

**Figure 10.5** The screen cage method for testing repellent formulations trial.

at room temperature. Percentage KD (immobilized) after 60 min and percentage mortality after 24 h were recorded (Figure 10.6).

The field test is the most expressive evidence to evaluate the effectiveness of treated textile. The field test is especially performed in such locations where abundant floodwater mosquitoes are present (Figure 10.7) [61]. The forearms are treated with 0.5 mL of repellent and exposed to mosquitoes, and the total number of bites is recorded.

### 10.6.4 Tunnel Test

The WHO cone test merely measures the killing effect. This ensures the requirement of an additional test method, so that the repellents cannot be eliminated on the basis of a single test of mortality for 3 min. The tunnel test is used to measure the mortality and blood-feeding success of host-seeking mosquitoes in an experimental chamber shown in Figure 10.8 [62]. The setup consists of a 60 cm (length, square section 25 cm × 25 cm) glass tunnel, with 25 cm square cages at both ends covered with polyester netting.

Figure 10.6 Cone used in the cone test.

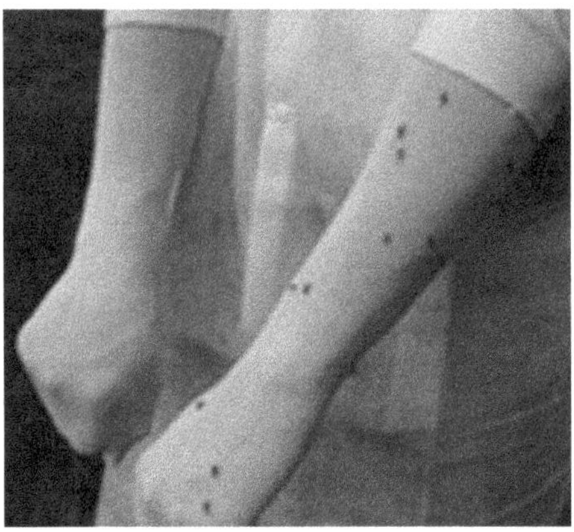

Figure 10.7 Field test.

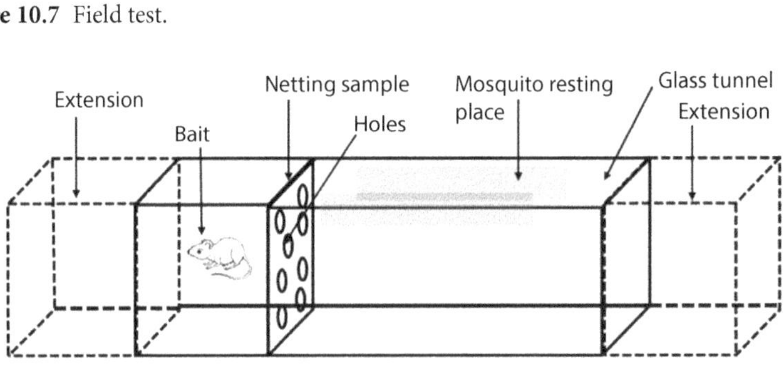


**Figure 10.8** A diagrammatic outline of the tunnel used for the study of effectiveness of long lasting insecticidal mosquito nets with extensions on either sides to fix mosquito cages.

A cardboard frame holding the repellent treated samples with nine holes (diameter 1 cm) is placed at one-third of the total tunnel length. On the shorter side, unmovable bait is fitted. In the whole process, an untreated sample is always used as a negative control.

### 10.6.5 USDA Laboratory Method

For the treated fabrics, Mosquito Repellency is calculated by USDA laboratory method which simulates the natural conditions of exposure. An artificially created tent is prepared in which laboratory reared strains of female culex species of mosquitoes are released in the range of 10 to 30. Hourly observations are made on their behaviour, mobility, etc. During a 3 h exposure, hourly data are recorded in terms of total KD; inability to fly on account of their loosing appendages is recorded [63].

## 10.7 Aroma Finish

Our sense of smell is the strongest of our five senses; it is the one sense which does not rest even when we sleep. Utilizing the sense of smell by adding fragrance into textile or garment can stimulate an influential and emotive sense [64]. Fragrance finishing has busted the textile industry with elated value. The addition of fragrance to textiles is a very common practice and started many years ago to impart fresh aroma to the textile. This can be carried out either as fabric conditioners in the wash or during tumble drying. The incorporation of different scents into fabrics, leads to the fabrication of scented fabrics and the psychology of acceptance of synthetic scents in textile goods.

Fragrance acts as a healing substance which is used in aromatherapy. Aromatherapy is a medicinal technique that helps people to heal with natural substance, by using pure fragrance compounds and EO. It enables a person to attempt for self-healing at home. It has been found that these fragrances have effects upon feelings, mood and emotions. They also have effects on several other human body functions like electrical activity in the brain, physiological parameters such as the heart rate and skin conductance, voluntary and involuntary behaviour, etc. [65].

However, the durability of fragrance effect remains a big challenge, which is not controllable by the technique used to impart the fragrance. Several attempts have been made at adding fragrance directly to the fibre and fabric, but all fail to endure the fragrance and thus vanished upon only one or two wash cycles. Only the fragrance imparted through microencapsulation technique gives high durable fragrance, which remains on the garment through a significant part of its lifetime [66]. The scent of various EO encapsulated into the fabric proved acted in a good way to achieve important psychological, emotional, and hygienic need. These kinds of fabrics are called aroma textiles [67].

Aroma incorporating micro encapsulated fragrance is typically used for:

- Relaxing or encouraging sleep by using fragrances for bedding and sleepwear,
- Muscle-easing when applied to after sports activity garments,
- Uplifting/head clearing when used in active sports wear,
- Confidence building when fragrance is used on formal business wears.

The main challenge to aromatic textile is to prepare fragrance compounds based microcapsules using EO without neglecting any of its ingredients to express their full effects. The application of polymer binder at low temperature to bind perfumed microcapsule to the textile surface is another important part of aromatic textile preparation. Along with these, the durability during laundering and a soft handle should be considered.

### 10.7.1 General Method of Application

In a general process meant for microcapsule application to textile materials, initially microcapsules can be introduced in the liquor prepared for textile treatment followed by addition of crosslinking agent to boost up the adhesion of microcapsules to the textile material. The application of microcapsules can also be done in the course of finishing process, which is described earlier. The majority of the commercially available microencapsulated fragrance systems in textile applications are highly based on formaldehyde systems (e.g. phenol-formaldehyde or melamine formaldehyde resins), which witness several restrictions under finishing of textile material for value addition. In this context, polyurethane–urea systems emerge as a more attractive and environment-friendly solution. Furthermore, this system is known as versatile polymer systems which can be fabricated from a wide range of raw materials to acquire the desired physical, chemical and mechanical properties. The only challenge with these systems is, the system must be optimized and designed as per the required characteristics of the active principle selected for encapsulation. The release rate of fragrance from various perfumed microcapsules depended on the nature of the fragrance oils themselves.

### 10.7.2 Green Methods: EO for Aroma Finish

Aromatic plants are a class of plants containing fragrant compounds or EO. The effective compounds of aromatic plants are classified into four main groups as the following: alkaloids, glycosides, EO or essences and other effective materials such as flavones, flavonoids, vitamins, carbohydrates, tannins and salicylic acid. The reason for which the above mentioned EO are showing fragrance with functional property is because of EO containing mixtures of volatile organic compounds. They are composed of hydrocarbons (terpenes and sesquiterpenes) and oxygenated compounds (aldehydes, alcohols, ethers, esters, lactones, ketones, phenols and phenol ethers). This aroma product also includes different alcohols, such as saturated alcohols (like ethanol, propylene glycol and di-propylene glycol), unsaturated alcohols (*cis*-3-hexenol (leaf alcohol), terpene alcohols and benzyl alcohols etc.) and also important phenolic compounds like guaiacol-4-carboxaldehyde, better known as Vanillin. Some acids such as acetic acid, phenyl acetic acid and trans-2-butenoic acid are also responsible for these properties. The other compound includes aliphatic aldehyde (acetaldehyde) and aromatic aldehyde (phenyl acetaldehyde); carotenoids and ketone based and few natural aroma chemicals are also present [54, 57, 68].

The microencapsulation is the most efficient method to store these fragrant ingredients against any odd conditions like temperature and self-evaporation. Various reports are available regarding modification of textiles for imparting fragrance using the microencapsulation technique [51, 69–72].

The durability of an encapsulated fragrant finish is an important aspect of the finished textiles which is affected by the type of binding agents, curing methods, and curing conditions used for fixing the finish of the fabric. The curing method of microcapsule treated fabrics plays an important role in retention of fragrance in treated fabrics. Li *et al.* [73] explained the mechanism of heating effect on application of microcapsules during finishing on fabric. The fabric was treated with an acrylic binder–encapsulated fragrant finish and cured separately using both conventional hot air stenter and infrared (IR) processes. Among the fixing agents employed, the acrylic binder demonstrated excellent fixing of the fragrant capsules for all curing methods. The conventional hot air stenter and the IR curing methods produced samples that could withstand at least 25 wash cycles. Although any thermal conditions could deplete the aroma from the capsules, microwave radiation was found to be the most severe curing process; because of its high penetrating ability it causes heating of both core (aroma) and wall of the capsule simultaneously, resulting in excessive loss of the fragrance through wall migration/evaporation of the fragrance and/or rupturing of the capsules. Therefore, microwave curing may not be suitable for curing encapsulated fragrant finishes.

Li *et al.* [74] also reported a method where Lemon fragrance capsules were applied on cotton fabric using UV resin and cured under the optimal conditions of UV light at low temperature. The aroma lost is limited during UV curing, which confirms higher durability of the finish, i.e. the aroma function can endure 50 wash cycles if applied under optimal conditions.

### 10.7.3 Evaluation of Aroma Finishes

The various methods can be utilized to confirm the modification of textile material with fragrance microcapsules as well as the efficacy of such finishes. Fourier Transform Infrared spectroscopy (FTIR) spectra can be compared to determine the composition of created perfume before and after encapsulation, and the finished fabric. In this analysis, we can see shift in methyl or methylene group as the major constituents of fragrance are based on these groups. If the microcapsules can be applied using carboxylic acid, then after application it can form ester which can easily be seen in this study. The particle size of microcapsule plays a dominant role in the efficacy of encapsulation and release of fragrance from prepared microcapsule. Scanning electron microscopy (SEM) can give the morphology and surface characteristics of the microcapsules as well as substrate finished with microcapsules. It indicates the general size and shape of the microcapsules, the evenness of their spread on textiles and the extent of their attachment to the textile material. Gas chromatography- flame ionization detector (GC-FID) headspace can also be used to determine the components of the fragrance. Aroma can be subjectively evaluated by using the Lewis method [75, 76].

## 10.8 Conclusion

Development of green methodologies for functional modification of textiles is the need of the hour. Huge research has been reported in the area of functional finishing of textiles using green methods; however, most of them either failed in their efficacy or the durability of such effects after washing. Commercialization of such methods also remains important as most of the research is available in the literature but no commercialization was reported. Important protective finishes like mosquito repellent, antimicrobial and flame retardance are coming into picture due to demand

and development of technical textiles. It is highly desirable to identify the safer chemicals and evaluate their efficacy as a finish for textiles. The opportunities for such products are endless and the market potential is huge.

## References


1. Gupta, D., Functional clothing – definition and classification. *Ind. J. Fib. Text. Res.*, 36, 321–326, 2011.
2. Shenai, V.A., *Technology of Finishing*, Sevak Publications, Mumbai, 1987.
3. Schindler, W.D. and Hauser, P.J., *Chemical Finishing of Textiles*, Woodhead Publishing Limited, Cambridge, 2004.
4. Ammayappan, L. and Moses, J., Study of antimicrobial activity of aloevera, chitosan, and curcumin on cotton, wool, and rabbit hair. *Fib. Polym.*, 10(2), 161–166, 2009.
5. Feng, X., Zheng, K., Wang, C., Chu, F., Chen, Y., Durable antibacterial cotton fabrics with chitosan based quaternary ammonium salt. *Fib. Polym.*, 17(3), 371–379, 2016.
6. Ganesan, P., Ramachandran, T., Karthik, T., PremAnand, V.S., Gowthaman, T., Process optimization of Aervalanata extract treated textile material for microbial resistance in healthcare textiles. *Fib. Polym.*, 14(10), 1663–1673, 2013.
7. Sharaf, S., Higazy, A., Hebeish, A., Propolis induced antibacterial activity and other technical properties of cotton textiles. *Int. J. Biol. Macromol.*, 59, 408–416, 2013.
8. Hong, K.H., Crosslinking phenolic compounds with cotton fabrics using succinic acid to develop functional clothing materials. *Fib. Polym.*, 17(5), 705–711, 2016.
9. Basak, S., Chattopadhyay, S.K., Samanta, K.K., Das, S., Narkar, R., Flame retardant and antimicrobial jute textile using sodium metasilicate nonahydrate. *Pol. J. Chem. Technol.*, 16, 106–113, 2014.
10. Horrocks, A.R., Flame retardant challenges for textiles and fibres: new chemistry versus innovatory solutions. *Polym. Degrad. Stab.*, 96, 377–392, 2011.
11. Alongi, J., Carletto, R.A., Blasio, A.D., Cuttica, F., Carosio, F., Bosco, F., Malucelli, G., Intrinsic intumescent like fire retardant properties of DNA treated cotton fabric. *Carbohydr. Polym.*, 96, 296–304, 2013.
12. Bosco, F., Carletto, R.A., Alongi, J., Marmo, L., Di Blasio, A., Malucelli, G., Thermal stability of flame resistance of cotton fabrics treated with whey proteins. *Carbohydr. Polym.*, 94, 372–377, 2012.
13. Malucelli, G., Bosco, F., Alongi, J., Carosio, F., Blasio, A.D., Mollea, C., Cuttica, F., Casale, A., Biomacromolecules as novel green flame retardant systems for textiles: an overview. *RSC Adv.*, 4, 46024–46039, 2014.

14. Shahid, M., Shahid-Ul-Islam, Mohammad, F., Recent advancements in natural dye applications: a review. *J. Clean. Prod.*, 53, 310–331, 2013.
15. Shahid-Ul-Islam and Mohammad, F., High-energy radiation induced sustainable coloration and functional finishing of textile materials. *Ind. Eng. Chem. Res.*, 54(15), 3727–3745, 2015.
16. Shahid-Ul-Islam and Sun, G., Thermodynamics, kinetics, and multifunctional finishing of textile materials with colorants extracted from natural renewable sources. *ACS Sustain. Chem. Eng.*, 5(9), 7451–7466, 2017.
17. USEPA. Registration Eligibility Decision (RED) DEET, EPA738-R-98-010, 1998.
18. Mount, G.A and Snoddy, E.L., Pressurized sprays of permethrin and deet on clothing for personal protection against the lone star tick and the American dog tick (Acari: Ixodidae). *J. Econ. Entomol.*, 76, 529–531, 1983.
19. Fradin, M.S., Mosquitoes and mosquito repellents: a clinician's guide. *Ann. Intern. Med.*, 128, 931–940, 1988.
20. WHO, World malaria fact sheet, http://www.who.int/malaria/publications/world_ malaria_report_2012/report/en/, 2012.
21. WHO, World Malaria Report, http://www.who.int/malaria/publications/world-malaria-report-2015/report/en/, 2015.
22. Pickett, J.A., Birkett, M.A., Logan, J.G., DEET repels ORNery mosquitoes. *Proc. Nat. Acad. Sci.*, 105(36), 13195–13196, 2008.
23. Alee, Y., Kim, S.H., Montell, C., Avoiding DEET through insect gustatory receptors. *Neuron*, 67(4), 555–561, 2010.
24. Akbarzadeh, A., Mokhtari, J., Kolkoohi, S., Sarli, M.A., Imparting insect repellency to nylon 6 fibers by means of a novel MCT reactive dye, *J. Appl. Polym. Sci.*, 126, 1097–1104, 2012.
25. Sabale, A.C., Teli, M.D., Adivarekar, R., Mosquito-repellent and fragrant polypropylene, *Melliand Int.*, 18(3), 158–160, 2012.
26. Romi, R., LoNostro, P., Bocci, E., Ridi, F., Baglioni, P., Bioengineering of a cellulosic fabric for insecticide delivery via grafted cyclodextrin. *Biotechnol. Prog.*, 21(6), 1724–1730, 2005.
27. Utunomiya, A., Hasegawa, K., Mori, Y., Analysis of pyrethroid pesticides, synergists and repellent in moth/mite-proofed household products and their mutagenicity. *Japan. J. Toxicol. Environ. Health*, 43, 366–375, 1997.
28. Fei, B. and Xin, J.H., N, N-diethyl-m-toluamide containing microcapsules for bio-cloth finishing. *Am. J. Trop. Med. Hyg.*, 77(1), 52–57, 2007.
29. Teli, M.D and Chavan, P.P., Modified application process on cotton fabric for improved mosquito repellency, *J. Text. Inst.*, 108, 915–921, 2017.
30. Teli, M.D and Chavan, P.P., Dyeing of cotton fabric for improved mosquito repellency. *J. Text. Inst.*, 109(4), 427–434, 2017.
31. Coppen, J.J.W., Flavours and fragrances of plant origin, *Non-Wood Forest Product*, pp. 1–6, FAO, Rome, 1995.
32. Nerio, L.S., Olivero-Verbel, J., Stashenko, E., Repellent activity of essential oils: a review. *Bioresour. Technol.*, 101, 372–378, 2010.

33. Trongtokit, Y., Rongsriyam, Y., Komalamisra, N., Apiwathnasorn, C., Comparative repellency of 38 essential oils against mosquito bites. *Phytother. Res.*, 19, 303–309, 2005.
34. Gopalakrishnan, D., Mosquito repellent fabrics. *Man-Made Text. India*, 50(1), 15–18, 2007.
35. Jaipura, L., Saini, M., Rangi, A., Development of herbal mosquito repellent textiles using essential oils of Pine, Guggul and Rosewood. *Colourage*, 63(1), 44–48, 2016.
36. Purev, N., Burgert, L., Prichystal, P., Hrdina, R., Kühn, J., Cerny, M., Oyuntulkhuur, J., Dyeing of leather with microencapsulated acid dye. *Color. Technol.*, 129(6), 412–417, 2013.
37. Christie, R.M and Bryant, I.D., An evaluation of thermochromic prints based on microencapsulated liquid crystals using variable temperature colour measurement. *Color. Technol.*, 121, 187–192, 2005.
38. Pratt, H.D. and Moore, C.G., Mosquitoes of public health importance and their control, Published by U.S. Department of Health and Human Services, Public Health Service, Centers for Disease Control and Prevention, Public Health Practice Program Office, Atlanta, Georgia, 1993.
39. Teli, M.D. and Chavan, P., Application of gelatine based microcapsules containing mosquito repellent/oils on cellulosic biopolymer. *J. Bionanosci.*, 10(5), 390–395, 2016.
40. Anitha, R., Ramachandran, T., Rajendran, R., Mahalakshmi, M., Microencapsulation of lemon grass oil for mosquito repellent finishes in polyester textiles. *Elixir Bio. Phys.*, 40, 5196–5200, 2011.
41. Specos, M.M.M., Garci, J.J., Gutierrez, A.C., Hermida, L.G., Application of microencapsulated biopesticides to improve repellent finishing of cotton fabrics. *J. Text. Inst.*, 108(8), 1454–1460, 2017.
42. Shahba, A.F., Halawa, O., Ragaei, M., Hashem, M., Development of longer-lasting insect repellence cellulosic based curtain fabrics. *Mater. Sci. Appl.*, 2(3), 200–208, 2011.
43. Abdel-Mohdy, F.A., Fouda, M.M.G., Rehan, M.F., Ali, A.S., Repellency of controlled-release treated-cotton fabrics based on permethrin and biollethrin against mosquitos. *J. Text. Inst.*, 100(8), 695–701, 2009.
44. Abdel-Mohdy, F.A., Fouda, M.M.G., Rehan, M.F., Ali, A.S., Repellency of controlled-release treated cotton fabrics based on cypermethrin and prallethrin. *Carbohydr. Polym.*, 73(1), 92–97, 2008.
45. Hebeish, A., Fouda, M.M.G., Hamdy, I.A., El-Sawy, S.M., Abdel-Mohdy, F.A., Preparation of durable insect repellent cotton fabric: Limonene as insecticide. *Carbohydr. Polym.*, 74(2), 268–273, 2008.
46. Hebeish, A., Hamdy, I.A., El-Sawy, S.M., Abdel-Mohdy, F.A., Prallethrin treatment–induced permanent insect repellency to cotton fabric. *Egypt. J. Text. Polym. Sci. Technol.*, 13(1), 35–48, 2009.
47. Gulrajani, M., Agarwal, A., Lohia, C., Preparation of mosquito repellent fabrics. *Asian Dyer*, 4(5), 53–55, 2007.

48. Frances, S.P., Watson, K., Constable, B.G., Comparative toxicity of permethrin- and bifenthrin-treated cloth fabric for Anopheles farauti and Aedes aegypti. *J. Am. Mosq. Control Assoc.*, 19(3), 275–278, 2003.
49. Holme, I., Innovative technologies for high performance textiles. *Color. Technol.*, 123, 59–73, 2007.
50. Moretti, M.D.L., Sanna-Passino, G., Demontis, S., Bazzoni, E., Essential oil formulations useful as a new tool for insect pest control. *AAPS Pharm. Sci. Tech.*, 3(2), 64–74, 2002.
51. Specos, M.M.M., Escobar, G., Marino, P., Puggia, C., Tesoriero, M.V.D., Hermida, L., Aroma finishing of cotton fabrics by means of microencapsulation techniques. *J. Ind. Text.*, 40(1), 13–32, 2010.
52. Ramya, K. and Maheshwari, V., Development of eco friendly mosquito repellent fabric finished with Andrographis Paniculata plant extracts. *Int. J. Pharm. Pharm. Sci.*, 6(5), 115–117, 2014.
53. Geethadevi, R. and Maheshwari, V., Long-lasting UV protection and mosquito repellent finish on bamboo/tencel blended fabric with microencapsulated essential oil. *Indian J. Fib. Text. Res.*, 40, 175–179, 2015.
54. Ghayempour, S. and Montazer, M., Micro/nanoencapsulation of essential oils and fragrances: focus on perfumed, antimicrobial, mosquito-repellent and medical textiles. *J. Microencapsul.*, 33(6), 497–510, 2016.
55. Versic, R.J., Flavor encapsulation: an overview, in: *Flavor Encapsulation*, S.A. Risch, G.A. Reineccius (Eds), vol. 370, pp. 1–6, US: ACS Symposium Series, American Chemical Society, Washington, DC, 1988.
56. Cheng, S.Y., Yuen, M.C.W., Kan, C.W., Cheuk, K.K.L., Chui, C.H., Lam, K,H., Cosmetic textiles with biological benefits: gelatin microcapsules containing Vitamin C, *Int. J. Mol. Med.*, 24, 411–419, 2009.
57. Gillij, Y.G., Gleiser, R.M., Zygadlo, J.A., Mosquito repellent activity of essential oils of aromatic plants growing in Argentina. *Bioresour. Technol.*, 99, 2507-2515, 2008.
58. Odalo, J.O., Omolo, M.O., Malebo, H., Angira, J., Njeru, P.M., Ndiege, I.O., Hassanali, A., Repellency of essential oils of some plants from the Kenyan coast against Anopheles gambiae. *Acta Trop.* 95, 210–218, 2005.
59. Tunon, H., Thorsell, W., Mikiver, A., Malander, I., Arthropod repellency, especially tick (*Ixodes ricinus*), exerted by extract from Artemisia abrotanum and essential oil from flowers of Dianthus caryophyllum. *Fitoterapia*, 77(4), 257–261, 2006.
60. Nerio, L.S., Jesus, O.V., Elena, S., Repellent activity of essential oils: a review. *Bioresour. Technol.* 101(1), 372–378, 2010.
61. Debboun, M., Frances, S.P., Strickman, D., *Insect Repellents: Principle, Methods and Uses*, CRC Press, Boca Raton, FL 33487, 2006.
62. http://apps.who.int/iris/bitstream/10665/44610/1/9789241501705_eng.pdf., (Accessed on 20th Nov. 2017)
63. Samson, R. D., McKinney, J. M., Russell, J., Insect repellent tent fabric, US Patent 5198287, 1993.

64. Bairagdar, S.M., Katkar, P.M., Microencapsulation in textiles. *Colourage*, 58(8), 41–44, 2009.
65. Harini, K., Leena, A., Giridev, V.R., Microencapsulation and its applications. *Synth. Fib.*, 36(1), 21–28, 2007.
66. Nelson, G., Microencapsulation in textile finishing. *Rev. Prog. Color.*, 31(3), 57–63, 2001.
67. Wang, C.X. and Chen, S.L., Aromachology and its application in the textile field. *Fib. Text. East. Eur.*, 13, no. 6 (54), 41–44, 2005.
68. Rowe, D.J., Aroma chemicals I: C, H, O compounds, in: *Chemistry and Technology of Flavors and Fragrances*, D.J. Rowe (Ed.), pp. 56–83, CRC Press, Boca Raton, FL, 2005.
69. Hipparagi, S.A., Srinivasa, T., Das, B., Naik, S.V., Purushotham, S.P., Studies on application of aroma finish on silk fabric. *J. Inst. India Ser. E*, 97(2), 159–165, 2016.
70. Silva, M., Martins, I.M., Barreiro, M.F., Dias, M.M., Rodrigues, A.E., Functionalized textiles with PUU/limonene microcapsules: effect of finishing methods on fragrance release. *J. Text. Inst.*, 108(3), 361–367, 2017.
71. Kumari, P., Rose, N.M., Jeet Singh, S.S., Microencapsulation of lime essential oil for fragrant textiles. *Ann. Agri Bio Res.*, 20(1), 152–157, 2015.
72. Li, Y., Ai, L., Yokoyama, W., Shoemaker, C.F., Wei, D., Ma, J., Zhong, F., Properties of chitosan-microencapsulated orange oil prepared by spray-drying and its stability to detergents. *J. Agric. Food Chem.*, 61(13), 3311–3319, 2013.
73. Li, S., Lewis, J.E., Stewart, N.M., Qian, L., Boyter, H., Effect of finishing methods on washing durability of microencapsulated aroma finishing. *J. Text. Inst.*, 99(2), 177–183, 2008.
74. Li, S., Boyter, H., Qian, L., UV curing for encapsulated aroma finish on cotton. *J. Text. Inst.*, 96(6), 407–411, 2005.
75. Teli, M.D., Finishing of carpets for value addition, in: *Advances in Carpet Manufacture, (Sec. Ed.) A volume in The Textile Institute Book Series*, K.K. Goswami (Ed.), pp. 175–211, Elsevier, UK, 2018.
76. Miyazawa, M., Nakahashi, H., Kashima, Y., Motooka, R., Hara, N., Nakagawa, H., Yoshii, T., Usami, A., Marumoto, S., Chemical composition and aroma evaluation of essential oils from skunk cabbage (Symplocarpus foetidus). *J. Oleo Sci.* 64(12), 1329–1336, 2015.

# 11

# Advances in Bio-Nanohybrid Materials

**Houda Saad[1], Pedro L. de Hoyos-Martinez[2,3], Ezzeddine Srasra[1] and Fatima Charrier-El Bouhtoury[2*]**

*[1]Laboratoire des Matériaux Composites et Minéraux Argileux (LMCMA), Centre National des Recherches en Sciences des Matériaux (CNRSM), Technopôle de Borj-Cédria, Soliman, Tunisia*
*[2]Université de Pau et des Pays de l'Adour, IPREM – UMR CNRS 5254, IUT des Pays de l'Adour, Mont de Marsan, France*
*[3]University of the Basque Country (UPV/EHU), Chemical and Environmental Engineering Department, Donostia-San Sebastian, Spain*

***Abstract***
This chapter presents an overview on the recent advances in research and development on bio-nanohybrid materials and focuses on the clays and polyphenols hybrids. Indeed, plant polyphenols have received a great deal of attention, in recent years, due to their diverse biological functions and their antioxidative, nutraceutical and dyeing properties. This chapter covers recent development and key challenges in the use of this kind of bio-nanohybrids and presents their preparation and properties. Their related applications in novel specific fields such as active packaging, functional coating, heavy metal removal and controlled delivery system are considered. It is also an opportunity to discuss challenges and future perspectives of polyphenols-based hybrids for diverse applications.

***Keywords:*** Bio nanohybrids, polyphenols, clays, hybrid organic–inorganic materials

## 11.1 Introduction

In recent years, the production and application of hybrid nanocomposites have attracted increasing research interest driven by the needs for replacing

**Corresponding author*: fatima.charrier@univ-pau.fr

Shakeel Ahmed and Chaudhery Mustansar Hussain (eds.) Green and Sustainable Advanced Materials, Vol. 1 (289–332) © 2018 Scrivener Publishing LLC

petroleum-based material and for tailoring material properties to fit the requirements of structural or advanced functional materials. Thus, hybrid nanocomposites have penetrated the fields of automotive, textile, packaging, construction, thermal and sound insulation. They are omnipresent in the fields of micro-optics, microelectronics and photonics. They are also emerging in the field of energy: they are found in the form of flexible cells, hybrid photovoltaic cells. Ecological concerns and sustainable development involves the elaboration of more environment friendly materials such as bio-nanocomposites that have various applications in different fields.

In this context, nano-structured hybrid materials derived from clays and natural substances attract more and more attention for many applications. They are considered among the most important topics in material science and engineering, as they are eco-friendly materials and their hybrid nature provides new properties to the involved organic and inorganic compounds, which surpass their original properties and even the properties of synthetic systems.

## 11.2 Inorganic/Organic Hybrids

Although an exact date of appearance of man-made hybrid materials is not known, they are said to have existed for a long period of time. Thus, the mixture of organic and inorganic compounds in one material to achieve improved features was reported from the very beginning of the mankind. This combination of compounds was most of the times driven by the necessity of some specific characteristics in the traditional materials, which could only be achieved by mixing two different substances. The combination of straw (organic fibre) and clay (inorganic matrix) for construction purposes, can be regarded as an example of that [1]. However, in the current scientific context this example would not count as hybrid material but more as a composite.

On the contrary, a nice example of a truly man-made hybrid material, extensively referred to in the literature [2–4] is *Maya blue*. It is normally related to the dawn of the history of organic–inorganic hybrid materials [5] and consisted of an organic pigment (indigo) and inorganic complex natural clay (palygorskite).

Despite these previous examples, it was not until the 20th century, with the availability of novel characterization methods and refined analytical techniques, that researchers were able to understand the nature and structure of hybrid materials [3]. This represented a great boost for investigations in the field. Ever since, there has been a sequence of breakthroughs

**Table 11.1** Timeline of the most relevant events in the development of hybrid materials during the 20th century.

| Years | Events | References |
|---|---|---|
| 1930s | Development of *sol–gel process* (preparation of multi-component systems using mild reaction conditions). | [7] |
| 1940s | The combination of organic and inorganic components took already part of the manufacturing technologies, e.g. nanopigments suspended in organic mixture, templated zeolites. | [8] |
| 1950s | Substantial contributions to the field of mixed organic–inorganic compounds, e.g. intercalation of organic compounds inside clays and inorganic lamellar compounds. | [9–10] |
| 1980s | Establishment of the *soft chemistry* by Livage, which opens the gate towards a variety of materials, i.e. hybrid materials. | [11] |
| 1990s | Appearance of the concept of hybrid materials. Arousal of two research directions: synthesis of periodically organized mesoporous materials and family of hybrids known as metal-organic-frameworks (MOFs). | [12–13] |

and advances in the subject (Table 11.1), which has led to an increase in the knowledge and the awareness of the potential of this sort of materials. Currently, hybrid materials chemistry is an established unifying interdisciplinary scientific domain, with considerable impact over various societal and industrial demands [6].

### 11.2.1 Definition, Classification and Synthetic Routes

Since the term *hybrid* is commonly used for describing different systems covering a broad number of materials, it becomes necessary to settle a proper definition. Several authors have dealt with this topic and therefore various definitions can be found along in the literature. Yamada *et al.* [14] defined these materials as mixtures of two or more materials with new properties created by new electron orbitals formed between each material. In the same direction, Makishima [15] stated that hybrid materials were mixtures of two or more materials with newly formed chemical bonds. Both authors highlighted in their definitions the formation of new linkages and therefore a new material; however, these definitions are said to be slightly vague and open to misunderstanding. Indeed, it is common that in material science the macroscopic composites were described as hybrid materials [16]. Nonetheless, from a strict chemical point of view both materials are clearly differentiated. Thus, hybrid materials would be related to a material with the structure of a composite but at molecular level [1]. Following to this, one current definition was given by the International Union of Pure and Applied Chemistry (IUPAC): *materials composed of an intimate mixture of inorganic components, organic components or both types in which these components usually interpenetrate on scales less than 1 μm* [17].

This definition has been completed by several authors, saying that *hybrid materials are composed of at least two components, which are usually organic and inorganic, and which are intimately mixed at molecular or nanoscopic scale* [3, 18, 19].

Despite the specificity of these definitions, there is not a clear borderline between hybrid materials and nanocomposites within the literature. Thus, nanocomposites, which are defined as materials in which at least one of its components should be in the nanoscopic scale [20], are often confused with hybrid materials. Nevertheless, a clear difference between both types of materials can be observed regarding their structural features [7]. Hence, the fact that in the hybrid material an intimate mixture between the components exists can change many properties compared to those of nanocomposites [1].

Taking into account the previous considerations, a clearer division can be observed between hybrids, composites and other materials (Figure 11.1).

The composition and structure of this type of materials present different possibilities regarding their main elements, namely matrix, building blocks and bonds between components. These possibilities, summarized in Table 11.2, show the heterogeneous nature of hybrid materials.

The diverse variations in composition and structure in hybrid materials can lead to different physicochemical properties, which are not just the simple combination of the characteristics of the single precursors. Hence, the interface and interactions between the components are a paramount parameter that can regulate important features of the material [6]. For this reason, the nature of the interface and interaction between organic and inorganic units, are employed to categorize hybrid materials [22]. Regarding this classification, two different types of hybrid materials can be encountered namely class I and class II hybrid materials (Figure 11.2). *Class I* is related to those hybrid materials where no covalent or weak chemical bonds are present. In this type, inorganic and organic units are linked via

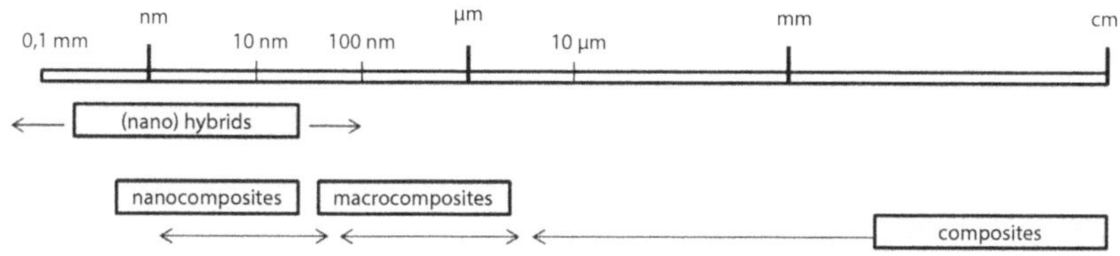


**Figure 11.1** Classification of materials at different scale levels, modified from the original figure by Nanko [21].

**Table 11.2** Variations of composition and structure in hybrid materials, modified from the original table by Kickelbick [7].

| Matrix nature | Building blocks types | Types of bonds between components |
|---|---|---|
| Crystalline or Amorphous | Molecules | Strong or Weak |
| | Macromolecules(polymers) | |
| Organic or Inorganic | Particles | |
| | Fibres | |

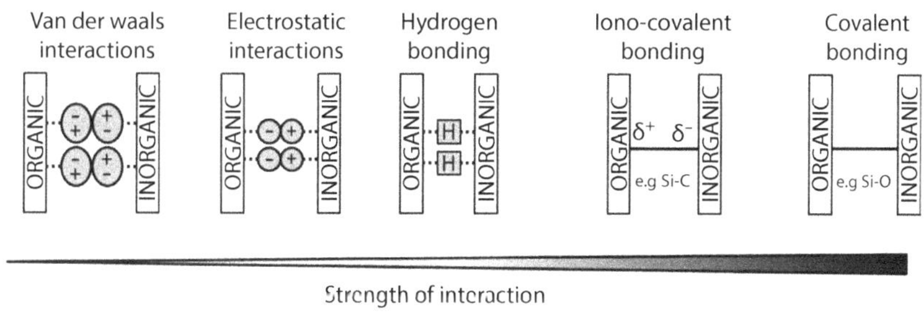


**Figure 11.2** Typical interactions present in hybrid materials and their relative strength.

Van der Waals or electrostatic interactions or hydrogen bonds [23]. On the other hand, *class II* is associated with hybrid materials that show strong interactions between its components, namely covalent or iono-covalent bonding. The iono-covalent bonds are referred to as covalent links with significant ionic character, usually found in binary or mixed metal oxides [24]. Within this sort of hybrid material, weak interactions characteristic of *class I* type can appear as well [25].

The performance of hybrid materials compared to the rest of materials is said to present several advantages since they combine the characteristics of both organic and inorganic components into one material. Thus, they allow the possibility of introducing functional groups and therefore to create multifunctional materials. Some examples are the introduction of inorganic cluster or nanoparticles with optical [26], electronic [27] or magnetic properties [28] into polymeric matrices. Another major advantage, which makes them an interesting type of material for many applications, is their processing. Thereby, opposite to pure inorganic solids, hybrid materials present a behaviour more similar to that of polymers in terms of handling. This is due to high organic content and to the formation of crosslinked inorganic structure from small precursors (polymerization-like reaction). This allows the production of these materials in a plethora of shapes and forms in bulk or film [7].

## General Synthetic Routes

Another important feature in the conformation of hybrid structures is related to the chemical routes used to design and synthesize hybrid materials. Several reviews can be found in the literature discussing in detail this topic [29–31]. Here the different paths to obtain inorganic–organic hybrids are summarized (Figure 11.3) and briefly presented.

- Path A: this pathway is associated with soft chemistry–based routes i.e. *conventional sol–gel process, bridge and multifunctional precursors* and *hydrothermal synthesis* including the synthesis of Metal Organic Frameworks (MOFs). The sol–gel method involves the hydrolysis and later condensation of organically modified metal alkoxides in an appropriate solvent with or without catalyst [32]. It is able to yield amorphous hybrid materials and it is interesting for its simplicity and low cost. On the other hand, the use of specific bridge precursors like *silsesquioxanes* ($X_3Si–R'–SiX_3$, where X = Cl, Br or OR and R′ = organic spacer) can yield homogeneous molecular hybrids [33]. Finally, the hydrothermal process is carried out with polar solvents, at moderate temperatures (20–200 °C) and in the presence of organic templates [34].
- Path B: this route is related to the hybridization of defined *nanobuilding blocks (NBB)* and is generally carried out via assembling or dispersion. The NBB are commonly clusters [35], organically functionalized nanoparticles [36], nanocoreshells [37] or layered compounds [38], which are able to intercalate organic constituents [39]. The main advantages

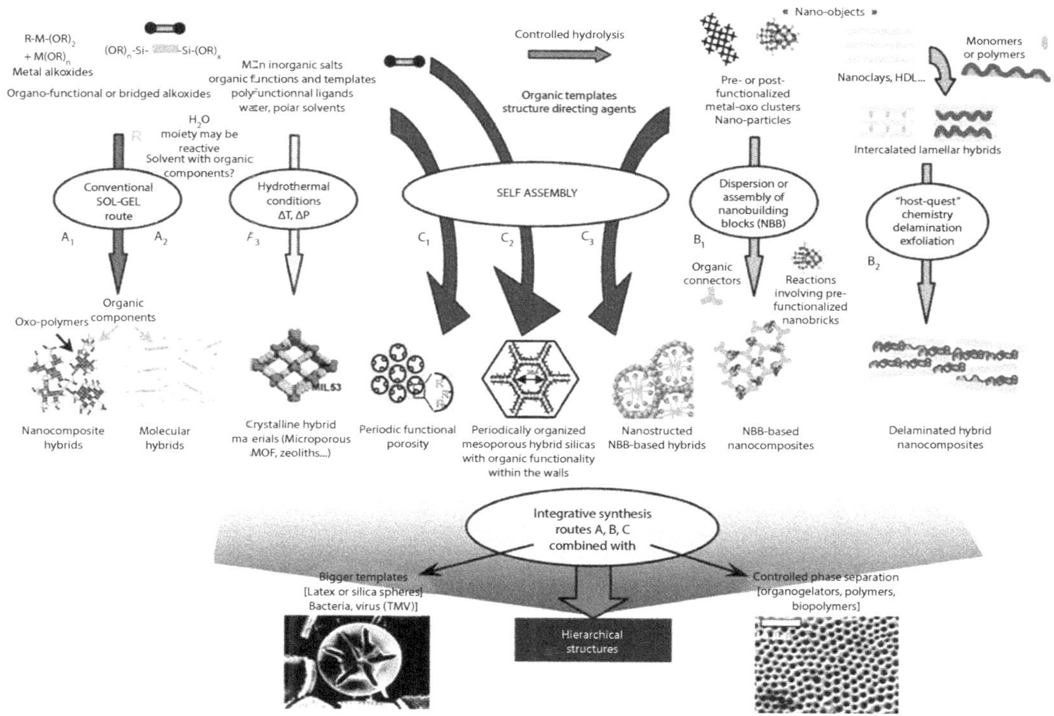


**Figure 11.3** Chemical routes for the synthesis of inorganic–organic hybrids [30] where $A_1$: sol–gel strategy, $A_2$: bridged precursors (silsesquioxanes), $A_3$: hydrothermal synthesis, $B_1$: assembling of NBB, $B_2$: dispersion of NBB, $C_1$: templated growth by organic surfactants, $C_2$: self-assembly via bridged precursors (silsesquioxanes), $C_3$: combination of self-assembly and NBB, D: integrative synthesis. (Reproduced from Sanchez *et al.* [30] with permission of the Royal Society of Chemistry via Copyright Clearance Center).

of the use of these units are their low reactivity towards hydrolysis or the attack of nucleophilic metal alkoxides and their nanometric and nano-dispersed nature with a well-defined structure. Besides, these units are able to maintain their integrity in the final hybrid.

- Path C: this strategy is based on the use of *organic templates able to self-assemble.* This route has been generally carried out via templated growth using organic surfactants [40]. Thus, a variety of ordered structures can be synthesized such as mesoporous structured hybrid networks [41]. Currently, another route, also employed, consists in the use of bridged precursors like silsesquioxanes, yielding a novel type of periodically ordered mesoporous hybrids with organic functionality [42]. The combination of the self-assembly and NBB strategies represents another possibility.
- Path D: this final route corresponds to an integrative synthesis coupling chemistry, physics and processing. Since the other strategies summarized above allow the control of hybrid materials only in the range of 1–500 Å, new pathways such as *micro-moulding methods* have been developed. Hence, the control of the shapes of complex objects in the micron scale is possible. In addition to that, the combination of this strategy with the previously reported (A, B, C) permit the creation of hierarchically organized materials as far as structure and functions are concerned [43].

### 11.2.2 Bio-Nanohybrid Materials

Bio-nanohybrids are commonly defined as materials in which molecular or polymeric species of biological origin are combined with inorganic nanostructured units via interactions on the nanometric scale [44]. However, the concept of biohybrid is not new since these types of materials have always been present in Nature. Hence, organism for several functions such as structural support, sensing ion storage or toxic waste removal [45], produces different natural bio-hybrids. Some of the most common natural hybrid examples [46] are shown in Table 11.3.

Currently, bio-nanohybrids have been prepared imitating these previous natural materials with the objective of developing novel biohybrids with improved mechanical properties, biocompatibility and in some cases functional behaviour. These materials have attracted interest in the last years, owing to their multiple applications in various important areas as

**Table 11.3** Examples of hybrid materials of natural origin and their main components.

| Natural hybrid materials | Components |
|---|---|
| Bone | Organic collagen fibres, inorganic hydroxyapatite crystals and bone matrix |
| Dentin | Apatite (calcium phosphate) and type I collagen fibrils |
| Nacre | Argonite (calcium carbonate) and chitin matrix |
| Ivory | Collagen matrix and hydroxyapatite mineral |
| Wood | Columns of cells supported by cellulose (polymer of glucose), matrix of lignin (polyphenolic resin) and hemicellulose |

different as regenerative medicine and new materials with improved structural and functional characteristics [47]. Besides, these sorts of materials appear as alternative to traditional polymer nanohybrids, since the use of molecules of natural origin assures that the final materials produced are renewable and environment friendly [46].

## 11.3 Bio-Nanohybrid Materials Based on Clay and Polyphenols

### 11.3.1 Clay Minerals and Organoclay

#### *11.3.1.1 Clay Minerals*

Natural clays come from the hydrothermal alteration of alkaline volcanic ash and rocks of the Cretaceous period (85–125 million years ago). Diverse suggestions have been presented concerning the mechanism of the ash to clay transformation. The most likely is that change started in marine water in reactions involving sufficient amounts of $Mg^{2+}$ and $Na^{+}$ [48].

According to the *Handbook of Clay Science*, clays are essentially composed of micro-particulated minerals, abundant in the Earth's crust, whose particle size is less than 2 μm. This family, called also 'Clay Minerals', is mostly composed of layered silicates (Phyllosilicates) of variable composition leading to diverse structural arrangements and properties [49].

A clay deposit usually contains non-clay minerals as impurities, namely quartz, sand, silt, feldspar, mica, chlorite, opal, volcanic dust, fossil fragments, heavy minerals, sulfates, sulfides, carbonate minerals, zeolites, and many other rock and mineral particles ranging in size from colloidal to pebbles [50].

#### 11.3.1.1.1 Structure of Clay Minerals

Most clay minerals are phyllosilicates that are also known as layer- or sheet silicate minerals. The terms layer- or sheet silicates are commonly used because the structural units of the minerals are ranged in long layers or sheets [51].

Clay minerals are hydrous layer silicates consisting of planes of atoms arranged in layers, their crystal habits and morphologies reflecting this arrangement so that the major are platy and have perfect basal cleavage. Each layer is composed of two key elements, which are:

- The tetrahedral sheet, where structural cations are coordinated in a four-fold tetrahedral manner by oxygen or hydroxyl anions.
- The octahedral sheet, where structural cations are coordinated in a six-fold octahedral manner by oxygen or hydroxyl anions [52].

Tetrahedral and octahedral sites are different. In fact, the space that can be occupied by a cation in a tetrahedral site is smaller than the space that can be occupied in an octahedral site. Smaller cations tend to take place in tetrahedral sites, while larger cations tend to take place in octahedral sites, and the largest cations must fit octahedral sites [53]. Table 11.4 recapitulates the structural sites in which cations tend to occur in clay minerals.

Variations among clay minerals and changes in their physical and chemical characteristics issue from the diverse combinations of octahedral and

**Table 11.4** Type of structural sites in which common cations tend to occur in phyllosilicate mineral structures. (Reproduced from Schulze [53] with permission of Elsevier Science &Technology Books via Copyright Clearance Center).

| Type of site | Cation |
|---|---|
| Tetrahedral only | $Si^{4+}$ |
| Tetrahedral or octahedral | $Al^{3+}$, $Fe^{3+}$ |
| Octahedral only | $Mg^{2+}$, $Ti^{4+}$, $Fe^{2+}$, $Mn^{2+}$ |
| Interlayer sites | $Na^{+}$, $Ca^{2+}$, $K^{+}$ |

tetrahedral sheets and the electrostatic effects that chemical substitutions have on the units [48].

- Tetrahedral sheets (T) are built of silica tetrahedrons. In each tetrahedron, a silicon atom is equidistant from four oxygens, or hydroxyls if needed to balance the structure, arranged in the form of a tetrahedron with the silicon atom at the centre. The silica tetrahedral groups are arranged to form a hexagonal network, which is repeated indefinitely to form a sheet of $Si_2O_6(OH)_4$ structure [54], and this via covalent bonding through shared oxygens. These shared oxygens form a basal plane of the tetrahedral groups, and the remaining apical oxygens are shared with another layer of cations [48].
- A side view of the tetrahedral sheet shows that it is formed of three parallel atomic planes, which are made of oxygens, silicones and oxygens respectively (Figure 11.4). The first plane is the basal plane composed of shared oxygen atoms. The second plane is composed of silicon atoms that fill the cavity at the junction of three oxygen atoms and so forming a hexagonal network. The third plane is constituted of hydroxyl atoms lying directly above the silicon at the tip of the tetrahedrons [54].
- Octahedral sheets (O), have aluminium, magnesium or iron cations that are coordinated with six oxygen atoms or hydroxyl groups, so producing an eight-sided structure called an octahedron (Figure 11.4). The octahedral units are horizontally linked via covalent bonds to form the octahedral sheet. The minerals brucite $Mg(OH)_2$ and gibbsite $Al(OH)_3$ are comparable to the octahedral sheets found in many clay minerals. Cations in the octahedral sites may exist in a divalent or trivalent state. When it comes to divalent cations ($Mg^{2+}$, $Fe^{2+}$), the layer exhibits a geometry identical to brucite, such that electrical neutrality is maintained. In this case, the ratio of divalent cations to oxygens is 1:2 and all three possible cation sites in the octahedron are occupied. This configuration shows a layer with trioctahedral arrangement. When the cations are trivalent ($Al^{3+}$, $Fe^{3+}$), the charge balance is maintained by leaving one of every three octahedral cation sites empty. With such composition, the ratio of trivalent cations to oxygens is 1:3 and the layer presents a gibbsite-like dioctahedral arrangement. A combination of

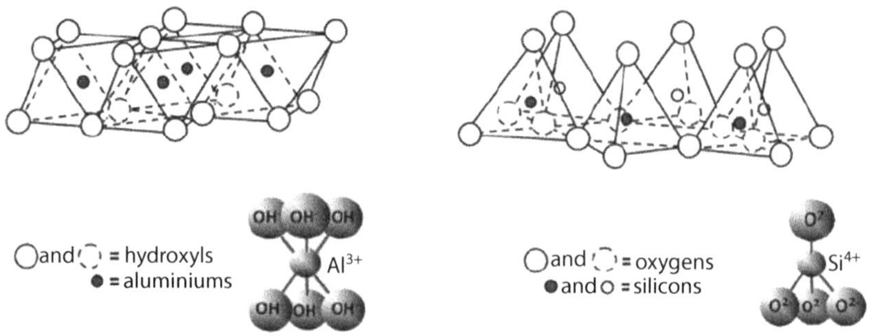


**Figure 11.4** Structure of an alumina octahedral sheet (left) and a silica tetrahedral sheet (right). (Reproduced from CalabiFloody *et al.* [55] with permission of Mineralogical Society of Great Britain and Ireland via Copyright Clearance Center).

tetrahedral and di- or trioctahedral sheets linked via shared oxygen atoms constitutes aluminosilicate layers that define the basic structural units of phyllosilicates [56].

The junction of tetrahedral and octahedral sheets together leads to the formation of two main clay mineral structures known as layers. The unit built by bonding one tetrahedral sheet and one octahedral sheet together is named a 1:1 layer or T-O layer [57]. The interconnection is accomplished by changing two out of every three anions of an octahedral sheet by the apical oxygens of a tetrahedral sheet. As a result, at the 'junction plane' the apical oxygens are shared jointly by tetrahedral and octahedral sheets. The remaining unshared anion of the octahedral sheet in the junction plane is a hydroxyl that lies in projection at the centre of the hexagonal ring of apical oxygens [52]. The second basic structure results from the similar linkage of a tetrahedral sheet to the other side of an octahedral sheet and the resulting unit is called 2:1 layer or T-O-T layer [57].

As it is described above, substitutions of cations by others of lower valence, such as $Al^{3+}$ for $Si^{4+}$ in tetrahedral sheets and $Mg^{2+}$ or $Fe^{2+}$ for $Al^{3+}$ in octahedral sheets are current. There is little or no substitution in 1:1 layers. Therefore, usually there is only a very small, overall layer charge or none at all. However, substitutions in 2:1 layers produce layers with a clear negative charge. The neutralization of this charge may be ensured by fixed cations, hydrated exchangeable cations, or by octahedrally coordinated hydroxyl groups or sheets, all of which reside between the layers known as the interlayer space [52].

The diverse combinations of tetrahedral (T) and octahedral (O) sheets as well as the electrostatic effects that chemical substitutions have on the units lead to a variety of clay minerals with various physical and chemical properties [48].

11.3.1.1.2 Classification of Clay Minerals

The classification of phyllosilicate clay minerals is based upon the following four main principles:

- The layer type (1:1 or 2:1). A third probable type that may exist is that there are two tetrahedral and two octahedral sheets. The combination being identical to the 2:1 structure with the difference that there is another octahedral sheet located between the 2:1 layers. This structure is defined as a 2:2 or 2:1:1 structure.
- The cation nature of the octahedral layers and especially whether it is dioctahedral or trioctahedral.
- The magnitude of the net negative layer charge (x) due to atomic substitutions in the layer structure.
- The cation content and type of the interlayer space [52, 58].

These various parameters are commonly used to classify phyllosilicate clay minerals. Nevertheless, another classification can be found based on the type of assembly and their morphology. Thereby the following classes of clays can be seen:

- Layered clays that are layered silicates of plate like morphology with structures based on layers built by tetrahedral and octahedral sheets (e.g. smectites, kaolinite and vermiculite).
- Microfibrous clay minerals, like palygorskite and spiolite, are tectosilicates. They are structurally developed by blocks constituted by octahedral sheets of magnesium or aluminium oxide/hydroxide loaded between two tetrahedral silica layers. The periodic inversion of $SiO_4$ tetrahedra induces a regular discontinuity of the sheets which is the origin of structural cavities (tunnels) extended along the c axis [9].

## The Layer 1:1 Type

### *Kaolin Minerals*

Clay minerals of the kaolin group are formed of dioctahedral 1:1 layer arrangements with a general formula of $Al_2Si_2O_5(OH)_4$. Kaolinite, dickite, nacrite and halloysite–endellite are polytypes belonging to the kaolin group [59]. Kaolinite is the most common member of this group. The kaolinite lattice consists of one sheet of tetrahedrally coordinated Si (with O) and one sheet of octahedrally coordinated Al (with O and OH). A layer of OH completes the charge requirements of the octahedral sheet. Adjacent cells are spaced about 7.1 Å across the (001) plane (Figure 11.5a) [50].

The isomorphous substitution is negligible in either the tetrahedral or octahedral sheets. As a consequence, the 1:1 layer type has little or no permanent charge, the cation exchange capacity is minor and the surface areas are low. Soils rich in kaolinite are commonly less fertile than soils in which 2:1 clay minerals overshadow [53].

Four names for clay materials are associated with kaolinite: ball clay, fireclay, fint clay and underclay. These terms come from the people that extract them and use them commercially. Ball clays generate a white product when fired. They have the plasticity and the green and dry strength for producing sanitary ware, ceramic tiles, etc. Fireclays are identical to ball clays. However, they are usually found directly beneath coals. When underclays have been referred to as fireclays, they are kaolinite-rich underclays. Flint clay can also be used to make refractory items. This term is more specific and applies to clay that is hard and smooth, breaks with conchoidal fracture, and does not present plasticity when ground up and wetted as other clays act, i.e. it does not slake in water. Kaolinite is major in diverse underclays, but underclays do not necessarily have kaolinite in

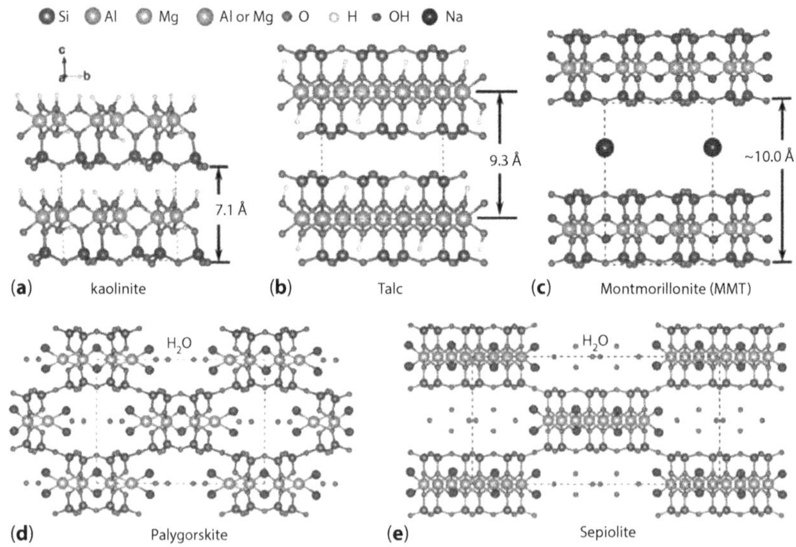


**Figure 11.5** Crystal structure of the pharmaceutical used clays; (a) kaolinite, (b) talc, (c) montmorillonite, (d) palygorskite and (e) sepiolite, where dashed lines represent the unit cell. (Reproduced from Yang *et al.* [60] with permission of Elsevier BV via Copyright Clearance Center).

them. Underclay is the unit directly below a coal bed that was the soil upon which the vegetation that formed the coal grew [48].

### *Serpentine Minerals*

The serpentine group is representative of the trioctahedral variant of the 1:1 layer type, with the general formula [$Mg_3Si_2O_5(OH)_4$] [55]. Subgroups are lizardite, antigorite, chrysotile, etc. [59]. Most serpentines present tubular or fibrous morphology. Chrysotile exists in both clino- and ortho-structures [50].

## The Layer 2:1 Type

### *Pyrophyllite-Talc Minerals*

The basic 2:1 structures with silicon in the tetrahedral sheets and aluminium or magnesium in the octahedral sheet, without any substitution of atoms, are named pyrophyllite and talc respectively [61]. Pyrophyllite (dioctahedral) and talc (trioctahedral), with general formulas [$Al_2Si_4O_{10}(OH)_2$] and [$Mg_2Si_4O_{10}(OH)_2$] respectively, present electrically neutral layers, and so no charge-balancing cation exists in the interlayer space (Figure 11.5b). Neighbouring layers are joined to each other by Van der Waals bonds [59]; the basal spacing in red by x-ray diffraction (XRD) is $d_{001}$ = 9.3 Å [60].

Since the layers do not expand in water, pyrophyllite has only an external surface area and typically no internal one [61].

### *Smectite Minerals*

Smectites, that are 2:1 phyllosilicates type, present a crystal structure which is formed of layers composed of two tetrahedrally coordinated silicon atoms fused to an edge-shared octahedral sheet of either aluminium or magnesium hydroxide. Stacking of the layers conducts to a regular Van der Waals gap between the layers called interlayer [62]. Smectite minerals are divided into two subclasses defined as:

- Dioctahedralsmectites, like montmorillonite, beidellite, nontronite, etc.
- And trioctahedralsmectites, like saponite, hectorite, sauconite, etc. [54].

Both subgroups exhibit the interesting property of being able to expand and contract their structures, while maintaining two-dimensional crystallographic integrity [48]. Isomorphic substitution within the layers occurs mainly in the octahedral layers, e.g. $Al^{3+}$ can be replaced by $Mg^{2+}$ or $Fe^{2+}$ and $Mg^{2+}$ can be substituted by $Li^{+}$. Consequently, when substitutions appear between ions of unlike charge, negative charge appears, and which is counterbalanced by hydrated cations (usually $Na^{+}$, $Ca^{2+}$, $K^{+}$ and $mg^{2+}$) in the interlayer space [50, 62]. These ions are hydrated due to the fact that, in nature, smectites are formed in aqueous environments. Because they are hydrated, these cations are only loosely held by the negatively charged clay layers. Smectites saturated with other cations dissociate in aqueous suspensions into exchangeable cations and tactoids, which are composed of several parallel TOT layers, held together by electrostatic forces by some of the exchangeable cations that remain in the interlayer space. Because they dissociate, the original cations are exchangeable by other inorganic and organic cations. In smectite tactoids, electrostatic forces keep the layers together. Parallel TOT layers are packed one above the other, and the exchangeable hydrated cations are located between the layers [48].

Montmorillonite and hectorite are considered as smectite clays of greatest commercial importance and value. Montmorillonite tends to have sheet morphology while a lath or strip morphology characterizes hectorite. Commercial availability of hectorite is limited whereas montmorillonite deposits are large and widely spread around the globe [50].

Montmorillonite is the result of the partial substitution of the trivalent Al-cation in the octahedral layer of the original pyrophyllite structure by the divalent Mg-cation (Figure 11.5c). In this case, the global negative charge is compensated by sodium and calcium ions in the interlayer. A special property of the resulting structure is that, water and other polar molecules can penetrate between the unit layers, leading to the expansion of the lattice [61]. The commonly used layered silicates in nanocomposite materials are of the 2:1 layers type (i.e. smectites, especially montmorillonite).

### *Mica Minerals*

Mica morphology results from the substitution of silicon in the tetrahedral sheet of the pyrophyllite structure by aluminium. Because of this substitution, micas are characterized by a negative surface charge, which is balanced by interlayer potassium cations ($K^{+}$). This gives an ideal formula of $KAl_2(AlSi_3)O_{10}(OH)_2$ for a mica mineral with Al in the octahedral sites. However, because the size of $K^{+}$ matches the hexagonal hole created by the Si/Al tetrahedral layer, it is able to fit very tightly between the layers. Consequently, the interlayers slump and the layers are held together by the electrostatic attraction between the negatively charged tetrahedral layer and the $K^+$ ions. That is why micas do not swell in water and, comparable to pyrophyllite, have no internal surface area [61]. The micas can be divided into the dioctahedral micas, like muscovite, illite, and phengite and the trioctahedral micas, like biotite [54].

### *Vermiculite Minerals*

Vermiculites have a 2:1 layer structure with a layer charge of 0.9–0.6 per formula unit and contains hydrated exchangeable cations, mainly Ca and Mg, in the interlayer. The high charge per formula unit accords vermiculites a high cation exchange capacity and it may exceed that of montmorillonite. That causes vermiculites to have a high affinity for weakly hydrated cations such as $K^+$, $NH^{4+}$ and $Cs^+$. Vermiculites are known to have high aspect ratio ($p \leq 2500$), exceeding that of montmorillonite by nearly one order of magnitude. The basal spacing of vermiculite varies from 14 to 15 Å with the nature of the interlayer cation and its hydration [50, 53].

### *Chlorite Minerals*

Like mica, chlorite minerals have a 2:1 layer structure with an excess of negative charge. Counter to mica, however, the excess charge is balanced by a positively charged interlayer hydroxide sheet, rather than $K^+$. The interlayer hydroxide sheet is an octahedral sheet and can be either di- or trioctahedral. The hydroxide sheet has a positive charge caused by substitution of higher-valence cations for lower-valence ones, as an example, $Mg_2Al(OH)_6^+$. Either octahedral sheet in the chlorite structure can be di- or trioctahedral, and can contain $Mg^{2+}$, $Fe^{2+}$, $Mn^{2+}$, $Ni^{2+}$, $Al^{3+}$, $Fe^{3+}$ and $Cr^{3+}$, which gives a broad number of different mineral species [53].

### *Sepiolite and Palygorskite*

Palygorskite (Figure 11.5d) and sepiolite (Figure 11.5e) are phyllosilicates characterized by the presence of continuous two-dimensional tetrahedral sheet and discontinuous octahedral sheets. Their morphology can be considered to consist of ribbons of 2:1 phyllosilicate structure, each ribbon being linked to the next by inversion of $SiO_4$ tetrahedra along a set of Si–O–Si bonds. Ribbons extend parallel to the X-axis and have an average width along Y of three linked pyroxene-like single chains in sepiolite and two linked chains in palygorskite; in this skeleton, rectangular channels run parallel to the X-axis between opposing 2:1 ribbons. As the octahedral sheet is interrupted at each inversion of tetrahedra, oxygen atoms in the octahedral at the edge of the ribbons are linked to cations on the ribbon side only, and coordination and charge balance are compensated along the channel by protons, coordinated water and a small number of exchangeable cations. A variable amount of zeolitic water exists in channels [63].

Sepiolite has eight possible octahedral positions per half unit cell, all are occupied and its structural formula is $[Si_{12}O_{30}Mg_8(OH)_4(OH_2)_2.4H_2O]$, while palygorskite has five octahedral positions and a structural formula $[Si_8O_{20}Mg_5(OH)_2(OH_2)_4.4H_2O]$. Yet, the five positions cannot be filled and a dioctahedral mineral with a structural formula $[Si_8O_{20}Al_2Mg_2(OH)_2(OH_2)_4.4H_2O]$ is accepted [64].

*Mixed Clays or Interstratified Clays*

Interstratification is the term used to define phyllosilicate minerals in which structural units of various kinds occur in the stacking sequence normal to the plane of the layers. Mixed clays appear in two ways. First, part of clay minerals (mainly smectites, vermiculites and halloysite) can sorb inorganic (e.g. water and cations) or organic (cations, alcohols, etc.) substance between the layers. This action may not occur equally in all the interlayers at the same time so more than one kind of interlayer separation in the same crystalline may result. Second, because the 2:1 and 1:1 layers are strongly bonded internally but relatively weakly bonded to each other and because the surface planes of the different sorts of layers and the hydroxide sheet of chlorites are geometrically identical, layers with different internal arrangements can stack together and articulate well at their interfaces [65].

The layer sequence is generally irregular. Infrequently, however, regular stacking can exist. Illite/montmorillonite and chlorite/vermiculite layers are the most known interstratified structures. Mixed clay usually disperse and expand in water more easily than other clay minerals, essentially when one constituent of the mixed-layer clay is expandable [48].

### *11.3.1.2 Surface Modification of Clay Minerals: Organoclays*

Surface modification of clay minerals is a concept of increasing interest due to the numerous and the creative applications that can offer in materials science. Modified clays are essential to develop polymer nanocomposites. They are also involved in other applications such as adsorbents for organic pollutants, rheological control agents, paints, cosmetics, etc. [66]. Montmorillonite and other phyllosilicate clays are naturally hydrophilic.

This makes them unwell suited to interacting with most hydrophobic compounds. In addition, the stacks of clay platelets are held tightly together by electrostatic forces. For these reasons, the clay must be modified before its use in nanocomposites or other fields [67]. There are various ways to modify 2:1 phyllosilicate type: (1) adsorption, (2) ion exchange with inorganic cations and cationic complexes, (3) ion exchange with organic cations, (4) binding of inorganic and organic anions, mainly at the edges, (5) grafting of organic compounds, (6) reaction with acids, (7) pillaring by different types of poly(hydroxo metal) cations, (8) interlamellar or intraparticle and interparticle polymerization, (9) dehydroxylation and calcination, (10) delamination and reaggregation of smectitic clay minerals, and (11) physical treatments such as lyophilization, ultrasound and plasma. Ion exchange with alkylammonium ions is a well-known method to make clay minerals and clays dispersible in inorganic solvents and to render them compatible with hydrophobic materials in compounding processes [68].

When alkali ammonium salts are used as an organic modifier, the clays are termed as 'organoclays'. The cation exchange reaction or the intercalation reaction usually occurs through the changing of $N^+$, $K^+$ or $Ca^{2+}$ ions with the host organic molecule usually quaternary ammonium cations, that are also called surfactants, with the general formula $(CH_3)_nNRX$ where R represents the aromatic or alkali hydrocarbon, n = 1–3 and X represents the $Cl^-$, $Br^-$ or $I^-$, etc. [69]. Initially, the interlamellar distance of the $d_{001}$ plane of the unmodified clay is relatively small, and the interlayer environment is hydrophilic. After intercalation of the surfactant between clay layers, the layers' basal spacing increases and they become hydrophobic (by decreasing the surface energy). This might facilitate the adsorption or the intercalation of hydrophobic organic compounds to the clay. Additionally, surfactants can provide functional groups that can react with polymer matrix for example [62, 67, 70].

Organoclays are classified into two classes on the basis of structure and sorption mechanism:

- Adsorptive clays, which are synthesized from the intercalation of quaternary ammonium ions, with $R \leq 12$, to clay interlayers. Due to their small size, these organic cations frequently act as 'pillars' in the interlamellar region of the clays, exposing clays surface to direct interaction with uncharged hydrophobic compounds [70].

  The current used organic cations for producing adsorptive clays are tetramethylammonium (TMA) and

trimethylphenylammonium (TMPA), or a number of primary to quaternary amines with different carbon chain lengths [70, 71].
- Organophilic clays, which result from exchange with relatively long-chain alkylammonium cations (R ≥ 12) that can form a hydrophobic partition medium within the clay interlayer, and behave similarly to a bulk organic phase such as octanol or hexane. The frequently used organic cations are hexadecyltrimethylammonium (HDTMA) and dioctadecyldimethylammonium (DODMA) [72].

The hydrophobicity of organic cations affects the sorption capacity of modified clays. For low hydrophobic compound, adsorptive clays function as better sorbent, while for the compounds with high hydrophobicity, organophilic clays play a better sorption behaviour. In addition, the sorption mechanism on both organoclays sorts is directly related to the structure of the organic modifier [69].

### *Organoclay Structures*

Depending on the layer charge (= interlayer cation density = packing density of the alkylammonium ions) of the clay mineral, the chain length of the organic ion and the cationic exchange capacity (CEC) of the clay mineral, different arrangements of the organic cations between the layers can be formed [57, 66, 72]. Generally, in adsorptive clays, the organic cations only form monolayers. While with organophilic clays, the organic cations may take different configurations (Figure 11.6) which can be identified by as they affect the basal spacing of organoclays [48, 66, 73] and described as follows:

- Monolayer structure: when the packing density of the chains is low, the chains are arranged with the C–C–C plane parallel to the layers, i.e. the chains will lie parallel to the silicate layers; the characteristics basal spacing is about 13.5 Å.
- Bilayer structure: bilayers of flat-lying alkylammonium ions are formed with a basal spacing of about 17.5 Å.
- Pseudotrimolecular layers structure: the polar end groups of the alkylammonium ions remain attached to the layers. By formation of kinks, the unipolar chain ends are shifted one above the other and the interlayer separation is determined by the thickness of three alkyl chains. The basal spacing is about 22 Å.

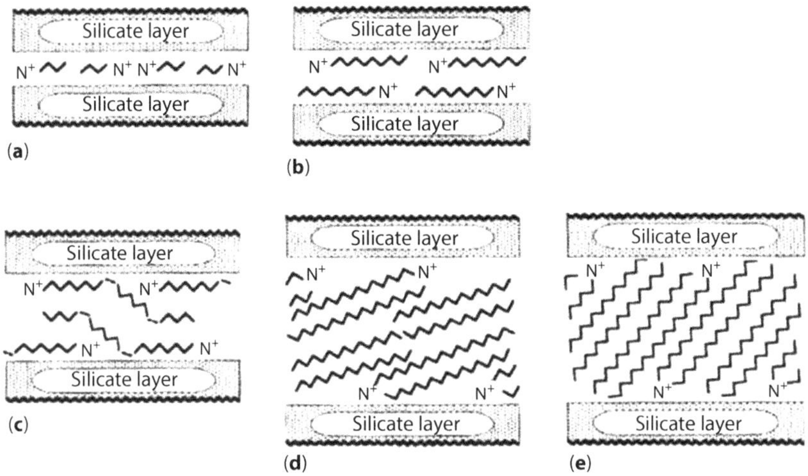


**Figure 11.6** Orientations of alkylammonium ions in the galleries of layered silicates: (a) monolayer, (b) bilayers, (c) pseudotrimolecular layers, and (d, e) paraffin-type arrangements of alkylammonium ions with different tilting angles of the alkyl chains. (Reproduced from De Paiva *et al.* [66] with permission of Elsevier BV via Copyright Clearance Center).

- Paraffin-type layers structure: in this case, the angles between the chain axes and the silicate layer increase with the layer charge. Therefore, the mean increase in basal spacing (>22 Å) with increasing chain length of the cation varies with the charge density.

## 11.3.2 Polyphenols as Natural Substances

Polyphenols are naturally occurring compounds defined by the presence of one to several hydroxyl groups and one or more aromatic rings within their structure [74]. This class of compounds is widely spread in nature and are said to be part of plants mainly as secondary metabolites. Thus, they are commonly involved in plant functions such as protection against radiation or pathogens [75].

Currently, more than 8000 different polyphenols have been identified in different plant species [76]. They occur generally as conjugated structures with one to various sugar moieties linked to the hydroxyl groups [77]. Nevertheless, direct bonds between aromatic carbons and sugars (mono- and polysaccharides) can also exist.

Although they are characterized by phenolic structures, this group of compounds is highly heterogeneous and therefore several subgroups or types can be distinguished. According to the number of phenol rings and the different structural parts binding between each other, four main classes can be found, namely phenolic acids, flavonoids, stilbenes and lignans [78]:

- Phenolic acids: they are abundantly found in foods and edible plants. Two main subclasses are differentiated, i.e. derivates of benzoic acid and derivates of Cinamic acid (Figure 11.7a).
- Flavonoids: this is said to be the most studied group of polyphenols. Their structure is defined by two aromatic rings linked by a three carbon–oxygen heterocycle (C6–C3–C6 backbone) as presented in Figure 11.7b. More than 4000 different species of this type of compounds have been identified so far [79].
- Stilbenes: these compounds are distinguished by the presence of two phenolic rings moieties interconnected by a two carbon methylene linkage (Figure 11.7c).

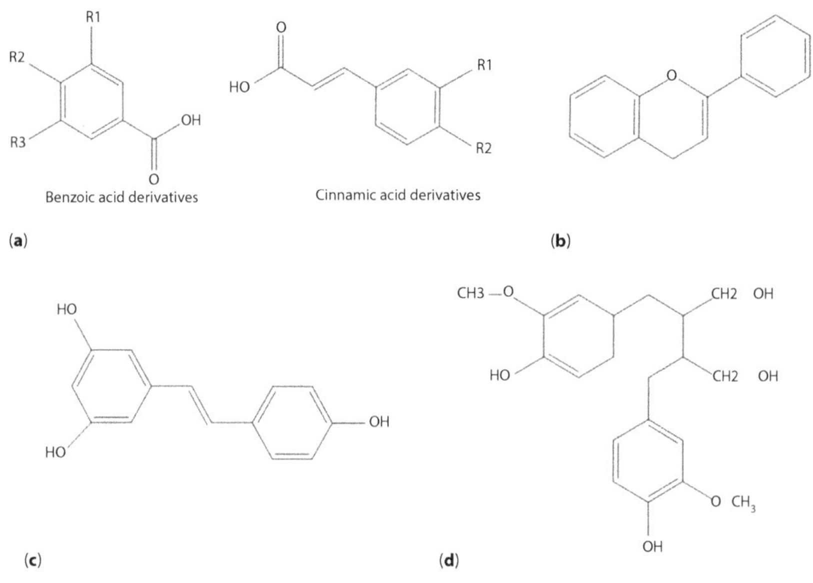


**Figure 11.7** Types of polyphenols based on structure: (a) phenolic acids, (b) flavonoids, (c) stilbenes and (d) lignans.

- Lignans: they are dimers of monolignols derived from hydroxycinnamic acids (p-coumaric, ferulic and sinapic acids) [80] and their structure is characterized by two 2,3-dibenzylbutane moieties (Figure 11.7d).

## 11.3.3 Clay/Polyphenols Hybrids

### *11.3.3.1 Techniques Used for Clay-Based Hybrids Characterization*

#### *X-ray Diffraction*

The advent of the uniaxial powder diffractometer along with using oriented clay samples has allowed for routine and reliable analysis of clay minerals. Prior to these developments, the small coherence domains in clay crystals (compounded by their sheet structure) and their small particle size (most are up to 10–20 times smaller than 2 μm) made them difficult to study. Diffraction effects are enhanced for the small crystallites by orienting them or by having them lie on the same plane by exploiting their sheet structure. As a results, the basal spacing (001) and higher orders of reflection (00l) are often enhanced relative to other diffraction planes (*hkl*, *hk0*), especially those without an *l* component. There are two standard polar molecules that are intercalated to determine the swelling properties of clays: water yields a 5 Å increase in basal spacing, and ethylene glycol expands layers by 7 Å. To enhance the (*hk*) planes, nonoriented clay specimens are employed, where the least physical pressure possible is used to ensure that the powders are not packed. The two most useful XRD regimes for clay characterization are the basal spacing region (2°–10° $2\theta$, Cu radiation) and the (060) reflection (near 60° $2\theta$, Cu radiation), which is diagnostic for several structural and compositional types [48]. Figure 11.8 illustrates the principle known as Bragg's law. The two paths, A and B, for the incident and diffracted beams differ in length by $2d\sin\theta$, where $d$ is the perpendicular distance between two adjacent parallel planes. The parameter $\theta$ is equal to the angle between the incident beam and each plane and is equal to the angle between the diffracted beam and each plane. For constructive interference to take place, this difference must be equal to a whole number of wavelengths. Bragg showed that diffraction takes place only if all the lattice points of the crystal are on the parallel planes.

$$n\lambda = 2dsin\theta$$

where $n$ is the degree of diffraction, $\lambda$ is the wavelength, $\theta$ is the measured diffraction angle, and $d$ refers to interlayer spacing [81].

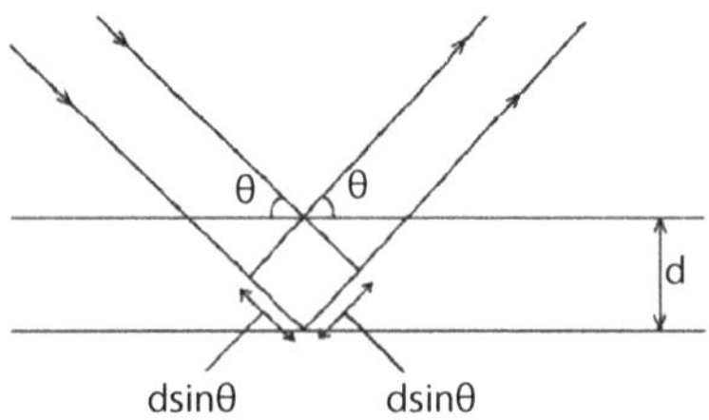


**Figure 11.8** Bragg's law.

## *Nuclear Magnetic Resonance Spectroscopy*

Nuclear magnetic resonance (NMR) spectroscopy has been used successfully in the study of clay minerals and clay organic interactions, providing information on both structural and dynamic aspects. NMR spectroscopy is sensitive to the local environment atoms, and information concerning crystallographic sites, coordination number, tetrahedral sheet distortions, and cation distribution has been obtained from $^{29}Si$ and $^{27}Al$ spectra [48]. $^{1}H$ and $^{13}C$ NMR have been used to study the orientation, interaction and mobility of species sorbed in the interlamellar space of clay minerals [82].

## *Thermal Analysis*

Several different types of water and hydroxyl groups are associated with a clay matrix that is liberated with increasing temperature. Thermogravimetric analysis (TGA) is often used to qualitatively determine the binding energies of these different types of water, which can then be used to characterize a mineral. In TGA, the mass of a sample is monitored versus temperature while it is heated according to a program in a specific atmosphere (oxidizing, reducing, inert). At lower temperatures, adsorbed water is lost from clay surfaces in defect sites or at broken bond sites; typical values for this are 1wt% at 80–90 °C. Zeolitic water is observed only for sepiolite–palygorskites, at 100–150 °C. Interlayer water, which is associated with interlayer cations in smectites as monolayers (2.5 Å) or bilayers (5 Å), is liberated between 100 and 200 °C. Dehydroxylation of the inner 2:1 or 1:1 lattice hydroxyl groups, also lost as water, is observed at 500–700 °C. Derivative thermogravimetry (DTG), where the first derivative of the TG curve is plotted against temperature, is often more useful.

In differential scanning calorimetry (DSC), the heat flow rate to a sample is monitored against temperature during programmed heating. Differential thermal analysis (DTA) allows for the observation of endothermic and exothermic phenomena. In DTA, the difference in temperature between a sample and a reference material is monitored against temperature, while the temperature of the sample in a specific atmosphere is programmed. Water losses are normally endothermic, and recrystallization is exothermic [48].

TGA and DTA have been widely been used to underscore the formation of clay/organic hybrids and analyse their stability and the possible organic changes that can occur and generate endo or exothermic phenomena [83–85].

### *Fourier Transformed Infrared Spectroscopy*

Much of the power of infrared (IR) techniques does not apply to clay minerals unless they are complexed with organic molecules, because IR is less sensitive to differences in clay structures than methods such as XRD and NMR. Only broad absorbances are observed for the Si–O and Si–O–Si tetrahedral stretching vibrations near 1000–1100 $cm^{-1}$, Al–O stretching near 600 $cm^{-1}$, and octahedral Fe, Al or Mg bending frequencies near 800–900 $cm^{-1}$. One of the most useful IR regions for pure clay minerals contains the OH vibrational modes from 3500 to 4000 $cm^{-1}$. Here the bonding energies of structural hydroxyl groups, and therefore octahedral site occupancy, are easily determined [48]. The Fourier transform infrared (FTIR) technique can be useful to highlight the clay/organic molecules hybrid formation. As FTIR technique is related to the vibrational energies of atoms or specific groups of atoms within a molecule and also to the rotational energies and by comparing the spectrum of the hybrid with that of the starting clay, it is possible to identify the presence or not of characteristic bands for the organic molecules [85–87].

An IR spectrum is commonly obtained by passing IR radiation through a sample and determining what fraction of the incident radiation is absorbed at a particular energy. The energy, at which any peak in an absorption spectrum appears, corresponds to the frequency of a vibration of a part of a sample molecule. For a molecule to show IR absorptions, it must possess a specific feature: an electric dipole moment of the molecule must change during the movement. The interactions of IR radiation with matter may be understood in terms of changes in molecular dipoles associated with vibrations and rotations. The atoms in molecules can move relative to one another, that is, bond lengths can vary or one atom can move out of its present plane. This is a description of stretching and bending movements, that is, vibrations. Vibrations can involve a change in either bond length (stretching) or bond angle (bending). Some bonds can stretch in phase (symmetrical stretching) or out of phase (asymmetric stretching). If a molecule has different terminal atoms, the amount of coupling will vary. Varying proportions of the stretching motion of each group are observed [48].

#### *Electron and Scanning Probe Microscopy*

In scanning electron microscopy (SEM), the flux of secondary and backscattered electrons from a clay sample bombarded by an electron beam is used to form an intensity image. The resulting three-dimensional maps are very useful for identifying textures and morphologies, with resolution on the order of 0.01 μm. Transmission electron microscopy (TEM) is used to image crystal shapes and sizes in essentially a two-dimensional plane. Clays are normally deposited on carbon-coated copper grids. In TEM, the electron beam is absorbed to a greater extent by the clay than the carbon, creating a shadow image of the clay particle. Resolution is greater than for SEM.

Atomic force microscopy (AFM) permits the observation of individual atoms on mineral surfaces, as well as the examination of the molecular-scale structure of cations, surfactants and organic molecules on various clay surfaces [48].

### *11.3.3.2 Applications of Clay–Polyphenols Hybrids*

#### *Active Packaging*

Active packaging is defined as a package system incorporating constituents that liberate or absorb substances into or from the packaged food or the environment surrounding the food to extend the shelf life or to maintain or improve the condition of the packaged food. Contrary to conventional packaging systems that play simply a passive barrier role to external harmful factors, active packaging plays an active role in food preservation and quality during the marketing process, as it interacts positively with product and environment to improve/preserve the quality of food longer [88–91]. Antioxidant/antimicrobial packaging is one of the most dynamic and promising concept of active packaging technologies. It is based on incorporating bioactive agents inside the packaging film. Thus, avoiding the direct addition of the active substances to the food product. The use of antimicrobial/antioxidant films permits the release of active compounds from the film, and continuous active effects take place on the food surface during the exposure time, increasing consumer safety because these substances are included in the packaging structure instead of being directly added to food and are released in modest amounts on the food surface [92, 93].

Synthetic additives with bioactive functions have been widely used in food industry but increasing health and ecological concerns due to the use of these chemical food ingredients have developed the consumer interest towards natural bioactive substances including plant and spices extracts, like polyphenols, flavonoids, flavonols, tannins and terpenoids, generally recognized as safe (GRAS) [94, 95].

In order to utilize plant extract in diverse food matrices, it is imperative to develop effective methodologies to preserve and deliver their bioactive functionality during aggressive conditions of handling, processing and storage of the final product [96].

Antioxidants and antimicrobials could be incorporated into clays, stand stable there for the needed time and finally be liberated from the film surface in a controlled way.

The direct incorporation of natural polyphenols in polymer/clay composite packaging film has been reported before [93, 97]. However, to the best of our knowledge the use of clay/natural polyphenols hybrid in active packaging making has not been well described and developed before. The only reference of clay/natural polyphenols hybrid utilization in active packaging is published by Busolo and Lagaron [85]. In this study, the authors have manufactured two grades of active composites of linear low density polyethylene (LLDPE) with the natural polyphenol resveratrol that was extracted from the *Polygonum Cuspidatum* root. The first composite (C1) was obtained by the direct addition of resveratrol to the polymer (1wt % resveratrol). The second composite (C2) was prepared by incorporation into the polymer of a suitable amount of the hybrid resveratrol/montmorillonite in order to achieve 1wt. % of the final content of resveratrol. The XRD was exploited to determine the way in which the hybrid was formed. The results showed that the basal peak of the natural clay (5.76°, 2θ) shifted to 6.05° and became less intense after hybridization with resveratrol. It was concluded that is an adsorption of resveratrol onto the clay surface rather than an intercalation in the interlayer space. The TGA showed a high thermal stability for resveratrol supported by the clay. The antioxidant capacity of the C1 sample was slightly less effective than its counterpart C2. According to authors, this difference is poor but significant. This can be possibly explained by the better distribution of the antioxidant across the matrix because the clay plays a dispersing vehicle role. So, the use of clay/antioxidant hybrid can own not only to transmit active properties to polymers but also to reinforce the activity of the agent. The antimicrobial activity evaluated in the active films C1 and C2 pointed out that there is no difference in the antimicrobial activity whether the resveratrol is in the composite or supported on the clay. As it is justified by the authors, the clay acts as an inert vehicle to distribute the active agent across the matrix. At the end of this study, the authors concluded that using natural antioxidant agents such as resveratrol immobilized on clay can maximize the shelf-life of food products prone to undergoing rapid oxidation, and can potentially play the role of additive for active food packaging designs and applications [98].

### *Biohybrid Pigments*

Maya blue is one of the oldest organic–inorganic nanohybrid materials and is the most famous example of biohybrid pigment. Maya blue was used in wall paintings, pottery and sculptures by the Mayas and other ancient Mesoamerican peoples. This ancient nano-structured material results from the assembling of indigo, a natural dye extracted from leaves of *Indigofera suffruticosa*, with palygorskite [4, 5, 99]. The exceptional thermal and chemical stability of 'Maya blue', as it persists to temperatures up to about 250 °C, organic solvents, and concentrated acids and alkalis [4], has attracted the attention of scientists to prepare clay/dye hybrid pigments and comprehend interactions between inorganic and organic components, with the hypothesis that there is a 'Maya Chemistry' no limited to Maya blue. This could lead to new hybrid pigments comparable to Maya blue [34, 100, 101]. Inspired by the 'Maya approach' and the need to use eco-friendly colourants for diverse applications (textile, printing inks, coatings, food industries, etc.), studies about the immobilization of biological dye on clays are emerging.

Ogawa *et al.* [102] studied the hybridization of purple sweet potato anthocyanin, a flavonoid compound, with saponite via cation exchange mechanism. The XRD analysis of the clay before and after the reaction with anthocyanin extract showed that the basal spacing of saponite did not change significantly before and after the reaction with the biological dye. According to the authors, this result could be due to hydration or to a possible immobilization of the anthocyanin on the external surface of saponite instead of their intercalation in the interlayer space. The authors also mentioned that saponite/anthocyanin hybrid revealed a utile colour change between blue and pink upon exposure to acidic (HCl) and basic ($NH_3$) vapours and the colour change was reversible and repeatable. This behaviour was explained by the structural change of anthocyanin molecules. The easy synthesis and the useful colour change of the present saponite/anthocyanin hybrid prompt its application as a 'smart pigment' [102].

Fournier *et al.* [99] reported the hybridization of carminic acid, a natural anthraquinonic dye, with montmorillonite. The XRD characterization did not show any intercalation of the organic molecules into the interlayer space. The adsorption of the dye onto clay surface was confirmed by TEM. The $^{13}C$ NMR was exploited to investigate the fixation of carminic acid on clay. The spectra showed the intervention of ketone and catechol functions in the immobilization of the dye leading to a stable lake pigment [99].

Doménech-Carbó *et al.* [103] investigated several clay/dye hybrid materials, which could be used as yellow pigments. These pigments were prepared using different natural flavonoid yellow dyes (*zacatlaxcalli*, fustic, marigold and cosmos), palygorskite and kaolinite clays [103].

Kohno *et al.* [104] prepared two kinds of hybrid pigments based on grape skins anthocyanin, montmorillonite and mica. The XRD analysis revealed the intercalation of anthocyanin into montmorillonite interlayer space, as the basal spacing increased with increasing amount of adsorbed anthocyanin. However, the basal reflection of mica remained unchanged, suggesting that anthocyanin was adsorbed on the external surface of the mineral particles. Anthocyanin intercalated into montmorillonite demonstrated a great stability against visible light irradiation and alkaline environment even pH 11 compared to anthocyanin simply adsorbed on mica external surface. According to the authors, the stability enhancement is associated with the electrostatic host/guest interaction and to steric protection from the attack of atmospheric oxygen [104].

Lima *et al.* [105] explored the hybridization of hibiscus *Sabdariffa L.* tropical flowers anthocyanins (known as the red Jamaica flowers) with different natural and synthetic inorganic matrices, including montmorillonite clay. Concerning the Jamaica flowers anthocyanins/montmorillonite, the thermal analysis confirmed the presence of organic substance strongly adsorbed in montmorillonite. The $^{27}Al$ MAS NMR spectra showed that the biological dye interacts solidly with the aluminium–oxygen groups of montmorillonite. In addition, it was pointed out that the interaction is preferentially with tetrahedral aluminium as it was the resonance most affected in the spectra regarding intensity and broadness [105].

### *Geomaterials*

Geomaterials for building are largely used in either traditional or modern construction. They are a mixture of clay and sand with diverse mineralogical composition and size distribution. Strength is governed by the binding properties of clays, which are reversible with water. Hence, constructions made of geomaterial bricks confront the risk of erosion. To ameliorate the strength in either dry or wet conditions, clay–sand mixtures can be stabilized using various binders, like cement, lime, bitumen and other synthetic products. Nevertheless, the problem posed by these stabilizers is the energy cost. Alternatively, natural stabilizers, used in traditional building techniques, could replace synthetic ones.

Sorgho *et al.* [106] and Keita *et al.* [107] proposed to use tannin compounds obtained from a decoction of pods of Néré trees (*Parkiabiglobosa*) with a mixture of clay–sand in order to enhance the macroscopic strength of the geomaterials. Clay samples were mined in Mali and Burkina Faso respectively. The comparison of the XRD patterns of clay/Néré complex with that of the clay alone revealed no significant differences in position and intensity of montmorillonite and kaolinite peaks, relatively to quartz that can be supposed to present low reactivity with tannin. Besides, the intensity of some illite peaks is reduced, at 3.737, 3.501 and 2.996 Å, suggesting an interaction process with tannins. However, the most important changes are the vanishing of goethites and hematite peaks at 2.987 and 3.685 Å in clay/Néré pattern, which proved the formation of tannin macromolecular complexes with iron from goethite on basal and interlayer surfaces of clay minerals and from hematite nanoparticles associated with clay minerals that lead to the bridging of clay layers. The FTIR characterization has supported the result of a chemical complex formation between tannin extract and clay–iron hydroxides. Mechanical characterization showed that tannin additives to clay–sand mixture increases the macroscopic strength of geomaterials and diminished micro-cracking [106, 107].

### *Drilling Process*

Drilling fluid plays imperative functions during the drilling process like hole cleaning, maintaining efficient lubrication between the bore-hole and drill string, cooling of the bit, and preserving convenient drilling pressure hence weight on bit. These operations have to be properly insured throughout the process regardless the operating conditions. The drilling fluid roles intently depend on the rheological properties of the drilling fluids, especially, viscosity, density and gel strength. Absence of any of these functions lead to serious drilling problems like lost circulation, high torque and drag, instability with changing conditions and stuck pipe events. Clays are used in drilling fluids. The drilling mud primarily provides the task of removing bit cuttings from the hole. In addition, it lubricates the bit, avoids hole sloughing and forms an impervious filter cake on the walls of the hole. The most important characteristic of clays for a drilling mud is the clay's ability to develop a suitable viscosity at relatively low solid levels and to keep the wished viscosity throughout the drilling operation [108].

Smectite clays have been widely used for this aim due to their small and plate-like particles, high surface area and high ability to hydrate and swell in water. The ability of smectite clay to realize the function mentioned above is solidly linked to the chemical environment in which clay is placed. When the clay is exposed to contaminants such as high salinity water formation, calcium chloride, cement, its capacity to hydrate, swell and therefore disperse in aqueous solution decreases largely. In fact, the presence of cations like calcium or a very high pH state induces the association of clay layers and the formation of an aggregated or flocculated structure, conducting to enormous fluctuation in rheological properties and great amount of fluid loss into the formation. To surround this problem, chemical additives known as 'deflocculant' are added into the drilling mud avoiding it from flocculating [109]. Nowadays, incorporating eco-friendly and cheap deflocculants becomes more and more interesting within the drilling sector.

Hafshejani *et al.* [109] reported the action of Oak seed extract (OSE) as a bio-based stabilizer for bentonite drilling mud. The main conclusions drawn from this study are as follows.

The OSE preserved the stability of bentonite in aqueous solution through prevention of bentonite to form a flocculated structure, reducing rheological parameters and fluid loss value. The plant extract also showed its high performance in decreasing plastic viscosity.

The particle size measurements revealed that OSE had a noticeable impact in diminishing particle size of contaminated bentonite muds. The SEM analysis demonstrated that there is no notable difference between the morphological features of bentonite with and without OSE, representing particle delamination in both cases and no inhibition property of OSE. These results approve that OSE can operate as a good deflocculant. The findings also indicated that the performance of OSE is more forceful for deflocculating cement-contaminated muds. The probable deflocculation mechanism proposed by the authors, is that tannins contained in the extract are responsible for deflocculating property of OSE. In fact, when OSE is in contact with water, hydrogen ions dissociate from the hydroxyl groups of tannin molecules. The resulted negative ions can adsorb on positive edges of Montmorillonite layers (dominant mineral in bentonite) and annul their ability to link together and so deflocculating bentonite mud [109].

Pérez *et al.* [110] examined the deflocculation effect of dividi tannins, extracted from fruit pods of *Caesalpinia coriaria* and present in tannic acid form. Satisfactory results were achieved [110].

Moslemizadeh *et al.* [111] investigated the effect of Henna extract (HE), containing Gallic acid, tannic acid and lawsone, on the swelling properties of sodium bentonite in aqueous solution for a potential use in drilling operations. Indeed, exposure of water sensitive shales (defined as sedimentary rocks with laminated layered characteristics and high clay content), mainly sodium saturated smectites, to the conventional water-based drilling fluids induces their hydration and swelling due to water adsorption leading to drilling problems. This study revealed that HE is able to reduce the swelling of sodium bentonite and has equivalent action compared to potassium chloride and polyamine, commonly clay stabilizers used for clay swelling inhibition. It has also been found that adsorption of HE increases hydrophobicity of sodium bentonite particles surface and that the plant extract has a deflocculating property at low concentrations. However, it has a satisfying inhibition performance for higher concentrations. The deflocculation mechanism proposed by Moslemizadeh *et al.* [111] is the same presented by Hafshejani *et al.* [109]. Yet, for the inhibition mechanism, the authors have justified the swelling inhibition by the interaction between the hydroxyl groups of HE constituents and oxygen atom attainable on silica surface of sodium bentonite through hydrogen bonding. Thus, HE is able to form inhibitive medium at concentrations much higher than that of deflocculating ones [111].

### *Therapeutical, Pharmaceutical and Bio-Medical Applications*

Clay minerals have long been used for therapeutic objectives since prehistoric times. They are commonly used in the pharmaceutical industry as excipients (lubricants, inert bases, emulsifiers) and as active principles administrated orally (antacids, gastrointestinal protectors, antidiarrhoeaics, osmotic oral laxatives, homoeostatic, antianaemic) and topically (antiseptics, disinfectants, dermatological protectors, local anaesthetics, keratolytic reducers, decongestive eye drops), notably due to their large specific area and their absorption/adsorption capacity, rheology, chemical inertness and small or no toxicity for the patient [112, 113].

In 1960s, it was discovered that oral absorption of certain drugs was diminished by the co-administration of clay-based intestinal adsorbents or by the presence of clay stabilizers in liquid formulations. It was soon realized that such behaviour in the associated administration of clay minerals with active ingredients could be valourized for new technological and biopharmaceutical interests. This was the beginning of the use of clays as drug delivery system [114–116]. Drug delivery systems have the benefit of decreasing the drug side effects due to sustained release and better patient compliance through shortened dose frequency. For many types of controlled release formulations, instantly upon placement in the release medium, a high initial bolus of drug is released wildly before it can reach a stable release rate, which can involve toxicity. Drug delivery system with controlled release rate can ensure the double target of reduced toxicity and improved compliance along with reinforced drug efficacy [117, 118].

Drug delivery systems are defined as hybrid materials where clays are proposed as host for drug encapsulation. So, these host–guest systems can also provide a high drug encapsulation efficiency, high drug protection against pH variations and light, drug solubility enhancement and easy passage through body tissue such as the skin [115, 119, 120, 121].

Clay/synthetic drug hybrid applied as drug delivery system in bio-medical and pharmaceutical fields has been widely studied over the past decade [115, 119, 122, 123, 124]. However, a narrow number of works dealing with clay/natural bioactive substances hybrid exist.

Rabiei *et al.* [125] examined montmorillonite/Gallic acid hybrid formation at two pH levels (pH 3 and pH 7) as a new controlled release system. Adsorption tests showed that Gallic acid loading amounts at pH 3 and pH 7 are about 33% and 22% respectively. XRD analysis demonstrated that intercalation process has taken place for Gallic acid charging into clay, as the basal reflection of montmorillonite ($d_{001}$) shifted for hybrids from 12.28 to 12.75 Å and 12.68 Å at pH 3 and pH 7 respectively.

FTIR and TGA analyses confirmed the high intercalation of Gallic acid at pH 3 compared to pH 7. The pattern of Gallic acid release from montmorillonite at PBS 0.1 M indicated that montmorillonite/Gallic acid hybrid could be exploited as a new controlled release system [125].

Dedzo *et al.* [126] reported the hybridization of Gallic acid with functionalized kaolinite (by a pyridinium-based ionic liquid). The results induced from this study showed an excellent rate of Gallic acid loading, in its anionic form, into modified kaolinite (with anion exchange properties). Compared to salicylic acid, Gallic acid was adsorbed in smaller amounts, due to the speciation in di-anionic form under the experimental conditions. According to the authors, this charge density conjoined with the presence of multiple hydroxyl groups in the Gallic acid structure justifies the more prolonged release time in a phosphate buffer solution at pH 7 [126].

Galbashy *et al.* [127] studied the loading of polyphenols, extracted from pomegranate peels, into montmorillonite clay for a possible use as a drug delivery vehicle. It was observed by XRD analysis that the basal distance of montmorillonite (12.3 Å) increased to 15.34 Å after interaction with pomegranate polyphenols, which confirms that montmorillonite/pomegranate polyphenols hybrid formation was based on the intercalation process.

FTIR and TGA analyses also confirmed the efficient intercalation of pomegranate polyphenols into montmorillonite interlayers. As proposed by the authors, the electrostatic forces, hydrogen bonding, ligand exchange and cation bridge may be predominant for the clay and polyphenols interaction [127].

Kim *et al.* [128] developed a novel nano-biohybrid based on engineered nanoclays and the antibacterial natural extract of *AgrimoniapilosaLedeb* (AP), containing flavonoids and tannins. According to the XRD analysis, the crystal structure of nanoclays was preserved after hybridization suggesting that the AP extract was attached on the surface of nanoclays. The SEM also confirmed this result by showing that nanoclays surface was covered with extract moiety with a few molecular layers. The authors pointed out that clay/AP extract biohybrid owned higher antibacterial activity than the natural extract itself. As it was argued, the hybrid form of the natural extract lets the antibacterial active components more exposed at the surface of nanoclays compared to the dried powder form of the extract [128].

Vergaro *et al.* [129] studied the encapsulation of resveratrol, known for having antioxidant and antineoplastic properties, inside halloysite. The clay/resveratrol hybrid showed slow 48 h release with a constant release rate with time. Halloysite tubes' external surface was coated with layer-by-layer polyelectrolyte multilayer to better control the release rate. The authors also demonstrated the potency of the polyphenolic compound on MCF-7 (breast cancer cell line) as apoptotic agent [129].

Teixeira-Neto *et al.* [130] investigated a hybrid material by combining clay minerals (montmorillonite and saponite) and anthocyanins from *Euterpe oleracea Mart.* (Açaí fruit) for nutraceutical supplements use. XRD, FTIR, Raman and electronic spectroscopic data revealed that flavylium cations were strongly loaded between clay interlayers. Thermal characterization showed an enhancement on the thermostability of the intercalated natural extract compared to the free one. This was argued by the strong interaction of anthocyanins with clay layers. The radical scavenging activity of immobilized anthocyanins was preserved after interaction. This was also attributed to the durable interaction between organic and inorganic parts. The authors concluded that clay/açaí anthocyanins hybrid is a promising material for antioxidant nutraceutical supplements uses [130].

Potarniche *et al.* [131] reported the formation of nanohybrids by modifying bentonite clay with quebracho tannins and methylene blue for later use in collagen-based nanocomposite membrane films development. TGA and XRD analyses proved the intercalation of active substances between the silicate layers. It was observed that modifying the layered silicate with biological ingredients affects the hydrophilicity of the final nanocomposite by reducing the water vapour adsorption. Although the methylene blue is a disinfectant, it negatively influences the growth and the development of the fibroblasts. Hence, it is inappropriate for use in case of an open lesion. On the contrary, quebracho tannins showed biocompatibility and biostimulating effect on the growth and development of the fibroblasts. The authors considered that bioactive films based on bentonite/quebracho tannins are good antiseptic and regenerating patches [131].

Madusanka *et al.* [132] presented a novel curcumin activated carboxymethylcellulose (CMC)–montmorillonite nanocomposite as a release system for curcumin, a turmeric derived water insoluble polyphenolic compound with antibacterial/anti-cancer properties.

CMC was used as an emulsifier for curcumin. FTIR data, TGA and XRD confirmed the interaction of curcumin and CMC with montmorillonite. It was observed that water solubility of encapsulated curcumin has significantly increased (60% release within 2 h and 30 min in distilled water at pH 5.4) compared to free curcumin. Based on these results the authors found that the elaborated nanocomposite could be a convenient curcumin carrier with enhanced release and structural properties [132].

*Other Applications*

Tian *et al.* [133] developed novel biohybrid materials based on clays and five Chinese traditional herbs extract (*Andrographispaniculata* Nees, *Gallachinensis*, *Punicagranatum* L., *Cortex phellodendrichinensis* and *Radix scutellariae*) for a potential use as algacides against two harmful algae, *Phaeocystisglobosa* and *Prorocentrumdonghaiense*. The study findings showed that *Gallachinensis* extract (containing mainly Gallic acid and Gallotannins) exhibited the strongest inhibitory effect, almost 100% during 96 h treatments. While functionalized clay with *Gallachinensis* extract remarkably enhanced the inhibitory effect, as it produced 95% of growth inhibition for both algae species at 24 h and kept the effect thereafter. These results led to conclude that clays modified with Chinese herbs extract might be an efficient method for mitigating and controlling harmful algae [133].

## 11.4 Conclusions and Perspectives

The examples presented in this chapter show that clay–polyphenols hybrids are of great interest due to the fields of their applications such as active packaging, functional coating, heavy-metal removal and controlled delivery system. The diversity of both polyphenols and clays suggests that there are numerous possibilities for hybridization, and therefore the production of a variety of nanocomposite materials with potential applications in important areas can be expected.

The development of these kinds of materials is an interdisciplinary topic because it involves Life Sciences, Material Sciences and Nanotechnology skills. Behind the development of Bio-nanohybrid materials and more, specifically the clay–polyphenols hybrids, there are several major areas of interest such as the potential economic impacts, environmental impacts, and the ability to meet material needs worldwide.

## References


1. Kickelbick, G., Hybrid materials – past, present and future. *Hybrid Mater.*, 1, 39, 2014.
2. Doménech, A., Doménech-Carbó, M.T., del Río, M.S., de Agredos Pascual, M.L.V., Lima, E., Maya Blue as a nanostructured polyfunctional hybrid organic–inorganic material: the need to change paradigms. *New J. Chem.*, 33(12), 2371, 2009.
3. Zhu, Y.P. and Yuan, Z.Y., History and classification of non-siliceous hybrid materials, in: *Mesoporous Organic–Inorganic Non-Siliceous Hybrid Materials*, SpringerBriefs in Molecular Science, pp. 7–23, Springer, Berlin Heidelberg, 2015.
4. Ouellet-Plamondon, C., Aranda, A., Favier, A., Habert, G., Van Dammed, H., Ruiz-Hitzky, E., The Maya blue nanostructured material concept applied to colouring geopolymers. *RSC Adv.*, 5, 98834, 2015.
5. Gómez-Romero, G. and Sanchez, G., Hybrid materials. Functional properties. From Maya Blue to 21st century materials. *New J. Chem.*, 29, 57, 2005.
6. Sanchez, C., Belleville, P., Popall, M., Nicole, L., Applications of advanced hybrid organic–inorganic nanomaterials: from laboratory to market. *Chem. Soc. Rev.*, 40(2), 696, 2011.
7. Kickelbick, G., Introduction to hybrid materials, in: *Hybrid Materials: Synthesis, Characterization, and Applications*, pp. 1–4, John Wiley & Sons, Darmstadt, Germany 2007.
8. Arkles, B., Commercial applications of sol-gel-derived hybrid materials. *MRS Bull.*, 26(5), 402, 2001.
9. Ruiz-Hitzky, E., Aranda, P., Darder, M., Rytwo, G., Hybrid materials based on clays for environmental and biomedical applications. *J. Mater. Chem.*, 20, 9306, 2010.
10. Darder, M., Aranda, P., Ruiz-Hitzky, E., Bionanocomposites: a new concept of ecological, bioinspired, and functional hybrid materials. *Adv. Mater.*, 19(10), 1309, 2007.

11. Livage, J., Henry, M., Sanchez, C., Sol-gel chemistry of transition metal oxides. *Prog. Solid State Chem.*, 18(4), 259, 1988.
12. Melde, B.J., Holland, B.T., Blanford, C.F., Stein, A., Mesoporous sieves with unified hybrid inorganic/organic frameworks. *Chem. Mater.*, 11(11), 3302, 1999.
13. 13.Yaghi, O.M., Li, H., Davis, C., Richardson, D., Groy, T.L., Synthetic strategies, structure patterns, and emerging properties in the chemistry of modular porous solids. *Acc. Chem. Res.*, 31(8), 474, 1998.
14. Yamada, A., Sasabe, H., Osada, Y., Shiroda, Y., Yamamoto, I., Concepts of hybrid materials, hybrid materials – concept and case studies. *ASM International, Ohio*, USA, 1989.
15. Makishima, A., Possibility of hybrid materials. *Ceram. Jap.*, 39(2), 90, 2004.
16. Ashby, M.F. and Brechet, Y.J.M., Designing hybrid materials. *Actamaterialia*, 51(19), 5801, 2003.
17. Alemán, J.V., Chadwick, A.V., He, J., Hess, M., Horie, K., Jones, R.G.,... Penczek, S., Definitions of terms relating to the structure and processing of sols, gels, networks, and inorganic-organic hybrid materials (IUPAC Recommendations 2007). *Pure Appl. Chem.*, 79(10), 1801, 2007.
18. Chujo, Y., Organic–inorganic nano-hybrid materials [Translated]. *KONA Powder Part. J.*, 25, 255, 2007.
19. Unterlass, M.M., Green synthesis of inorganic–organic hybrid materials: state of the art and future perspectives. *Eur. J. Inorg. Chem.*, 8, 1135, 2016.
20. Camargo, P.H.C., Satyanarayana, K.G., Wypych, F., Nanocomposites: synthesis, structure, properties and new application opportunities. *Mater. Res.*, 12(1), 1, 2009.
21. Nanko, M., Definitions and categories of hybrid materials. *Adv. Tech. Mater. Mater. Process. J.*, 11, 1, 2009.
22. Judeinstein, P., Rault, J., Alonso, B., Sanchez, C., Macroscopic–microscopic mechanical relaxation behavior of hybrid organic–inorganic materials. *J. Polym. Sci. Part B: Polym. Phys.*, 39(6), 645, 2001.
23. Liu, L.H., Métivier, R., Wang, S., Wang, H., Advanced nanohybrid materials: surface modification and applications. *J. Nanomater.*, 2012, 1, 2012.
24. Lenglet, M., Iono-covalent character of the metal–oxygen bonds in oxides: a comparison of experimental and theoretical data. *Active Passive Electron. Comp.*, 27(1), 1, 2004.
25. Judeinstein, P. and Sanchez, C., Hybrid organic–inorganic materials: a land of multidisciplinarity. *J. Mater. Chem.*, 6(4), 511, 1996.
26. Ma, H.Y., Wang, T.L., Chang, P.Y., Yang, C.H., High refractive organic–inorganic hybrid films prepared by low water sol-gel and UV-irradiation processes. *Nanomaterials*, 6(3), 44, 2016.
27. Da Costa Júnior, A.E., Mota, J.P.F., Pontes, S.M.A., Maia, F.J.N., da Silva Clemente, C., Fechine, P.B.A., Mele, G., A self-assembly of graphene oxide@ Fe3O4/metallo-phthalocyaninenanohybrid materials: synthesis, characterization, dielectric and thermal properties. *J. Mater. Sci.*, 52(16), 9546, 2017.

28. Wang, J., Zhang, W., Yue, X., Yang, Q., Liu, F., Wang, Y., Wang, J., One-pot synthesis of multifunctional magnetic ferrite–MoS 2–carbon dot nanohybrid adsorbent for efficient Pb (ii) removal. *J. Mater. Chem. A*, 4(10), 3893, 2016.
29. Sanchez, C., Soler-Illia, G.D.A., Ribot, F., Lalot, T., Mayer, C.R., Cabuil, V., Designed hybrid organic–inorganic nanocomposites from functional nanobuilding blocks. *Chem. Mater.*, 13(10), 3061, 2001.
30. Sanchez, C., Julián, B., Belleville, P., Popall, M., Applications of hybrid organic–inorganic nanocomposites. *J. Mater. Chem.*, 15(35–36), 3559, 2005.
31. Ribot, F. and Sanchez, C., Organically functionalized metallic oxo-clusters: structurally well-defined nanobuilding blocks for the design of hybrid organic-inorganic materials. *Comments Inorg. Chem.*, 20(4–6), 327, 1999.
32. Mackenzie, J.D. and Bescher, E.P., Chemical routes in the synthesis of nanomaterials using the sol–gel process. *Acc. Chem. Res.*, 40(9), 810, 2007.
33. Mori, H., Design and synthesis of functional silsesquioxane-based hybrids by hydrolytic condensation of bulky triethoxysilanes. *Int. J. Polym. Sci.*, 2012, 1, 2012.
34. Paul, L., Dolai, M., Panja, A., Ali, M., Hydrothermal synthesis of two supramolecular inorganic–organic hybrid phosphomolybdates based on Ni (ii) and Co (ii) ions: structural diversity and heterogeneous catalytic activities. *New J. Chem.*, 40(8), 6931, 2016.
35. Rozes, L. and Sanchez, C., Titanium oxo-clusters: precursors for a Lego-like construction of nanostructured hybrid materials. *Chem. Soc. Rev.*, 40(2), 1006, 2011.
36. Samiey, B., Cheng, C.H., Wu, J., Organic-inorganic hybrid polymers as adsorbents for removal of heavy metal ions from solutions: a review. *Materials*, 7(2), 673, 2014.
37. Bourgeat-Lami, E., Organic–inorganic nanostructured colloids. *J. Nanosci. Nanotechnol.*, 2(1), 1, 2002.
38. Hussein, M.Z., Nazarudin, N.F.B., Sarijo, S.H., Yarmo, M.A., Synthesis of a layered organic-inorganic nanohybrid of 4-chlorophenoxyacetate-zinc-layered hydroxide with sustained release properties. *J. Nanomater.*, 2012, 9pp, 2012.
39. Vaia, R.A. and Giannelis, E.P., Polymer nanocomposites: status and opportunities. *MRS Bull.*, 26(5), 394, 2001.
40. Hoffmann, F., Cornelius, M., Morell, J., Fröba, M., Silica-based mesoporous organic–inorganic hybrid materials. *Angewandte Chemie International Edition*, 45(20), 3216, 2006.
41. Shimojima, A., Kuge, H., Kuroda, K., Synthesis of mesostructured silica from monoalkyl-substituted double five-ring units. *J. Sol-Gel Sci. Technol.*, 57(3), 263, 2011.
42. Van Der Voort, P., Vercaemst, C., Schaubroeck, D., Verpoort, F., Ordered mesoporous materials at the beginning of the third millennium: new strategies to create hybrid and non-siliceous variants. *Phys. Chem. Chem. Phys.*, 10(3), 347, 2008.

43. Soler-Illia, G.J.D.A., Sanchez, C., Lebeau, B., Patarin, J., Chemical strategies to design textured materials: from microporous and mesoporous oxides to nanonetworks and hierarchical structures. *Chem. Rev.*, 102(11), 4093, 2002.
44. Acevedo, M. and Alexander, R., Design and synthesis of novel bio-nanohybrid materials: catalytic applications in reactions of interest to the fine-chemical/pharmaceutical industries, *Doctoral Thesis Dissertation*. University Rovira and Virgili, Tarragona (Spain), 2012.
45. Currie, H.A., Patwardhan, S.V., Perry, C.C., Roach, P., Shirtcliffe, N.J., Natural and artificial hybrid biomaterials, in: *Hybrid Materials – Synthesis, Characterization and Applications*. pp. 255–299, G. Kickelbick (Ed.), Wiley-VCH, Weinheim, 2007.
46. Ruiz-Hitzky, E., Darder, M., Aranda, P., An introduction to bio-nanohybrid materials, in: *Bio-Inorganic Hybrid Nanomaterials Strategies, Syntheses, Characterization and Applications*, pp. 1–40, Wiley-VCHVerlagGmbH& Co. KGaA, Weinheim, Germany, 2008.
47. Palocci, C. and Chronopoulou, L., Hybrid systems biomolecule-polymeric nanoparticle: synthesis, properties and biotechnological applications, in: *Advances in Macromolecules*, pp. 219–259, Springer, Netherlands, 2010.
48. Gürses, A., *Introduction to polymer-clay nanocomposites*, pp. 88–106, Pan Stanford Publishing Pte. Ltd, Singapore, 2015.
49. Bergaya, F. and Lagaly, G., General introduction: clays, clay minerals, and clay science, in: *Handbook of Clay Science*, F. Bergaya, B.K.G. Theng, G. Lagaly (Eds), pp. 1–18, Elsevier, Amsterdam, Netherlands, 2006.
50. Utraki, A.L., *Clay-containing polymeric nanocomposites (volume 1)*, Rapra Technology, Shawbury, Shrewsbury, Shropshire, UK, 2004.
51. Jacobs, P.M., Soil and clay mineralogy, in: *The International Encyclopedia of Geography*, D. Richardson., N. Castree *et al.* (Eds), pp. 1–8, John Wiley and Sons, Ltd, Chichester, UK; Hoboken, NJ, 2017.
52. Wilson, M.J., *Rock-forming minerals: Clay minerals, Sheet silicates (volume 3C)*, pp. 2–5, The Geological Society, London, 2013.
53. Schulze, D.G., Clay minerals, in: *Encyclopedia of Soils in The Environment (Volume 2)*, D. Hillel and J.L. Hatfield, pp. 246–254, Elsevier, Amsterdam, Netherlands, 2005.
54. Obaje, S.O., Omada, J.I., Dambatta, U.A., Clays and their industrial applications: Synoptic Review. *Int. J. Sci. Technol.*, 3, 264, 2013.
55. CalabiFloody, M., Theng, B.K.G., Reyes, P., Mora, M.L., Natural clays: applications and future trends – a Chilean perspective. *Clay Miner.*, 44, 161, 2009.
56. Barton, C.D. and Karathanasis, A.D., Clay minerals, in: *Encyclopedia of Soil Science*, R. Lal (Ed), pp. 187–192, Marcel Dekker, Inc., NY, 2002.
57. Lee, S.M. and Tiwari, D., Organo and inorgano-modified clays in the remediation of aqueous solutions: an overview. *Appl. Clay Sci.*, 59–60, 84, 2012.
58. Mott, C.J.B., Clay minerals – an introduction. *Catal. Today*, 2, 199, 1988.

59. Brigatti, M.F., Galan, E., Theng, B.K.G., Structures and mineralogy of clay minerals, in: *Handbook of Clay Science*, F. Bergaya., B.K.G. Theng., G. Lagaly (Eds), pp. 19–48, Elsevier, Amsterdam, Netherlands, 2006.
60. Yang, J.H., Lee, J.H., Ryu, H.J., Elzatahry, A.A., Alothman, Z.A., Choy, J.H., Drug-clay nanohybrid as sustained delivery systems. *Appl. Clay Sci.*, 130, 20, 2016.
61. Pavlidou, S. and Papaspyrides, C.D., A review on polymer-layered silicate nanocomposites. *Prog. Polym. Sci.*, 33, 1119, 2008.
62. Ray, S.S. and Okamoto, M., Polymer/layered silicate nanocomposites: a review from preparation to processing. *Prog. Polym. Sci.*, 28, 1539, 2003.
63. Galan, E., Properties and applications of palygorskite-sepiolite clays. *Clay Miner.*, 31, 443, 1996.
64. Garcia-Romero, E. and Suárez, M., On the chemical composition of sepiolite and palygorskite. *Clays Clay Miner.*, 58, 1, 2010.
65. Borwn, G., Crystal structures of clay minerals and related phyllosilicates. *Phil. Trans. R. Soc. Lond.*, A311, 221, 1984.
66. De Paiva, L.B., Morales, A.R., Valenzuela Díaz, F.R., Organoclays: properties, preparation and applications. *Appl. Clay Sci.*, 42, 8, 2008.
67. Singla, P., Mehta, R., Upadhyary, S.N., Clay modification by the use of organic cations. *Green Sustain. Chem.*, 2, 21, 2012.
68. Bergaya, F. and Lagaly, G., Surface modification of clay minerals. *Appl. Clay Sci.*, 19, 1, 2001.
69. Nafees, M. and Waseem, A., Organoclays as sorbent material for phenolic compounds: A review. *Clean-Soil, Air, Water*, 42, 1500, 2014.
70. Jaynes, W.F. and Boyd, S.A., Hydrophobicity of siloxane surfaces in smectites as revealed by aromatic hydrocarbon adsorption from water. *Clays Clay Miner.*, 39, 428, 1991.
71. Theng, B.K.G., Grennland, D.J., Quik, J.P., Adsorption of alkylammonium cations by montmorillonite. *Clay Miner.*, 7, 1, 1967.
72. Bonczek, J.L., Harris, W.G., Nkedi-Kizza, P., Monolayer to bilayer transitional arrangements of hexadecyltrimethylammonium cations on Na-montmorillonite. *Clays Clay Miner.*, 50, 11, 2002.
73. Sarkar, B., Xi, Y., Megharaj, M., Krishnamurti, G.S.R., Bowman, M., Rose, H., Naidu, R., Bioreactiveorganoclay: a new technology for environmental remediation. *Crit. Rev. Environ. Sci. Technol.*, 42, 435, 2012.
74. Dai, J. and Mumper, R.J., Plant phenolics: extraction, analysis and their antioxidant and anticancer properties. *Molecules*, 15(10), 7313, 2010.
75. Beckman, C.H., Phenolic-storing cells: keys to programmed cell death and periderm formation in wilt disease resistance and in general defence responses in plants?. *Physiol. Mol. Plant Pathol.*, 57(3), 101, 2000.
76. Cheynier, V., Polyphenols in foods are more complex than often thought. *Am. J. Clin. Nutr.*, 81(1), 223S, 2005.
77. Pandey, K.B. and Rizvi, S.I., Plant polyphenols as dietary antioxidants in human health and disease. *Oxidat. Med. Cell. Longevity*, 2(5), 270, 2009.

78. Spencer, J.P., El Mohsen, M.M.A., Minihane, A.M., Mathers, J.C., Biomarkers of the intake of dietary polyphenols: strengths, limitations and application in nutrition research. *Brit. J. Nutr.*, 99(1), 12, 2008.
79. Tsao, R., Chemistry and biochemistry of dietary polyphenols. *Nutrients*, 2(12), 1231–1246, 2010.
80. Peterson, J., Dwyer, J., Adlercreutz, H., Scalbert, A., Jacques, P., McCullough, M.L., Dietary lignans: physiology and potential for cardiovascular disease risk reduction. *Nutr. Rev.*, 68(10), 571, 2010.
81. Reventós, M.M., Rius, J., Amigó, J.M., Mineralogy and geology: the role of crystallography since the discovery of X-ray diffraction in 1912. *Revista de la Sociedad Geológica de España.*, 25, 133, 2012.
82. Fournier, F., De Viguerie, L., Balme, S., Janot, J.M., Physico-chemical characterization of lake pigments based on montmorillonite and carminic acid. *Appl. Clay Sci.*, 130, 12, 2016.
83. Tenci, M., Rossi, S., Aguzzi, C., Carazo, E., Sandri, G., Bonferoni, M.C., Grisoli, P., Viseras, C., Caramella, C.M., Ferrzri, F., Carvacrol/clay hybrids loaded into *in situ* gelling films. *Int. J. Pharmaceut.*, 531(2), 676, 2017.
84. Xi, Y., Martens, W., He, H., Frost, R.L., Thermogravimetric analysis of organoclaysintercalated with the surfactant Octadecyltrimethylammonium bromide. *J. Therm. Anal. Cal.*, 81, 91, 2005.
85. Busolo, M.A. and Lagaron, J.M., Antioxidant polyethylene films based on a resveratrol containing clay of interest in food packaging applications. *Food Packag. Shelf Life*, 6, 30, 2015.
86. Blanchart, P., Dembelé, A., Demblé, C., Pléa, M., Bergström, L., Granet, R., Sol, V., Gloaguen, V., Degot, M., Krausz, P., Mechanism of traditional Bogolan dyeing technique with clay on cotton fabric. *Appl. Clay Sci.*, 50, 455, 2010.
87. Kaya, K., Onganer, Y., Tabak, A., Preparation and characterization of 'green' hybrid clay-dye nanopigments. *J. Phys. Chem. Solids*, 78, 95, 2015.
88. Abreu, A.S., Oliveira, M., De Sá, A., Rodriguez, R.M., Cerqueira, M.A., Antimicrobial nanostructured starch based films for packaging. *Carbohydr. Polym.*, 129, 127, 2015.
89. Gómez-Estaca, J., López-de-Dicastillo, C., hernández-Muñoz, P., Catalá, R., Gavara, R., Advances in antioxidant active foodpackaging. *Trends Food Sci. Technol.*, 35, 42, 2014.
90. Barbosa-Pereira, L., Angulo, I., Lagarón, L.M., Paseiro-Losada, P., Cruz, J.M., Development of new active packaging films containing bioactive nanocomposites. *Innov. Food. Sci. Emerg. Technol.*, 26, 310, 2014.
91. Ranjan, S., Dasgupta, N., Chakraborty, A.R., Samuel, S.M., Ramalingam, C., Shanker, R., Kumar, A., Nanoscience and nanotechnologies in food industries: opportunities and research trends. *J. Nanopart. Res.*, 16, 2464, 2014.
92. Rodriguez, F., Sepulveda, H.M., Bruna, J., Guarda, A., Galotto, M.J., Development of cellulose eco-nanocomposites with antimicrobial properties oriented for food packaging. *Packag. Technol. Sci.*, 26, 149, 2013.

93. Campos-Requena, V.H., Riva, B.L., Pérez, M.A., Garrido-Miranda, C.A., Pereira, E.D., Polymer/clay nanocomposite films as active packaging material: modeling of antimicrobial release. *Eur. Polym. J.*, 71, 461, 2015.
94. Contini, C., Katsikogianni, M.G., O'Neill, F.T., O'Sullivan, M., Dowling, D.P., Monahan, F.J., Development of active packaging containing natural antioxidants. *Procedia Food Sci.*, 1, 224, 2011.
95. Valdés, A., Mellinas, A.C., Ramos, M., Burgos, N., Jiménez, A., Garrigós., M.C., Use of herbs, spices and their bioactive compounds in active food packaging. *RSC Adv.*, 5, 40324, 2015.
96. Wang, S., Marcone, M.F., Barbut, S., Lim, L.T., Fortification of dietary biopolymers-based packaging material with bioactive plant extracts. *Food Res. Int.*, 49, 80, 2012.
97. Lim, G.O., Jang, S., Bing Song, K., Physical and antimicrobial properties of *Gelidiumcorneum*/nano-clay composite film containing grapefruit seed extract or thymol. *J. Food Eng.*, 98, 415, 2010.
98. Busolo, M.A. and Lagaron, J.M., Antioxidant polyethylene films based on a resveratrol containing clay of interest in food packaging applications. *Food Packag. Shelf Life*, 6, 30, 2015.
99. Fournier, F., De Viguerie, L., Balme, S., Janot, J.M., Walter, P., Jaber, M., Physico-chemical characterization of lake pigments based on montmorillonite and carminic acid. *Appl. Clay Sci.*, 130, 12, 2016.
100. Giustetto, R. and Wahyudi, O., Sorption of red dyes on palygorskite: synthesis and stability of red/purple Mayan nanocomposites. *Microporous Mesoporous Mater.*, 142, 221, 2011.
101. Giustetto, R., Seenivasan, K., Pellerej, D., Ricchiardi, G., Bordiga, S., Spectroscopic characterization and photo/thermal resistance of a hybrid palygorskite/methyl red Mayan pigment. *Microporous Mesoporous Mater.*, 155, 167, 2012.
102. Ogawa, M., Takee, R., Okabe, Y., Seki, Y., Bio-geo hybrid pigment: clay-anthocyanin complex which changes colour depending on the atmosphere. *Dyes Pigm.*, 139, 561, 2017.
103. Doménech-Carbó, A., Doménech-Carbó, M.T., Osete-Cortina, L., Valle-Algarra, F.M., Buti, D., Isomerization and redox tuning in 'Maya yellow' hybrids from flavonoid dyes plus palygorskite and kaolinite clays. *Microporous Mesoporous Mater.*, 194, 135, 2014.
104. Kohno, Y., Kinoshita, R., Ikoma, S., Yoda, K., Shibata, M., Matsushima, R., Tomita, Y., Mmaeda, Y., Kobayashi, K., Stabilization of natural anthocyanin by intercalation into montmorillonite. *Appl. Clay Sci.*, 42, 519, 2009.
105. Lima, E., Martinez-Ortiz, M.J., Fregoso, E., Mendez-Vivar, J., Capturing natural chromophores on natural and synthetic aluminosilicates. *Stud. Surf. Sci. Catal.*, 170, 2110, 2007.
106. Sorgho, B., Zerbo, L., Keita, I., Dembele, C., Plea, M., Sol, V., Gomina, M., Blanchart, P., Strength and creep behavior of geomaterials for building with tannin addition. *Mater. Struct.*, 47, 937, 2014.

107. Keita, I., Sorgho, B., Dembele, C., Plea, M., Zerbo, L., Guel, B., Ouedraogo, R., Gomina, M., Blanchart, P., Ageing of clay and clay-tannin geomaterials for building. *Construct. Build. Mater.*, 61, 114, 2014.
108. Abdo, K. and Haneef, M.D., Clay nanoparticles modified drilling fluids for drilling of deep hydrocarbon wells. *Appl. Clay Sci.*, 86, 76, 2013.
109. Hafshejani, K.S., Moslemizadeh, A., Shahbazi, K., A novel bio-based deflocculant for bentonite drilling mud. *Appl. Clay Sci.*, 127–128, 23, 2016.
110. Pérez, M.A., Rengifo, R., Pereira, C., Hernández, V., Dividi tannins: an ecological product for water-based drilling fluids. *Environ. Dev. Sustain.*, 19, 1815, 2017.
111. Moslemizadeh, A., Shadizadeh, S.R., Moomenie, M., Experimental investigation of the effect of henna extract on the swelling of sodium bentonite in aqueous solution. *Appl. Clay Sci.*, 105–106, 78, 2015.
112. Carretero, M.I., Clay minerals and their beneficial effects upon human health: a review. *Appl. Clay Sci.*, 21, 155, 2022.
113. Carretero, M.I. and Pozo, M., Clay and non-clay minerals in the pharmaceutical and cosmetic industries: part II. Active ingredients. *Appl. Clay Sci.*, 47, 171, 2010.
114. Aguzzi, C., Cerezo, P., Viseras, C., Caramella, C., Use of clays as drug delivery systems: possibilities and limitations. *Appl. Clay Sci.*, 36, 22, 2007.
115. Apps, M.G., Ammit, A.J., Gu, A., Wheate, N.J., Analysis of montmorillonite clay as a vehicle in platinum anticancer drug delivery. *Inorg. Chim. Acta*, 421, 513, 2014.
116. Viseras, C., Cerezo, P., Sanchez, R., Salcedo, I., Aguzzi, C., Current challenges in clay minerals for drug delivery. *Appl. Clay Sci.*, 48, 291, 2010.
117. Rodriguez, L.A.S., Figueiras, A., Veiga, F., Freitas, R.M., Nunes, L.C.C., Silva Filho, E.C., Silva Leite, C.M., The systems containing clays and clay minerals from modified drug release: A review. *Colloids Surf. B*, 103, 642, 2013.
118. Kim, M.H., Choi, G., Elzatahry, A., Vinu, A., Choy, Y.B., Choy, J.H., Review of clay-drug hybrid materials for biomedical applications: administration routes. *Clays Clay Miner.*, 64, 115, 2016.
119. Gârea, S.A., Mihai, A.I., Ghebaur, A., Nistor, C., Sârbu, A., Porous clay heterostructures: a new inorganic host for 5-fluorouracil encapsulation. *Int. J. Pharm.*, 491, 299, 2015.
120. De Oliveira, J.L., Ramos Campos, E.V., Bakshi, M., Abhilash, P.C., Fraceto, L.F., Application of nanotechnology for the encapsulation of botanical insecticides for sustainable agriculture: prospects and promises. *Biotechnol. Adv.*, 32, 1550, 2014.
121. Liu, Y. and Feng, N., Nanocarriers for the delivery of active ingredients and fractions extracted from natural products used in traditional Chinese medicine (TCM). *Adv. Colloid Interface Sci.*, 221, 60, 2015.
122. Ambrogi, V., Nocchetti, M., Latterini, L., Promethazine–montmorillonite inclusion complex to enhance drug photostability. *Langmuir*, 30, 14616, 2014.

123. Bonina, F.P., Giannossi, M.L., Medici, L., Puglia, C., Summa, V., Tateo, F., Adsorption of salicylic acid on bentonite and kaolin and release experiments. *Appl. Clay Sci.*, 36, 77, 2007.
124. El-Nahhal, Y., Nir, S., Margulies, L., Rubin, B., Reduction of photodegradation and volatilization of herbicides in organo-clay formulations. *Appl. Clay Sci.*, 14, 105, 1999.
125. Rabiei, M., Sabahi, H., Rezayan, A.H., Gallic acid-loaded montmorillonite nanostructure as a new controlled release system. *Appl. Clay Sci.*, 119, 236, 2016.
126. Dedzo, G.K. and Detellier, C., Intercalation of two phenolic acids in an ionic liquid-kaolinite nanohybrid material and desorption studies. *Appl. Clay Sci.*, 97–98, 153, 2014.
127. Galbashy, M., Sabahi, H., Allahdadi, I., Nazokdast, H., Hosseini, M., Synthesis the montmorillonite–pomegranate (*Punicagranatum L.*) peel polyphenols nanostructure as a drug delivery vehicle. *Biomed. Pharmacol. J.*, 9, 385, 2016.
128. Kim, H.J., Kim, T.H., Kim, H.M., Hong, I.K., Kim, E.J., Choi, A.J., Choi, H.J., Oh, J.M., Nano-biohybrids of engineered nanoclays and natural extract for antibacterial agents. *Appl. Clay Sci.*, 134, 19, 2016.
129. Vergaro, V., Lvov, Y.M., Leporatti, S., Halloysite clay nanotubes for resveratrol delivery to cancer cells. *Macromol. Biosci.*, 12, 1265, 2012.
130. Teixeira-Neto, A.A., Izumi, C.M.S., Temperini, M.L.A., Ferreira, A.M., Constantino, V.R.L., Hybrid materials based on smectite clays and nutraceutical anthocyanins from the açaí fruit. *Eur. J. Chem.*, 2012, 5411, 2012.
131. Potarniche, C.G., Vuluga, Z., Radovici, C., Serban, S., Vuluga, D.M., Ghiurea, M., Purcar, V., Trandafir, V., Iordáchescu, D., Albu, M.G., Nanocomposites based on collagen and Na-montmorillonite modified with bioactive substances. *Mater. Plast.*, 47, 267, 2010.
132. Madusanka, N., Silva, K.M.N., Amaratunga, G., A curcumin activated carboxymethyl cellulose-montmorillonite clay nanocomposite having enhanced curcumin release in aqueous media. *Carbohydr. Polym.*, 134, 695, 2015.
133. Tian, F., Zhou, J., Sun, Z., Cai, Z., Xu, N., An, M., Duan, S., Inhibitory effects of Chinese traditional herbs and herb-modified clays on the growth of harmful algae, *Phaeocystisglobosa* and *Prorocentrumdonghaiense*. *Harmful Algae*, 37, 153, 2014.

12

# Green and Sustainable Selenium Nanoparticles and Their Biotechnological Applications

**Meryam Sardar*** **and Hammad Alam**

*Department of Biosciences, Jamia Millia Islamia, New Delhi, India*


## *Abstract*

Nanotechnology is the branch of science that deals in tailoring and manufacturing of materials at the nanoscale that is having a dimension smaller than 100 nm. The inorganic nanoparticles are of great interest due to their unique optical, catalytic, and electronic properties. These unique properties have enhanced their applications in various industries such as cosmetics, coatings, packaging, medicine, and biotechnology. The properties of the nanoparticles depend upon their shape and size. At present, the environment-friendly methods of synthesis are preferred over other methods which employ harsh conditions and toxic chemicals. These green methods employ biological materials for synthesis which includes plant/plant extracts, enzymes/proteins, biomolecules, and microorganism. Selenium is an essential trace element found in various enzymes and proteins. Selenium is toxic at higher concentration, but it is necessary in dietary constituent for selenocysteine and enzymes like glutathione etc. In addition, selenium has antibacterial and anticancerous properties also. Therefore, there is a need to develop methods/protocols to synthesize selenium nanoparticles which are green and economical. In this chapter, the biological synthesis of selenium nanoparticles has been discussed in detail. The applications of selenium nanoparticles mainly the biomedical applications are also highlighted in this chapter.


***Keywords:*** Selenium nanoparticles, biocompatible, antimicrobial, anticancerous

**Corresponding author*: msardar@jmi.ac.in

Shakeel Ahmed and Chaudhery Mustansar Hussain (eds.) Green and Sustainable Advanced Materials, Vol. 1 (333–354) © 2018 Scrivener Publishing LLC

## 12.1 Introduction

A bulk material has constant physico-chemical properties regardless of its size, but when tailored at nanoscale they acquire some novel properties due to their very high aspect ratio. These characteristic properties make these nanomaterials important for humans with their wide range of applications in electronics [1], cosmetics [2], coatings [3], packaging [4], imaging [5], agriculture [6], as catalysts [7], and in medicine [8]. These nanomaterials are produced naturally or synthetically and can be classified into four general categories:

1. Carbon-based: mostly made up of carbon, e.g. carbon nanotubes and fullerenes.
2. Metal-based: includes quantum dots, metal, and metal oxide nanoparticles (NPs).
3. Dendrimers: a highly branched polymer at nanoscale dimensions.
4. Composites: 'A Nanocomposite is a multiphase solid material where one of the phases has one, two or three dimensions of less than 100 nm, or structure having nano-scale repeat distance between the different phases that make up the material'.

So far, a number of reviews and books have been published on the different types of nanomaterials and their applications [1, 2]. Among the various types of nanomaterials, selenium nanoparticles (SeNPs) have been extensively studied for their biomedical applications and therapeutic use as they are less toxic compared to other nanomaterials. In this chapter, synthesis of SeNPs, mainly green synthesis, has been discussed at length.

Selenium (Se) is a trace element found in soil; the bioavailability of Se compounds varies depending on the source and form of Se. Our primary source of Se intake is through food, which includes grains, animal products, and dietary supplements [9]. It is present in environment in various oxidation states such as $Se^{-2}$, $Se^{+4}$, and $Se^{+6}$ [10].

It is an essential micronutrient for living beings and is present in many proteins as the amino acid selenocysteine. The human genome consists of 25 different genes for seleno-proteins [11]. It is also present in various enzymes like glutathione peroxidase, formate dehydrogenase, xanthene dehydrogenase, glycine reductase, and thioredoxin reductase [11]. These enzymes/proteins are useful in redox reactions for preventing cellular damage, scavengers of free radicals, and also helps in thyroid regulation and to boost the immune system [11]. Most of the seleno-proteins help regulate antioxidant reactions and can thus participate in the protection of cells against oxidative stress. Some studies also suggest a strong correlation of seleno-proteins with human diseases like cancer, abnormalities in immune responses, and thyroid hormone function.

It has been reported that elemental Se shows good anticancerous/antibacterial activity but at very high concentration; such a high dose, however, is considered lethal to the living organism [12]. Thus, research is required to develop nanostructures of Se. Many reports are available in the literature on synthesis of SeNPs having good anticancerous activity at low dose [12–17].

Synthesis of SeNPs can be achieved by physical, chemical, and biological methods. The physical methods comprise laser ablation [13], UV radiation [14], and hydrothermal techniques [15]. The chemical synthesis of SeNPs includes catalytic reduction, acid decomposition, and precipitation using ascorbic acid [16], glucose [17], sodium dodecyl sulphate [18], etc. These methods often employ toxic chemicals or high temperature and high pressure which further pollute the environment. The large-scale synthesis requires the complex instrumentation or specific chemicals which may be time consuming and costly. To overcome the effect of toxic chemicals, biological protocols are generally followed. Moreover, the NPs produced by biological agents are more stable as compared to the chemically synthesized ones since the former is stabilized by natural polymers. The biological methods employ biomolecules, microbes, and plants for the synthesis of SeNPs. Till date, around 50 different species of microorganisms are reported to reduce colorless selenite and selenate to red elemental Se of various size and shapes. Numerous bacterial strains are known for the reduction of selenite/selenate into SeNPs in various environments even in sewage and sludge under both anaerobic and aerobic conditions. The synthesis of SeNPs can be easily monitored by UV-VIS (Ultraviolet Visible) spectroscopy; their characteristic surface Plasmon Resonance occurs in the range of 250 to 400 nm. The shapes and morphology of synthesized NPs can be determined by XRD (x-ray diffraction), DLS (Differential light scattering), TEM (Transmission electron microscopy) and FESEM (field emission scanning electron microscopy).

## 12.2 Synthesis of SeNPs

The aqueous solution of either Sodium Selenite or Selenous acid/Selenium chloride is generally taken as the raw material or precursor of Se to convert it into its nano form.

$$H_2SeO_3 \,/\, Na_2SeO_3 \xrightarrow{\text{Reducing agents}} Se^0$$

### 12.2.1 Physical Methods of Synthesis of SeNPs

Physical methods generally involve high temperature and use of radiation to initiate the reaction. Gamma irradiations generate hydrated electrons which reduces the Se ions to SeNPs. Tan *et al.* (2009) synthesized SeNPs using sodium alginate, selenious acid, and ascorbic acid by passing ultrasonic radiations; synthesis was completed under 2 h [19]. Microwave assisted synthesis of SeNPs using selenium tetrachloride ($SeCl_4$) and hydrazine has also been reported [20]. Guisbiers *et al.* (2016) synthesized SeNPs by Liquid Phase–Pulsed Laser Ablation (LP-PLA) in de-ionized water [21]. Glu-SeNPs were also synthesized by autoclaving the glucose solution with sodium selenite at 115 °C for 15 min. The synthesized NPs were biologically active against cancerous cells, with low toxicity against normal cell line [17].

### 12.2.2 Chemical Methods for Synthesis of SeNPs

Under chemical methods, the solution phase approach is the most common one which is generally used in labs to fabricate the NPs of different sizes and morphologies. The synthesis can be achieved by precipitation, acid decomposition, and catalytic reduction. Panahi-Kalamuei *et al.* (2015) synthesized SeNPs using selenium tetrachloride as a Se source and hydrazine, potassium borohydride, and thioglycolic acid (TGA) as reducing reagents via four different synthesis routes: co-precipitation, hydrothermal, microwave, and sonication [22]. The nature and concentration of reducing agent solvent used, and the type of method applied affect the morphology and size of the NPs. The synthesis of SeNPs using organic acids like ascorbic acid, oxalic acid, acetic acid, benzoic acid, and gallic acid as reducing agents has been described by Dwivedi *et al.* (2011) [23]. They reported that the carboxyl group of the acid reduces the Se salt to NPs. SeNPs of spherical shape and of size about 35–70 nm have been synthesized using different organic acids and polyvinyl alcohol as the stabilizing agent [23]. The synthesis of SeNPs using aldehydic group of glucose has also been described [17].

The size and shape of the NPs is directly related to its activity/applications which is dictated by several parameters like reducing agents, concentration, pH, temperature of incubation, time of incubation, and the stabilizing agents used for capping the nanomaterials.

### 12.2.3 Microbial Synthesis of SeNPs

Microbes play an important role in the synthesis of SeNPs. When the microbes are subjected to grow in the presence of Se salt, they utilize the detoxification mechanism to reduce the Se into the nano form. The nanomaterials of different morphologies and sizes can be obtained using the microbes, e.g. nanowires and nanorods have been synthesized from *Rhizobium selenireducens sp.* [24], *Bacillus sp.* [25], *Pseudomonas sp.* [26], *Pantoea sp.* [27], *Enterobactorsp.* [28], *Thaurea sp.* [29], *Sulfurospirillium sp.* [25], *Dugnellasp.* [30], and *Shewanella sp.* [31]. The shape and size of the nanomaterials is governed by the nature of reducing agents, concentration of the reactants, and the time of incubation. Temperature and pH are also known to be size and shape modulating agents. The synthesis of nanomaterials can be carried out inside or outside the microbial cell; the latter is the preferred method since the extraction step is eliminated in this case. SeNPs synthesized from different strains of bacteria and fungi are listed in Table 12.1.

Zhang *et al.* (2011) studied that $SeO_3^{2-}$ is reduced to elemental Se by the protein excreted by the *Pseudomonas alcaliphila* [26]. The microbes like *Enterobacter cloacae* SLD1a-1 [32], *Thauera selenatis* [37], and *Escherichia coli* K-12 [46] were studied for their potential and mechanism of SeNP synthesis. Fungi like *Yarrowia lipolytica* [55], *Alternaria alternata* [56], *Sacchromyces cerevisiae* [57], *Aspergillus terreus* [58], and *Cordyceps sinensis* [59] and algae *Gracilaria lamaneiform* [60] also reduced sodium selenite and form nano Se. Wadhwani *et al.* (2016) suggested the mechanism for reduction of Se by microbes [61]. Their process for selenate reduction suggested firstly catalyzing selenate to selenite by selenate reductase and then further catalyzing the product from the first step to insoluble elemental Se using nitrite and sulphite reductases.

Figure 12.1 shows the proposed diagram for reduction of Se with microbes. Selenate reductase is a membrane bound enzyme and is the key enzyme for the reduction of selanate [61].

### 12.2.4 Plant Based Synthesis of SeNPs

Plant based synthesis of NPs is a cost-effective and eco-friendly approach. Various plants and plant metabolites have been explored to synthesize nanomaterials. The SeNPs synthesized using different plants are shown in Table 12.2.

The biologically produced SeNPs are considered less toxic than the chemically produced ones because here the natural biomolecules are used for reduction and stabilization of NPs. The studies reported that the secondary metabolites present in plants mainly polyphenols, flavonoids,

**Table 12.1** Synthesis of SeNPs from microorganisms.

| Microorganism | Size of SeNPs (nm) | Shape of SeNPs | References |
|---|---|---|---|
| **Bacteria** | | | |
| *Pseudomonas alcaliphila* | 50–200 | Spherical | [26] |
| *Pantoea agglomerans* | 30–300 | Spherical | [27] |
| *Enterobacter cloacae and Escherichia coli* | – | – | [32] |
| *Streptomyces bikiniensis* | 17 nm | Spherical | [33] |
| *Bacillus subtilis* | 50–400 | Spherical | [34] |
| *Bacillus cereus* | 150–200 | Spherical | [35] |
| *Klebsiella pneumoniae* | 100–550 | – | [36] |
| *Thauera selenatis* | – | – | [37] |
| *Zooglea ramigera* | 30–150 | Spherical | [38] |
| *Bacillus licheniformis JS-2* | 40–180 | Spherical | [39] |
| *Lactobacillus plantarum and Lactobacillus johnsoni* | 25–250 | Spherical | [40] |
| *Pseudomonas aeruginosa strain JS-11* | 21 | Spherical | [41] |
| *Bacillus sp. MSh-1* | 80–220 | Spherical | [42] |
| *Lactobacillus casei; Streptococcus thermophilus;*<br>*Bifidobacterium; Lactobacillus acidophilus; Lactobacillus helveticus and Klebsiella pneumonia* | 50–500<br>400–500<br>50–500<br>100–500 | Spherical | [43] |
| *Stenotrophomonas maltophilia and Ochrobactrum sp.* | 76.2 and 221–345 | Spherical | [44] |
| *Enterobacter faecalis* | 29–195 | Spherical | [45] |
| *Escherichia coli K-12* | <50 | Spherical | [46] |
| *Stenotrophomonas maltophilia* | – | – | [47] |
| *Bacillus mycoides* | 50–400 | Spherical | [48] |
| *Azoarcus sp.* | 123 ± 35 | Spherical | [49] |
| *Tetrahymena thermophila* | 50–500 | Spherical | [50] |

(*Continued*)

**Table 12.1** Cont.

| Microorganism | Size of SeNPs (nm) | Shape of SeNPs | References |
|---|---|---|---|
| *Bacillus licheniformis JS-2* | 110 | Spherical | [51] |
| *Acinetobacter sp. SW30* | 78 | Spherical | [52] |
| *Streptomyces minutiscleroticus* | 10–250 | Spherical | [53] |
| *Citrobacter fruendii* | 45–70 | Spherical | [54] |
| **Fungi** | | | |
| *Yarrowia lipolytica* | – | – | [55] |
| *Alternaria alternata* | 30–150 | Spherical | [56] |
| *Sacchromyces cerevisiae* | 30–100 | Spherical | [57] |
| *Aspergillus terreus* | 47 | Spherical | [58] |
| *Cordyceps sinensis* | 80–125 | – | [59] |
| **Algae** | | | |
| *Gracilaria lamaneiform* | 50–123 | Spherical | [60] |

sugars, and proteins are responsible in green synthesis [62, 63–67]. Sharma *et al.* (2014) have synthesized SeNPs using dried extract of raisin (*Vitis vinifera*), which were characterized by UV-VIS spectroscopy and its size determined by TEM (transmission electron microscopy) [62]. The study reported that phenols and proteins were mainly responsible for the formation of NPs using FTIR (Fourier Transform infrared spectroscopy) analysis. Furthermore, there were some other peaks of medium intensity of $CH_3$ and $OCH_3$ which are associated with biopolymers present in *Vitis vinifera*. Tentatively, the phenols and reducing sugars are oxidized to aldehydes and ketones, thereby reducing the Se ions into NPs [62]. The synthesis of α-Se NPs using *Capsicum annum L.* extract have been reported and the study suggested that mainly phenols, proteins, and polysaccharides are responsible for the synthesis [66]. The comparative analysis of FTIR of only *Capsicum annum L.* extract and NPs showed a shift in the peaks of 1652, 1542, and 1241 $cm^{-1}$ which corresponds to amide I, II, III, respectively. Furthermore, the x-ray photoemission spectroscopy (XPS) confirms that these reducing agents present in *Capsicum annum* extract help in the formation of NPs [66]. The SeNPs have also been synthesized using leaves of *Undaria pinnatifida* (a sea vegetable) [67], *Citrus limon* [68], and *Terminalia arjuna* extract and it was reported that biomolecules

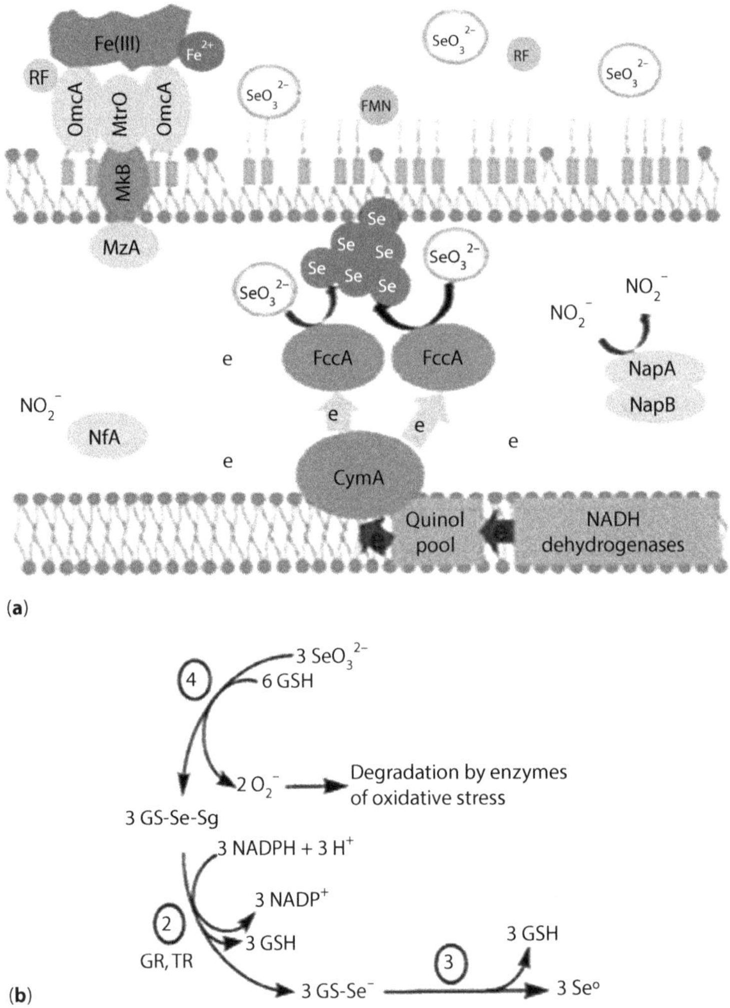


**Figure 12.1** Schematic diagrams showing the proposed mechanisms for reduction of selenate/selenite by microorganisms. (a) Selenite reduction and anaerobic respiration in *Shewanella oneidensis* MR-1; (b) reduction of selenite in *R. rubrum* and *E. coli* by glutathione (GSH). Reprinted with permission from ref. no. [61], Sweety A. Wadhwani, Utkarsha U. Shedbalkar, Richa Singh, and Balu A. Chopade, Journal of Applied Microbiology and Biotechnology. 100(6), 2555–2566 (2016). Copyright@Springer.

in the extract are alcohol, aldehydes, phenols, etc. which are getting oxidized to carboxylic acid, ketone, and phenone [63]. An exact mechanistic approach is needed to find out the biomolecules responsible for reduction and capping. Figure 12.2 shows the electron micrograph images of SeNPs synthesized using different microbes and plants.

Table 12.2 Synthesis of SeNPs using various plants.

| Plants | Part | Size of SeNPs (nm) | Shape of SeNPs | References |
|---|---|---|---|---|
| *Vitis vinifera* | Seed | 3–14 | Spherical | [62] |
| *Capsicum annum L.* | Leaf | 200–500 | Ellipsoidal | [66] |
| *Undaria pinnatifida* | Sea vegetable | 44–94 | Spherical | [67] |
| *Citrus limon* | Leaf | 60–80 | Spherical | [68] |
| *Termanalia arjuna* | Leaf | 10–80 | Spherical | [63] |
| *Bougainvillea spectabilis* | Flowers | 24.24–2.95 | Spherical and hollow | [69] |
| *Citrus reticulata* | Peel | 70 | Spherical | [70] |
| *Allium sativum* | Garlic cloves | 40–100 | Spherical | [71] |
| *Zingiber officinale* | Rhizomes | 10–20 | Spherical | [72] |
| *Acacia senegal* | Gum | 34.9 | Spherical | [73] |
| *Brassica oleracea* | Vegetable | 50–150 | Spherical | [74] |
| *Leucas lavandulifolia* | Leaf | 56–75 | Spherical | [76] |
| *Clausena dentata* | Leaf | 46.32–78.88 | Spherical | [77] |

Thus, SeNPs of different shapes and sizes can be obtained using microbial/plant extract; conditions need to be optimized in each case. The methods are simple and can be performed at room temperature without using toxic chemicals and sophisticated instruments. India has a great diversity of plants; this methodology can be adopted to synthesize nano Se at large scale.

## 12.3 Biotechnological Applications of SeNPs

Se is an important semiconductor metalloid, and has been used in photocells, photographic exposure meters, and solar cells [78]. In addition, Se is beneficial for the human health because of its anti-oxidant properties [79]. Some of the biomedical applications of SeNPs are described below.

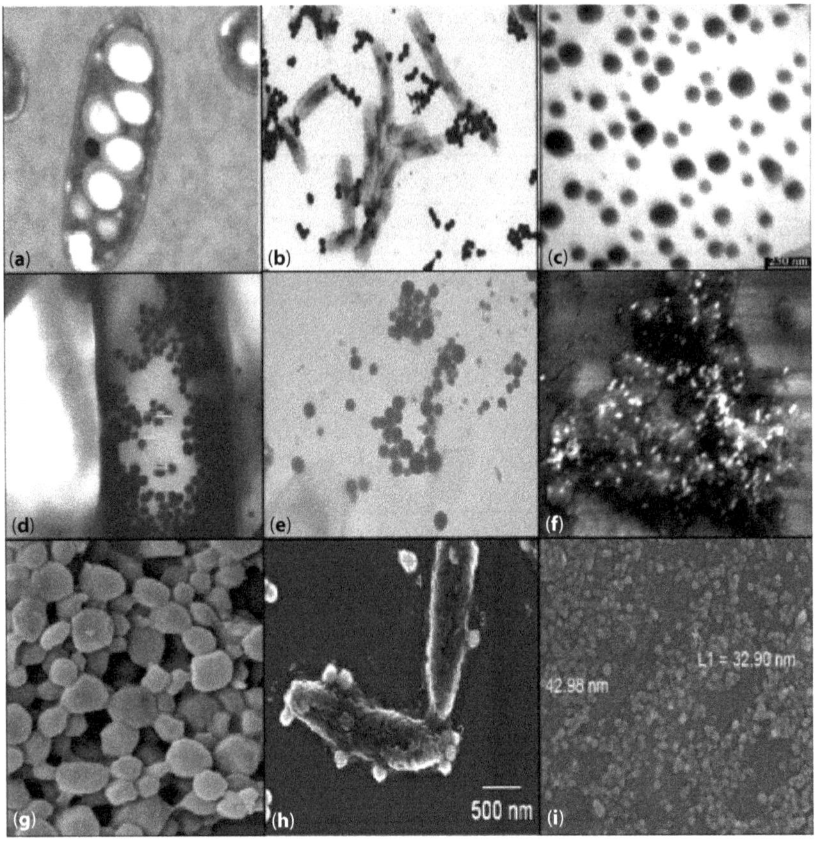


**Figure 12.2** Electron micrographs of SeNPs synthesized by microorganisms and plants. (a) TEM image of (a) intracellular SeNPs synthesized by *Thauera selenatis*, (b) extracellular SeNPs by *B. selenitreducens*, (c) SeNPs synthesized by *Bacillus sp.* MSh-1 a, (d) SeNPs synthesized by hyphae of *Lentinula edodes*, (e) SeNPs synthesized by *Terminalia arjuna* leaf extract. (f) FESEM micrographs of SeNPs on *Bacillus cereus* cell surface, (g) FESEM micrograph of SeNPs synthesized by *Capsicum annuum L.* SEM image of SeNPs synthesized by (h) *Duganella*, (i) culture supernatant of *A. terrus* BZ1. Reprinted with permission from ref. no. [61], Sweety A. Wadhwani, Utkarsha U. Shedbalkar, Richa Singh, and Balu A. Chopade, Journal of Applied Microbiology and Biotechnology. 100(6): 2555–2566 (2016). Copyright@Springer.

## 12.3.1 Anticancerous Activity

The SeNPs have been widely explored as antitumour and anticancerous agents because of their low cytotoxicity. Yazdi *et al.* (2012) synthesized SeNPs using gram-positive *Lactobacillus plantarum* bacteria [80]. The biosynthesized SeNPs were spherical with the diameter of approximately 250 nm. These SeNPs were orally administered (100 μg/day) to BALB/c (albino mouse model) every day for 2 weeks before and after tumor induction. The results indicated reduction in the rate of tumor growth in treated mouse as compared to untreated mouse. This effect was a result of enhancement of cellular immunity triggered by administering SeNPs, which was confirmed by assessing the levels of pro-inflammatory cytokines [80]. Chen *et al.* (2008) shows the *in vitro* cytotoxicity of SeNPs against various human cancer cell lines in comparison to normal cell lines. The SeNPs show broad spectrum inhibition against A375 (human skin malignant melanoma cells), CNE-2 (Cellosaurus cell line), HepG2 (liver hepatocellular carcinoma cell lines), and MCF-7 (human breast cancer cell line) cancer cells with $IC_{50}$ (50% inhibitory concentration) values ranging from 3.0 to 14.1 μM whereas the SeNPs show very less toxicity against normal cell lines [67]. Similarly, Yang *et al.* (2012) investigated the anticancer activity of SeNPs against A375 melanoma cells, human cervical carcinoma HeLa-229 cells, MCF-7 breast adenocarcinoma cells, osteosarcoma MG-63 (osteosarcoma cell lines), and human normal kidney HK-2 cells (human kidney cell lines) [81]. The anticancer activity of SeNPs may be due to (1) enhanced oxidative stress, carcinogen detoxification, and immune surveillance; (2) induction of cellular and mitochondria-mediated apoptosis; (3) inhibition of tumor cell invasion and angiogenesis; (4) cell cycle arrest at S-phase (synthesis phase); (5) metastasis prevention by inhibition of matrix metalloproteinases expression; and (6) mobilization of endogenous copper [33, 76–78]. Researchers have also reported that conjugation of SeNPs with organic moieties and drugs prevents aggregation of NPs, enhances cellular uptake, and enhances their anticancer activity. The SeNPs were functionalized with *Spirulina* polysaccharide (SPS-SeNPs) and these SPS-SeNPs are reported to have higher cytotoxic effect than SeNPs without polysaccharide against cancer cell lines [81]. The polysaccharide modified SeNP can easily enter the cell/lipid bilayer, and thus increases the anticancerous activity. SeNPs with good anticancerous activity have also been synthesized from a polysaccharide extracted from *Gracilaria lamaneiform* which is an algal species [60]. Alkylating anticancer drug Cisplatin which causes oxidative stress and DNA (deoxyribonucleic acid) cross-linking was conjugated to SeNPs and this Cisplatin conjugated SeNPs can improve the quality of sperms and improve spermatogenesis [82]. Figure 12.3 depicts the mechanism of anticancer activity of glucose synthesized SeNPs.

### 12.3.2 Antioxidant Activity

The antioxidant activity of SeNPs was studied by Torres *et al.* (2012) in human umbilical vein endothelial cells [27]. The SeNPs lead to less intracellular

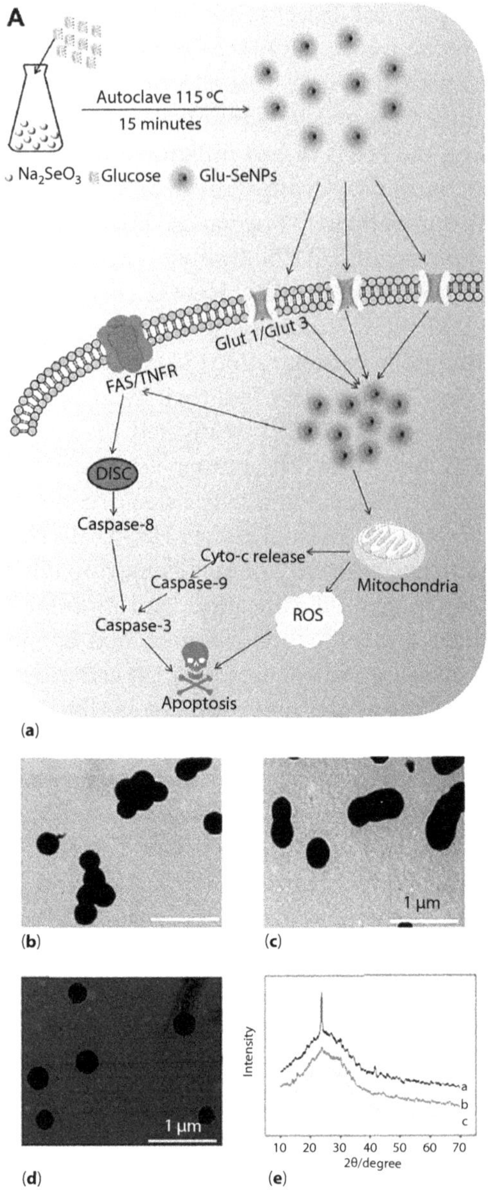


**Figure 12.3** TEM images and XRD patterns of SeNPs with different concentrations of glucose. (A) Synthetic scheme for the Glu–SeNPs and its general mechanism of anticancer effect; (B) SeNPs with 0.25 M glucose concentration; (C) SeNPs with 0.5 M glucose concentration; (D) SeNPs with 1.0 M glucose concentration; (E) XRD patterns of Glu–SeNPs. From top to bottom: 0.25 M (a), 0.50 M (b), and 1.0 M (c) glucose. Reprinted with permission from ref. no. [17], Tianqi Nie, Hualian Wu, Ka-Hing Wong, and Tianfeng Chen, Journal of Materials Chemistry B, 4, 2351 (2016). Copyright@Royal Society of Chemistry.

accumulation of reactive oxygen species (ROS) and thus, reduction in oxidative stress [27]. Recently Vyas and Rana (2017) reported that Aloe Vera mediated SeNPs prevent the oxidation of essential biological macromolecules by inhibiting the propagation of the oxidizing chain reaction and hence can be used as chemo-preventive agent in cancer diagnosis [75].

### 12.3.3 Antidiabetic Effect

Kumar *et al.* (2014) explored the therapeutic potential of SeNPs in preventing diabetic nephropathy. The diabetic male Sprague-Dawley rats were treated with SeNPs having a diameter of 40–90 nm [83]. These SeNPs were highly effective in lowering the blood urea, nitrogen, fibronectin, creatinine, and collagen and increased the albumin levels. The SeNPs treated kidney of diabetic rats showed increased level of longevity protein SIRT1 (a protein NAD-dependent deacetylase sirtuin-1 in human encoded by SIRT1), heat shock protein (HSP-70), and balanced apoptotic proteins Bax and Bcl-2 (gene regulate apoptosis). Hence, results suggest the prevention of diabetic nephropathy by not only quenching oxidative stress but also by activating cyto-protective protein HSP-70 and longevity protein SIRT1 [83].

### 12.3.4 Wound Healing

Ramya *et al.* (2015) have reported the wound healing properties of SeNPs as shown in Figure 12.4. Wistar Albino rats were selected for the studies and wound healing was evaluated *in vitro* [53]. At low and high concentrations of SeNPs along with ointment, the artificially created wound was healed in 21 and 18 days, respectively, whereas the wound treated with standard antibiotic lyramycin was healed in 21 days and in group of control, i.e. without any treatment, the wound was completely healed in 30 days. The SeNPs treated rats showed resemblance to normal skin, less hypertrophic scarring, normal hair growth in the wounded area, hence it was inferred that SeNPs based ointments are more efficient medication than lyramycin [53].

### 12.3.5 Antibacterial Activity

The bacterial infections are the most prevailing infections in the world and the occurrence of resistance among bacteria for the common antibacterial therapies makes the condition more severe. Therefore, researchers are continuously seeking for new antibacterial agents. SeNPs have been reported as antibacterial agents against gram-positive and gram-negative bacteria.

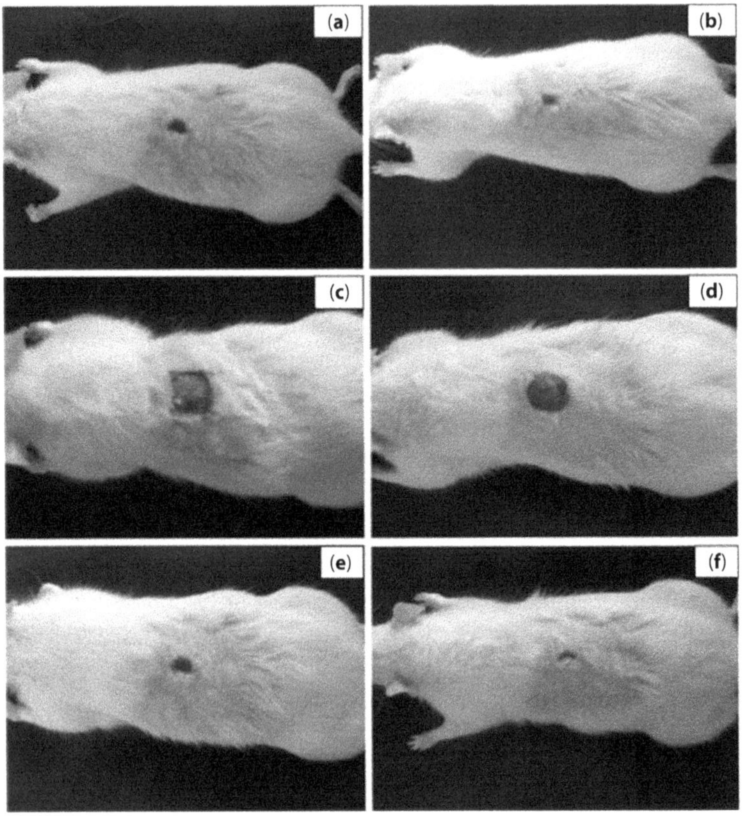


**Figure 12.4** (a and b) Wound treatment with higher dose of selenium nanoparticles, (c and d) wound treatment with induced model, and (e and f) wound treated with standard antibiotic lyramycin. Reprinted with permission from ref. no. [53], S. Rmaya, T. Shanmugasundaram, and R. Balagurunathan, Journal of Trace Element in Medicine and Biology. 32, 30–39 (2015). Copyright@Elsevier.

Kirupagaran *et al.* (2016) synthesized SeNPs using *Leucas lavandulifolia* leaf extract using selenious acid solution and tested their efficacy against four gram-positive bacterial strains *Bacillus subtilis*, *Enterococcus faedalis*, *Staphylococcus aureus*, *Staphylococcus epidermidis*, and three gram-negative strains *Salmonella typhii*, *Klebsiella pneumonia*, and *Shigella boydii* [76]. The clear zone of inhibition shows that *Leucas lavandulifolia* leaf extract mediated SeNPs are efficient against all tested microbes [76]. The high antibacterial activity of SeNPs is due to their large surface area, which provides better contact with microorganisms. The SeNPs have also been reported to disrupt the microbial biofilms [44].

### 12.3.6 Antilarvicidal Activity

Mosquitoes are the major vectors for the transmission of many diseases, like chikungunya, malaria, dengue, zika, etc. Nearly 100 countries are prevalent to mosquito borne disease, infecting 70 million people world-wide and 4 million peoples in India. The nanotechnology is proven a boon to such conditions as chemical insecticides are proven to be environmental as well as human health hazards. The SeNPs possess the larvicidal activity against *Anopheles stephensi*, *Culex quinquefasciatus*, and *Aedes aegypti* [77]. The green synthesized *Christella dentata* mediated SeNPs show high mortality with very low concentrations of 240.714, 104.13, and 99.602 mg/L for *Anopheles stephensi*, *Culex quinquefasciatus*, and *Aedes aegypti*, respectively [77]. The larvicidal activity may be due to the denaturation of sulfur-containing proteins or phosphorous-containing compounds like DNA that leads to denaturation of organelles and enzymes and thus reduces the cellular permeability and loss of cellular function [51, 77].

### 12.3.7 Biosensors

The SeNPs are known to have electro-catalytic activity. Very few reports have been published on the electrochemical nature of SeNPs for the detection of hydrogen peroxide ($H_2O_2$) [16, 34]. Wang *et al.* (2010) have deposited the Se nanomaterials on glassy carbon electrode with HRP (horseradish peroxidase) and studied the oxidation/reduction reactions [34]. The study reported that Se coated electrodes have more peak currents than the electrode alone as SeNPs possess good adhesive properties and are biocompatible, hence can be used in HRP-biosensor (horseradish peroxidase based biosensor) [34]. The SeNPs based biosensor exhibits good electro-catalytic activity toward the reduction of $H_2O_2$ having the detection limit of $8 \times 10^{-8}$ M. Thus, they can be exploited for the development of biosensors [34].

## 12.4 Conclusion

This chapter highlights the applications and synthesis of SeNPs. The green method of synthesis is described in detail, which involves natural biomolecules as the reducing and capping agents. These biomolecules can be obtained either from micro-organisms or from plants as secondary metabolites. Mainly the carbohydrates, amino acids, flavonoids which have the potential to reduce the Se act as reducing agents and the polymeric substances like polysaccharides, proteins serve as the stabilizing agents.

The methodology is quite simple, and renewable source is used for production of nano Se. There are numerous applications of these synthesized NPs especially in medicine as they are considered less toxic as compared to chemically synthesized ones. The anticancerous activity of the SeNPs has been studied in detail by several investigators. The toxicity of these NPs is the major concern when used in therapy. Future research on the mechanism of biosynthesis can give an insight on the exact reducing and capping agents which can synthesize the size-controlled NPs, thus this technology is feasible for large-scale synthesis.

## Acknowledgments

The authors acknowledge the Indian Council of Medical Research, Government of India, for providing the financial support.

## References

1. Lee, Y., Choi, J.R., Lee, K.J., Stott, N.E., Kim, D., Large-scale synthesis of copper nanoparticles by chemically controlled reduction for applications of inkjet-printed electronics, *Nanotechnology*, 19, 415604–415610, 2008.
2. Mu, L. and Sprando, R.L., Application of nanotechnology in cosmetics, *Pharm. Res.* 27, 1746–1749, 2010.
3. Salata, O.V., Applications of nanoparticles in biology and medicine, *J. Nanobiotechnol.*, 2, 3–8, 2004.
4. Espitia, P.J.P., Soares, N.F.F., dos Reis Coimbra, J.S., de Andrade, N.J., Cruz, R.S., Medeiros, E.A.A., Zinc oxide nanoparticles: synthesis, antimicrobial activity and food packaging applications, *Food Bioproc. Tech.*, 5, 1447–1464, 2012.
5. Gindy, M.E. and Prud'homme, R.K., Multifunctional nanoparticles for imaging, delivery and targeting in cancer therapy, *Expert Opin. Drug Deliv.*, 6, 865–878, 2009.
6. Lai, F., Wissing, S.A., Muller, R.H., Fadda, A.M., Artemisia arborescens L essential oil-loaded solid lipid nanoparticles for potential agricultural application: preparation and characterization, *Aaps PharmSciTech*, 7, E1-E10, 2006.
7. Mishra, A. and Sardar, M., Alpha amylase mediated synthesis of gold nanoparticles and their application in the reduction of nitroaromatic pollutants, *Energy Environ. Focus*, 3, 179–184, 2014.
8. Yildiz, I., Shukla, S., Steinmetz, N.F., Applications of viral nanoparticles in medicine, *Curr. Opin. Biotechnol.*, 22, 901–908, 2011.
9. Hatfield, D.L., Berry, M.J., Gladyshev, V.N., *Selenium: Its Molecular Biology and Role in Human Health*, Springer Science & Business Media, USA, 2011.

10. Papp, L.V., Lu, J., Holmgren, A., Khanna, K.K., From selenium to selenoproteins: synthesis, identity, and their role in human health, *Antioxid. Redox Signal.*, 9, 775–806, 2007.
11. Gladyshev, V.N., Arner, E.S., Berry, M.J., Brigelius-Flohe, R., Bruford, E.A., Burk, R.F., Carlson, B.A., Castellano, S., Chavatte, L., Conrad, M., Selenoprotein gene nomenclature, *J. Biol. Chem.*, 291, 24036–24040, 2016.
12. Sonkusre, P., Nanduri, R., Gupta, P., Cameotra, S.S., Improved extraction of intracellular biogenic selenium nanoparticles and their specificity for cancer chemoprevention, *J. Nanomed. Nanotechnol.*, 5, 1–9, 2014.
13. Quintana, M., Haro-Poniatowski, E., Morales, J., Batina, N., Synthesis of selenium nanoparticles by pulsed laser ablation, *Appl. Sci. Res.*, 195, 175–186. 2002.
14. Yang, L.B., Shen, Y.H., Xie, A.J., Liang, J.J., Zhang, B.C., Synthesis of Se nanoparticles by using TSA ion and its photocatalytic application for decolorization of cango red under UV irradiation, *Mater. Res. Bull.*, 43, 572–582, 2008.
15. Shin, Y., Blackwood, J.M., Bae, I.-T., Arey, B.W., Exarhos, G.J., Synthesis and stabilization of selenium nanoparticles on cellulose nanocrystal, *Mater. Lett.*, 61, 4297–4300, 2007.
16. Zhang, S.Y., Zhang, J., Wang, H.Y., Chen, H.Y., Synthesis of selenium nanoparticles in the presence of polysaccharides, *Mater. Lett.*, 58, 2590–2594, 2004.
17. Nie, T., Wu, H., Wong, K.-H., Chen, T., Facile synthesis of highly uniform selenium nanoparticles using glucose as the reductant and surface decorator to induce cancer cell apoptosis, *J. Mater. Chem.*, B.4, 2351–2358, 2016.
18. Lin, Z.H. and Wang, C.R.C., Evidence on the size-dependent absorption spectral evolution of selenium nanoparticles, *Mater. Chem. Phys.*, 92, 591–594, 2005.
19. Tan, L., Jiang, X., Zhang, Y., Tang, H., Yao, S., Xie, Q., *In vitro* study on the individual and synergistic cytotoxicity of adriamycin and selenium nanoparticles against Bel7402 cells with a quartz crystal microbalance, *Biosens. Bioelectron.*, 24, 2268–2272, 2009.
20. Panahi-Kalamuei, M., Salavati-Niasari, M., Hosseinpour-Mashkani, S.M., Facile microwave synthesis, characterization, and solar cell application of selenium nanoparticles, *J. Alloys Compd.*, 617, 627–632, 2014.
21. Guisbiers, G., Wang, Q., Khachatryan, E., Mimun, L.C., Mendoza-Cruz, R., Larese-Casanova, P., Webster, T.J., Nash, K.L., Inhibition of E. coli and S. aureus with selenium nanoparticles synthesized by pulsed laser ablation in deionized water, *Int. J. Nanomed.*, 11, 3731–3736, 2016.
22. Panahi-Kalamuei, M., Salavati-Niasari, M., Zarghami, Z., Mousavi-Kamazani, M., Taqriri, H., Mohsenikia, A., Synthesis and characterization of Se nanostructures via co-precipitation, hydrothermal, microwave and sonochemical routes using novel starting reagents for solar cells, *J. Mater. Sci. Mater. Electron.*, 26, 2851–2860, 2015.

23. Dwivedi, C., Shah, C.P., Singh, K., Kumar, M., Bajaj, P.N., An organic acid-induced synthesis and characterization of selenium nanoparticles, *J. Nanotechnol.*, 2011, 1–6, 2011.
24. Hunter, W.J. and Kuykendall, L.D., Reduction of selenite to elemental red selenium by *Rhizobium sp.* strain B1, *Curr. Microbiol.*, 55, 344–349, 2007.
25. Oremland, R.S., Herbel, M.J., Blum, J.S., Langley, S., Beveridge, T.J., Ajayan, P.M., Sutto, T., Ellis, A.V., Curran, S., Structural and spectral features of selenium nanospheres produced by Se-respiring bacteria, *Appl. Environ. Microbiol.*, 70, 52–60, 2004.
26. Zhang, W., Chen, Z., Liu, H., Zhang, L., Gao, P., Li, D., Biosynthesis and structural characteristics of selenium nanoparticles by Pseudomonas alcaliphila, *Colloids Surf. B Biointerfaces*, 88, 196–201, 2011.
27. Torres, S.K., Campos, V.L., Leon, C.G., Rodriguez-Llamazares, S.M., Rojas, S.M., Gonzalez, M., Smith, C., Mondaca, M.A., Biosynthesis of selenium nanoparticles by *Pantoea agglomerans* and their antioxidant activity, *J. Nanopart. Res.*, 14, 1–9, 2012.
28. Losi, M.E. and Frankenberger, W.T., Reduction of selenium oxyanions by *Enterobacter cloacae* SLD1a-1: isolation and growth of the bacterium and its expulsion of selenium particles, *Appl. Environ. Microbiol.*, 63, 3079–3084, 1997.
29. Debieux, C.M., Dridge, E.J., Mueller, C.M., Splatt, P., Paszkiewicz, K., Knight, I., Florance, H., Love, J., Titball, R.W., Lewis, R.J., A bacterial process for selenium nanosphere assembly, *Proc. Natl Acad. Sci.*, 108, 13480–13485, 2011.
30. Bajaj, M., Schmidt, S., Winter, J., Formation of Se (0) nanoparticles by *Duganella sp.* and *Agrobacterium sp.* isolated from Se-laden soil of North-East Punjab, India, *Microb. Cell Fact.*, 11, 1–14, 2012.
31. Li, D.B., Cheng, Y.Y., Wu, C., Li, W.W., Li, N., Yang, Z.C., Tong, Z.H., Yu, H.Q., Selenite reduction by *Shewanella oneidensis* MR-1 is mediated by fumarate reductase in periplasm, *Sci. Rep.*, 4, 3735–3741, 2014.
32. Ma, J., Kobayashi, D.Y., Yee, N., Role of menaquinone biosynthesis genes in selenate reduction by *Enterobacter cloacae SLD1a-1* and *Escherichia coli K12*, *Environ. Microbiol.*, 11, 149–158, 2009.
33. Ahmad, M.S., Yasser, M.M., Sholkamy, E.N., Ali, A.M., Mehanni, M.M., Anticancer activity of biostabilized selenium nanorods synthesized by *Streptomyces bikiniensis* strain Ess_amA-1, *Int. J. Nanomed.*, 10, 3389–3401, 2015.
34. Wang, T., Yang, L., Zhang, B., Liu, J., Extracellular biosynthesis and transformation of selenium nanoparticles and application in $H_2O_2$ biosensor, *Colloids Surf. B Biointerfaces*, 80, 94–102, 2010.
35. Dhanjal, S., Cameotra, S.S., Aerobic biogenesis of selenium nanospheres by *Bacillus cereus* isolated from coalmine soil, *Microb. Cell. Fact.*, 9, 1–11, 2010.
36. Kazempour, Z.B., Yazdi, M.H., Rafii, F., Shahverdi, A.R., Sub-inhibitory concentration of biogenic selenium nanoparticles lacks post antifungal effect for

*Aspergillus niger* and *Candida albicans* and stimulates the growth of *Aspergillus niger*, *Iran. J. Microbiol.*, 5, 81–85, 2013.
37. Butler, C.S., Debieux, C.M., Dridge, E.J., Splatt, P., Wright, M., Biomineralization of selenium by the selenate-respiring bacterium *Thaueraselenatis*, *Biochem. Soc. Trans.*, 40, 1239–1243, 2012.
38. Srivastava, N. and Mukhopadhyay, M., Biosynthesis and structural characterization of selenium nanoparticles mediated by *Zooglearamigera*, *Powder Technol.*, 244, 26–29, 2013.
39. Sonkusre, P. and Cameotra, S.S., Biogenic selenium nanoparticles induce ROS-mediated necroptosis in PC-3 cancer cells through TNF activation, *J. Nanobiotechnol.* 15, 1–12, 2017.
40. Kheradmand, E., Rafii, F., Yazdi, M.H., Sepahi, A.A., Shahverdi, A.R., Oveisi, M.R., The antimicrobial effects of selenium nanoparticle-enriched probiotics and their fermented broth against *Candida albicans*, *DARU J. Pharm. Sci.*, 22, 1–6, 2014.
41. Dwivedi, S., AlKhedhairy, A.A., Ahamed, M., Musarrat, J., Biomimetic synthesis of selenium nanospheres by bacterial strain JS-11 and its role as a biosensor for nanotoxicity assessment: a novel Se-bioassay, *PLoS One*, 8, 1–10, 2013.
42. Forootanfar, H., Adeli-Sardou, M., Nikkhoo, M., Mehrabani, M., Amir-Heidari, B., Shahverdi, A.R., Shakibaie, M., Antioxidant and cytotoxic effect of biologically synthesized selenium nanoparticles in comparison to selenium dioxide, *J. Trace Elem. Med. Biol.*, 28, 75–79, 2014.
43. Sasidharan, S. and Balakrishnaraja, R., Comparison studies on the synthesis of selenium nanoparticles by various microorganisms, *Int. J. Pure* Appl. *Biosci.*, 2, 112–117, 2014.
44. Zonaro, E., Lampis, S., Turner, R.J., Qazi, S.J.S., Vallini, G., Biogenic selenium and tellurium nanoparticles synthesized by environmental microbial isolates efficaciously inhibit bacterial planktonic cultures and biofilms, *Front. Microbiol.*, 6, 1–11, 2015.
45. Shoeibi, S. and Mashreghi, M., Biosynthesis of selenium nanoparticles using *Enterococcus faecalis* and evaluation of their antibacterial activities, *J. Trace Elem. Med. Biol.*, 39, 135–139, 2017.
46. Dobias, J., Suvorova, E.I., Bernier-Latmani, R., Role of proteins in controlling selenium nanoparticle size, *Nanotechnology*, 22, 1–9, 2010.
47. Lampis, S., Zonaro, E., Bertolini, C., Cecconi, D., Monti, F., Micaroni, M., Turner, R.J., Butler, C.S., Vallini, G., Selenite biotransformation and detoxification by *Stenotrophomonas maltophilia* SeITE02: novel clues on the route to bacterial biogenesis of selenium nanoparticles, *J. Hazard. Mater.*, 324, 3–14, 2017.
48. Lampis, S., Zonaro, E., Bertolini, C., Bernardi, P., Butler, C.S., Vallini, G., Delayed formation of zero-valent selenium nanoparticles by *Bacillus mycoides* SeITE01 as a consequence of selenite reduction under aerobic conditions, *Microb. Cell. Fact.*, 13, 1–14, 2014.

49. Fernandez-Llamosas, H., Castro, L., Blazquez, M.L., Diaz, E., Carmona, M., Biosynthesis of selenium nanoparticles by *Azoarcus sp.* CIB, *Microb. Cell. Fact.*, 15, 1–10, 2016.
50. Cui, Y.H., Li, L.L., Zhou, N.Q., Liu, J.H., Huang, Q., Wang, H.J., Tian, J., Yu, H.Q., *In vivo* synthesis of nano-selenium by *Tetrahymena thermophila* SB210, *Enzyme Microb. Technol.*, 95, 185–191, 2016.
51. Sonkusre, P., Nanduri, R., Gupta, P., Cameotra, S.S., Improved extraction of extracellular biogenic selenium nanoparticle and their specificity for cancer chemoprevention, *J. Nanomed. Nanotechnol.*, 5, 1–9, 2014.
52. Wadhwani, S.A., Gorain, M., Banerjee, P., Shedbalkar, U.U., Singh, R., Kundu, G.C., Chopade, B.A., Green synthesis of selenium nanoparticles using *Acinetobacter sp.* SW30: optimization, characterization and its anti-cancer activity in breast cancer cells, *Int. J. Nanomed.*, 12, 6841–6855, 2017.
53. Ramya, S., Shanmugasundaram, T., Balagurunathan, R., Biomedical potential of actinobacterially synthesized selenium nanoparticles with special reference to anti-biofilm, anti-oxidant, wound healing, cytotoxic and anti-viral activities, *J. Trace Elem. Med. Biol.*, 32, 30–39, 2015.
54. Samant, S., Naik, M., Parulekar, K., Charya, L., Vaigankar, D., Selenium reducing *Citrobacter fruendii* strain KP6 from Mandovi estuary and its potential application in selenium nanoparticle synthesis, *Proc. Natl Acad. Sci., India Sect. B: Biol. Sci.*, 8, 1–8, 2016.
55. Hamza, F., Vaidya, A., Apte, M., Kumar, A.R., Zinjarde, S., Selenium nanoparticle-enriched biomass of *Yarrowia lipolytica* enhances growth and survival of *Artemia salina*, *Enzyme Microb. Technol.*, 106, 48–54, 2017.
56. Sarkar, J., Dey, P., Saha, S., Acharya, K., Mycosynthesis of selenium nanoparticles, *Micro Nano Lett.*, 6, 599–602, 2011.
57. Hariharan, H., Al-Dhabi, N.A., Karuppiah, P., Rajaram, S.K., Microbial synthesis of selenium nanocomposite using *Saccharomyces cerevisiae* and its antimicrobial activity against pathogens causing nosocomial infection, *Chalcogenide Lett.*, 9, 509–515, 2012.
58. Zare, B., Babaie, S., Setayesh, N., Shahverdi, A.R., Isolation and characterization of a fungus for extracellular synthesis of small selenium nanoparticles, *Nanomed. J.*, 1, 13–19, 2013.
59. Xiao, Y., Huang, Q., Zheng, Z., Guan, H., Liu, S., Construction of a *Cordyceps sinensis* exopolysaccharide-conjugated selenium nanoparticles and enhancement of their antioxidant activities, *Int. J. Biol. Macromol.*, 99, 483–491, 2017.
60. Jiang, W., Fu, Y., Yang, F., Yang, Y., Liu, T., Zheng, W., Zeng, L., Chen, T., *Gracilaria lemaneiformis* polysaccharide as integrin-targeting surface decorator of selenium nanoparticles to achieve enhanced anticancer efficacy, *ACS Appl. Mater. Interfaces*, 6, 13738–13748, 2014.
61. Wadhwani, S.A., Shedbalkar, U.U., Singh, R., Chopade, B.A., Biogenic selenium nanoparticles: currentstatus and future prospects, *Appl. Microbiol. Biotechnol.*, 100, 2555–2566, 2016.

62. Sharma, G., Sharma, A.R., Bhavesh, R., Park, J., Ganbold, B., Nam, J.S., Lee, S.S., Biomolecule-mediated synthesis of selenium nanoparticles using dried Vitis vinifera (raisin) extract, *Molecules*, 19, 2761–2770, 2014.
63. Prasad, K.S. and Selvaraj, K., Biogenic synthesis of selenium nanoparticles and their effect on As(III)-induced toxicity on human lymphocytes, *Biol. Trace Elem. Res.*, 157, 275–283, 2014.
64. Rajasree, R.S.R. and Gayathri, S., Extracellular biosynthesis of Selenium nanoparticles using some species of *Lactobacillus*, *Indian J. Mar. Sci.*, 44, 766–775, 2015.
65. Shakibaie, M., Shahverdi, A.R., Faramarzi, M.A., Hassanzadeh, G.R., Rahimi, H.R., Sabzevari, O., Acute and subacute toxicity of novel biogenic selenium nanoparticles in mice, *Pharm. Biol.*, 51, 58–63, 2013.
66. Li, S., Shen, Y., Xie, A., Yu, X., Zhang, X., Yang, L., Li, C., Rapid, room-temperature synthesis of amorphous selenium/protein composites using *Capsicum annuum L* extract, *Nanotechnology*, 18, 405101–405110, 2007.
67. Chen, T., Wong, Y.S., Zheng, W., Bai, Y., Huang, L., Selenium nanoparticles fabricated in *Undaria pinnatifida* polysaccharide solutions induce mitochondria-mediated apoptosis in A375 human melanoma cells, *Colloids Surf. B Biointerfaces*, 67, 26–31, 2008.
68. Prasad, K.S., Patel, H., Patel, T., Patel, K., Selvaraj, K., Biosynthesis of Se nanoparticles and its effect on UV-induced DNA damage, *Colloids Surf. B Biointerfaces*, 103, 261–266, 2013.
69. Deepa, B. and Ganesan, V., Biogenic synthesis and characterization of selenium nanoparticles using the flower of *Bougainvillea spectabilis* wild, *Int. J. Sci. Res.*, 4, 690–695, 2015.
70. Santanu, S., Sowmiya, R., Rengaraju, B., Biosynthesis of selenium nanoparticles using *Citrus reticulata* peel extract, *World J. Pharm. Res.*, 4, 1322–1330, 2015.
71. Ezhuthupurakkal, P.B., Polaki, L.R., Suyavaran, A., Subastri, A., Sujatha, V., Thirunavukkarasu, C., Selenium nanoparticles synthesized in aqueous extract of *Allium sativum* perturbs the structural integrity of Calf thymus DNA through intercalation and groove binding, *Mater. Sci. Eng. C*, 74, 597–608, 2016.
72. Zahran, W.E., Elsonbaty, S.M., Moawed, F.S.M., Selenium nanoparticles with low-level ionizing radiation exposure ameliorate nicotine-induced inflammatory impairment in rat kidney, *Environ. Sci. Pollut. Res.*, 24, 19980–19989, 2017.
73. Kong, H., Yang, J., Zhang, Y., Fang, Y., Nishinari, K., Phillips, G.O., Synthesis and antioxidant properties of gum arabic-stabilized selenium nanoparticles, *Int. J. Biol. Macromol.* 65, 155–162, 2014.
74. Kapur, M., Soni, K., Kohli, K., Green synthesis of selenium nanoparticles from broccoli, characterization, application and toxicity, *Adv. Tech. Biol. Med.*, 5, 1–7, 2017.
75. Vyas, J. and Rana, S., Antioxidant activity and biogenic synthesis of selenium nanoparticles using the leaf extract of *Aloe vera*, *Int. J. Curr. Pharm. Res.*, 9, 147–152, 2017.

76. Kirupagaran, R., Saritha, A., Bhuvaneswari, S., Green synthesis of selenium nanoparticles from leaf and stem extract of *Leucas lavandulifolia Sm.* and their application, *J. Nanosci. Nanotechnol.*, 2, 224–226, 2016.
77. Sowndarya, P., Ramkumar, G., Shivakumar, M.S., Green synthesis of selenium nanoparticles conjugated *Clausena dentata* plant leaf extract and their insecticidal potential against mosquito vectors, *Artif. Cells Nanomed. Biotechnol.*, 45, 1490–1495, 2017.
78. Luo, H., Wang, F., Bai, Y., Chen, T., Zheng, W., Selenium nanoparticles inhibit the growth of HeLa and MDA-MB-231 cells through induction of S phase arrest, *Colloids Surf. B Biointerfaces*, 94, 304–308, 2012.
79. Battin, E.E. and Brumaghim, J.L., Antioxidant activity of sulfur and selenium: a review of reactive oxygen species scavenging, glutathione peroxidase, and metal-binding antioxidant mechanisms, *Cell Biochem. Biophys.*, 55, 1–23, 2009.
80. Yazdi, M.H., Mahdavi, M., Kheradmand, E., Shahverdi, A.R., The preventive oral supplementation of a selenium nanoparticle-enriched probiotic increases the immune response and lifespan of 4T1 breast cancer bearing mice, *Arzneimittelforschung*, 62, 525–531, 2012.
81. Yang, F., Tang, Q., Zhong, X., Bai, Y., Chen, T., Zhang, Y., Li, Y., Zheng, W., Surface decoration by *Spirulina* polysaccharide enhances the cellular uptake and anticancer efficacy of selenium nanoparticles, *Int. J. Nanomed.*, 7, 835–844, 2012.
82. Li, Y., Li, X., Wong, Y.S., Chen, T., Zhang, H., Liu, C., Zheng, W., the reversal of cisplatin-induced nephrotoxicity by selenium nanoparticles functionalized with 11-mercapto-1-undecanol by inhibition of ROS-mediated apoptosis, *Biomaterials*, 32, 9068–9076, 2011.
83. Kumar, G.S., Kulkarni, A., Khurana, A., Kaur, J., Tikoo, K., Selenium nanoparticles involve HSP-70 and SIRT1 in preventing the progression of type 1 diabetic nephropathy, *Chem. Biol. Interact.*, 223, 125–133, 2014.

# Index

Lightning Source UK Ltd.
Milton Keynes UK
UKHW051211311018
331463UK00005B/3/P

9 781119 407041